D1596113

Mechanical Engineering Series

Frederick F. Ling
Editor-in-Chief

Mechanical Engineering Series

Makoto Ohsaki and Kiyohiro Ikeda, **Stability and Optimization of Structures: Generalized Sensitivity Analysis**

A.C. Fischer-Cripps, **Introduction to Contact Mechanics, 2nd ed.**

W. Cheng and I. Finnie, **Residual Stress Measurement and the Slitting Method**

J. Angeles, **Fundamentals of Robotic Mechanical Systems: Theory Methods and Algorithms, 3rd ed.**

J. Angeles, **Fundamentals of Robotic Mechanical Systems: Theory, Methods, and Algorithms, 2nd ed.**

P. Basu, C. Kefa, and L. Jestin, **Boilers and Burners: Design and Theory**

J.M. Berthelot, **Composite Materials: Mechanical Behavior and Structural Analysis**

I.J. Busch-Vishniac, **Electromechanical Sensors and Actuators**

J. Chakrabarty, **Applied Plasticity**

K.K. Choi and N.H. Kim, **Structural Sensitivity Analysis and Optimization 1: Linear Systems**

K.K. Choi and N.H. Kim, **Structural Sensitivity Analysis and Optimization 2: Nonlinear Systems and Applications**

G. Chryssolouris, **Laser Machining: Theory and Practice**

V.N. Constantinescu, **Laminar Viscous Flow**

G.A. Costello, **Theory of Wire Rope, 2nd ed.**

K. Czolczynski, **Rotordynamics of Gas-Lubricated Journal Bearing Systems**

M.S. Darlow, **Balancing of High-Speed Machinery**

W. R. DeVries, **Analysis of Material Removal Processes**

J.F. Doyle, **Nonlinear Analysis of Thin-Walled Structures: Statics, Dynamics, and Stability**

J.F. Doyle, **Wave Propagation in Structures: Spectral Analysis Using Fast Discrete Fourier Transforms, 2nd ed.**

P.A. Engel, **Structural Analysis of Printed Circuit Board Systems**

A.C. Fischer-Cripps, **Introduction to Contact Mechanics**

A.C. Fischer-Cripps, **Nanoindentation, 2nd ed.**

(continued after index)

Makoto Ohsaki
Kiyohiro Ikeda

Stability and Optimization of Structures

Generalized Sensitivity Analysis

Makoto Ohsaki
Kyoto University
Kyoto, Japan

Kiyohiro Ikeda
Tohoku University
Sendai, Japan

Series Editor:
Frederick F. Ling

Library of Congress Control Number: 2007923779

ISBN 978-0-387-68183-2 e-ISBN 978-0-387-68184-9

Printed on acid-free paper.

9 8 7 6 5 4 3 2 1

springer.com

Mechanical Engineering Series

Frederick F. Ling
Editor-in-Chief

The Mechanical Engineering Series features graduate texts and research monographs to address the need for information in contemporary mechanical engineering, including areas of concentration of applied mechanics, biomechanics, computational mechanics, dynamical systems and control, energetics, mechanics of materials, processing, production systems, thermal science, and tribology.

Series Preface

Mechanical engineering, and engineering discipline born of the needs of the industrial revolution, is once again asked to do its substantial share in the call for industrial renewal. The general call is urgent as we face profound issues of productivity and competitiveness that require engineering solutions, among others. The Mechanical Engineering Series is a series featuring graduate texts and research monographs intended to address the need for information in contemporary areas of mechanical engineering.

The series is conceived as a comprehensive one that covers a broad range of concentrations important to mechanical engineering graduate education and research. We are fortunate to have a distinguished roster of series editors, each an expert in one of the areas of concentration. The names of the series editors are listed on page v of this volume. The areas of concentration are applied mechanics, biomechanics, computational mechanics, dynamic systems and control, energetics, mechanics of materials, processing, thermal science, and tribology.

Preface

In our modern world, best structures with specified shape, stiffness, strength, stability, frequency, and so on, can be designed with the assistance of computer-aided methodologies including sensitivity analysis, reliability-based design, inverse engineering, optimization, and anti-optimization. Buckling is an extremely important design constraint for structures with slender members such as latticed domes and frames; buckling of geometrically nonlinear structures is a well developed field of research. Nevertheless, because of the complexity of nonlinear buckling behavior, optimization of nonlinear structures has come to be conducted only recently despite its importance.

It is possible to consider structural optimization as a straightforward application of mathematical programming and operations research, as well as heuristics and evolutionary approaches. Such, however, is not the case for optimization of structures that undergo buckling. As cautioned by the "danger of naive optimization [287]," optimized structures often become more imperfection-sensitive and their buckling loads are reduced sharply because of the inevitable presence of initial imperfections that arise from errors in manufacturing processes, material defects, and other causes. It is certainly ironic that dangerous structures are produced in the attempt to optimize their performance. In any search for the best structure, the imperfection sensitivity of optimized structures must be investigated.

This book offers an introduction to "optimization of geometrically nonlinear structures under stability constraint," which is an exciting and fast-growing branch of application of structural and mechanical engineering, and also necessarily involves applied mathematics. The premise of this book is that a thorough and profound knowledge of nonlinear buckling behaviors is crucial, via proper problem setting, as a step toward the successful design of the best structure.

Some optimized structures are shown to be safe, and readers are encouraged to carry out optimization-based design with confidence.

In Part I, design sensitivity analysis and imperfection sensitivity analysis are introduced as systematic tools to perform stability design of structures. The influence of design parameters on structural performance is to be expressed as parameter sensitivity. Design sensitivity analysis is implemented into the gradient-based algorithm for structural optimization in Part II. Imperfection sensitivity laws are introduced to evaluate the influence of initial imperfections on buckling loads quantitatively. In this book, design sensitivity analysis and imperfection sensitivity analysis, which have been addressed independently up to now, are described in a synthetic manner based on the general theory of elastic stability [166]. This theory, which once was an established means to describe the buckling of structures, is thus given a new role in the computer age. Part I is organized as follows.

- The overview of design sensitivity analysis and its theoretical backgrounds are presented in Chapter 1.
- Numerical methods of design sensitivity analysis are provided in Chapter 2.
- Imperfection sensitivity analysis is presented in the framework of modern stability theory in Chapter 3.

In Part II, based on the synthetic description of sensitivity analyses presented in Part I, we introduce state-of-the-art optimization methodologies of geometrically nonlinear finite-dimensional structures under stability constraints. These optimization methodologies are reinforced on the one hand by the stability theory and on the other hand by finite element method and mathematical programming with ever-increasing computing power. Design of compliant mechanisms is highlighted as an engineering application of shape and topology optimization with extensive utilization of snapthrough buckling. Part II is organized as follows.

- In Chapter 4, general formulation for optimization under stability constraints is provided. An optimized truss dome is shown to be less imperfection-sensitive than a non-optimal one.
- Optimal structures with snapthrough are investigated in Chapter 5 to pave the way for shape design of compliant mechanisms using snapthrough behavior in Chapter 6.
- Optimal frames with coincident buckling loads are investigated in Chapter 7.
- Imperfection sensitivity of hilltop branching points with simple, multiple, and degenerate bifurcation points are investigated in Chapters 8–10.

In Part III, in order to ensure the performance of optimized structures, we introduce two design methodologies:

- optimization via the worst imperfection, and
- probabilistic analysis via random imperfections.

In particular, imperfection sensitivity laws are extended to be applicable to many imperfection variables and, in turn, to deal with the probabilistic variation of the buckling loads of structures. Part III is organized as follows.

- The asymptotic theory on the worst imperfection is formulated in Chapter 11.
- An anti-optimization problem is formulated in Chapter 12 to minimize the lowest eigenvalue of the tangent stiffness matrix, and a design methodology is presented for a laterally braced frame.
- The worst imperfection is defined and investigated for a stable-symmetric bifurcation point in Chapter 13.
- The theory on random imperfections is presented in Chapter 14, and is applied to steel specimens with hilltop branching in Chapter 15.
- The theory is extended to the second-order imperfections in Chapter 16.

In the Appendix, derivations of several formulations and details of numerical examples are presented. In particular, the derivation of imperfection sensitivity laws by the power series expansion method is an important ingredient for readers who are interested in stability theory.

This book consequently offers a wide and profound insight into optimization-based and computer-assisted stability design of finite-dimensional structures in a readable and illustrative form for graduate students of engineering and applied mathematicians. General methodology is emphasized instead of studies of particular structures. Historical developments are outlined with many references to assist readers' further studies.

The authors are grateful to Dr. J. S. Arora for his support of an optimization program. The suggestion of Dr. K. K. Choi was vital for the publication of this book. The authors thank for the comments of Drs. K. Murota and Y. Kanno. For the realization of this book, the authors owe much to Drs. K. Uetani, K. Terada, S. Okazawa, S. Nishiwaki, J. Takagi, K. Oide, and Mr. J. Y. Zhang. The support of C. Simpson, E. Tham and K. Stanne was indispensable for the publication of this book. The authors conclude the preface with many thanks to Dr. I. Elishakoff for his encouragement.

February 2007

Makoto Ohsaki
Kiyohiro Ikeda

Contents

II Optimization Methods for Stability Design 59

Part I: Generalized Sensitivity of Nonlinear Elastic Systems

1

Introduction to Design Sensitivity Analysis

1.1 Introduction

Sensitivity analysis is conducted to evaluate the dependence of structural performances on design or imperfection parameters. As stated in the Preface, dependent on parameters to be employed, sensitivity analysis in structural stability can be classified as follows:

- In the design sensitivity analysis, employed as parameters are design variables, such as member stiffnesses and geometrical variables. The sensitivity (differential) coefficients of structural responses, such as displacements, stresses and buckling loads, with respect to these parameters are obtained. These coefficients, in turn, are put to use in gradient-based optimization algorithms [223, 227, 262].
- In the imperfection sensitivity analysis for structures subjected to buckling, employed as parameters are initial imperfections, such as errors in manufacturing process and material defects [20, 285].

Different notations and terminologies are used in the two sensitivity analyses presented above for the same physical properties. Because the design sensitivity and imperfection sensitivity are mathematically equivalent, especially for a limit point load, the two sensitivity analyses are formulated in Part I in a synthetic manner under the name of "parameter sensitivity." Part I serves as a theoretical foundation of the optimization methodologies presented in Part II.

Many branches of parameter sensitivity analysis exist, as shown in Fig. 1.1, and methods of sensitivity analysis to be employed vary with these branches. Sensitivity analysis is classified dependent on whether the governing equation is linear or nonlinear. Nonlinear sensitivity analysis is further classified according

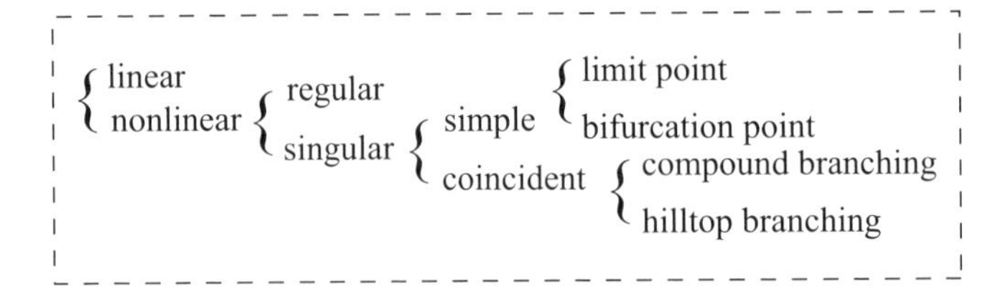

Fig. 1.1 Branches of parameter sensitivity analysis.

to whether the structure in question is in a regular state or in a singular state. Nonlinear sensitivity analysis for a singular state has sub-branches that are dependent on whether the singular state is associated with a limit point, a simple bifurcation point, a coincident critical point, and so on.

As a prologue to the main contents of this part, Chapters 2 and 3, we present in this chapter general framework of elastic stability to provide fundamental tools for design sensitivity analysis to compute the differential coefficients of structural responses with respect to the design parameters. We introduce sensitivity analysis for simple cases, including:

- linear elastic response,
- linear buckling load as a weakly nonlinear case,
- regular state for nonlinear response, and
- limit point load for nonlinear response.

This chapter is organized as follows. General framework of elastic stability is introduced in Section 1.2. Design parameterization is presented in Section 1.3. A simple introduction to design sensitivity analysis for linear and nonlinear responses is given in Sections 1.4 and 1.5, respectively. A historical review of studies on design sensitivity analysis is presented in Section 1.6.

1.2 General Framework of Elastic Stability

General framework of elastic stability of a conservative system that has the total potential energy is briefly introduced to define variables, basic equations, and critical points.

1.2.1 Governing equations and stability

Consider an elastic structure discretized by the finite element method and subjected to quasi-static nodal loads, which are parameterized by a load parameter or load factor Λ. The deformation of the structure is expressed in terms of an n-dimensional nodal displacement vector $\mathbf{U} = (U_i) \in \mathbb{R}^n$. The stationary condition of the total potential energy[1] $\Pi(\mathbf{U}, \Lambda)$ at a specified value of Λ leads to the following set of equilibrium equations:

$$S_{,i} = 0, \quad (i = 1, \ldots, n) \tag{1.1}$$

where

$$S_{,i} \equiv \frac{\partial \Pi}{\partial U_i}, \quad (i = 1, \ldots, n) \tag{1.2}$$

is the partial differentiation of Π with respect to U_i. The set of equilibrium points in $(\mathbf{U}, \Lambda)$-space is called an *equilibrium path*, and a path that contains the undeformed initial state is called the *fundamental equilibrium path* or the fundamental path.

In the description of stability, we refer to the tangent stiffness matrix or the stability matrix

$$\mathbf{S} = [S_{,ij}] \tag{1.3}$$

where $S_{,ij} \equiv \partial \Pi^2 / \partial U_i \partial U_j$ $(i, j = 1, \ldots, n)$. Note that $\mathbf{S} \in \mathbb{R}^{n \times n}$ is symmetric owing to existence of potential, i.e.,

$$S_{,ij} = S_{,ji}, \quad (i, j = 1, \ldots, n) \tag{1.4}$$

The rth eigenvalue λ_r and the associated eigenvector $\mathbf{\Phi}_r = (\phi_{ri}) \in \mathbb{R}^n$ of $\mathbf{S}$ are defined by

$$\sum_{j=1}^{n} S_{,ij} \phi_{rj} = \lambda_r \phi_{ri}, \quad (i = 1, \ldots, n) \tag{1.5}$$

Since $\mathbf{S}$ is symmetric, all eigenvalues of $\mathbf{S}$ are real, and are ordered such that

$$\lambda_1 \leq \lambda_2 \leq \cdots \leq \lambda_n \tag{1.6}$$

The eigenvectors $\mathbf{\Phi}_1, \ldots, \mathbf{\Phi}_n$ are ortho-normalized by

$$\sum_{i=1}^{n} \phi_{ri} \phi_{si} = \delta_{rs}, \quad (r, s = 1, \ldots, n) \tag{1.7}$$

where δ_{rs} is the Kronecker delta, being equal to 1 for $r = s$ and 0 for $r \neq s$.

The stability of an equilibrium state is classified as

$$\begin{cases} \textit{stable:} & \text{all eigenvalues of } \mathbf{S} \text{ are positive } (\lambda_1 > 0) \\ \textit{unstable:} & \text{at least one eigenvalue of } \mathbf{S} \text{ is negative } (\lambda_1 < 0) \end{cases} \tag{1.8}$$

Thus the stability depends on the sign of λ_1.

1.2.2 Critical state

A critical state is defined as an equilibrium state at which at least one eigenvalue is zero. The first critical state on the fundamental path that is defined by $\lambda_1 = 0$ is of most engineering interest as λ_1 defines the stability by (1.8).

[1] The total potential energy function $\Pi(\mathbf{U}, \Lambda)$ is assumed to be sufficiently smooth. The contact problem, for example, is out of scope of this book.

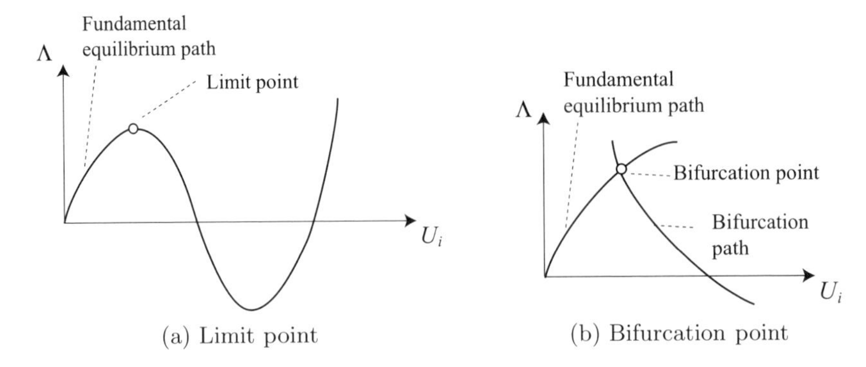

(a) Limit point (b) Bifurcation point

Fig. 1.2 Equilibrium paths for simple critical points.

Load factor, eigenvalue and eigenvector at the critical state are denoted by Λ^{c}, λ_r^{c} and $\mathbf{\Phi}_r^{\mathrm{c}}$, respectively. The superscript $(\,\cdot\,)^{\mathrm{c}}$ indicates a value at a critical state throughout this book.

Critical points are classified as follows:

- **simple critical point**: a critical point is called *simple* or *distinct* if only one eigenvalue λ_r becomes zero. A simple critical point is classified to a *limit point* and a *bifurcation point.*
- **coincident critical point**: a critical point with m (≥ 2) zero eigenvalues is called *coincident*[2]. The coincident critical point consisting of several bifurcation points is called a *compound branching point.* The coincident point with bifurcation point(s) at a limit point is called a *hilltop branching point.*

A more detailed classification of critical points will be given in Section 3.3.

Equilibrium paths for simple critical points are illustrated in Fig. 1.2 in terms of the relation between a displacement U_i and Λ. At a limit point, Λ reaches a maximum or minimum in the $(\mathbf{U}, \Lambda)$-space, while the uniqueness of the equilibrium state is lost at a bifurcation point.

Remark 1.2.1 An equilibrium path can be traced by the load control method, for which Λ is chosen as the path parameter t. Namely, the equilibrium equations (1.1) are solved for a given $t = \Lambda$. However, the path near the limit point should be traced by the displacement increment method or the arc-length method [255], for which the path parameter $t = t(\mathbf{U}, \Lambda)$ is generally defined as a function of $\mathbf{U}$ and Λ. □

Remark 1.2.2 Many numerical methods for the computation of critical points are available [66, 90, 156, 311, 312]. For example, a simple critical point can be pinpointed accurately by iteratively solving the extended system formulated by (1.1) and (1.5) with (1.7) for $r = 1$, $\lambda_1 = 0$ considering $\mathbf{U}$, $\mathbf{\Phi}_1$ and Λ as variables. In numerical examples of this book, critical points are computed within

[2] *Coincident*, *compound*, *multiple* and *repeated* have a similar meaning, and *coincident* is mainly used in this book.

sufficiently good accuracy so as not to influence the accuracy of sensitivity coefficients. □

1.2.3 Proportional loading

For a proportional loading, most common in structural analysis, the nodal load vector $\mathbf{P} \in \mathbb{R}^n$ is given as the product of the load factor Λ and the specified load pattern vector $\mathbf{p} = (p_i) \in \mathbb{R}^n$, namely,

$$\mathbf{P} = \Lambda \mathbf{p} \tag{1.9}$$

Then the total potential energy $\Pi(\mathbf{U}, \Lambda)$ is given as

$$\Pi(\mathbf{U}, \Lambda) = H(\mathbf{U}) - \Lambda \mathbf{p}^\top \mathbf{U} \tag{1.10}$$

where $H(\mathbf{U})$ denotes the strain energy corresponding to the deformation $\mathbf{U}$ and $(\,\cdot\,)^\top$ means the transpose of the associated vector or matrix. The equilibrium equations are written as

$$S_{,i} = H_{,i} - \Lambda p_i = 0, \quad (i = 1, \dots, n) \tag{1.11}$$

where $H_{,i}$ is the partial differentiation of H with respect to U_i, and expresses the equivalent nodal load in the direction of U_i. From (1.11), we have $\mathbf{S} = [H_{,ij}]$.

1.3 Design Parameterization

We define a design parameter in structural optimization (or an imperfection parameter in stability analysis). Let $\mathbf{v} \in \mathbb{R}^\nu$ denote a ν-dimensional vector representing the mechanical properties of the structure. The vector $\mathbf{v}$ may correspond, e.g., to

- sizing design parameter such as cross-sectional areas and plate thickness,
- shape design parameter such as nodal coordinates, and
- material property design parameter such as Young's modulus and Poisson's ratio.

It is possible to choose topology (member location) and external loads also as design parameters. Among many design parameters, we focus mainly on cross-sectional areas of members and nodal coordinates in this book. See [49] for more issues on design parameters.

Indicate by $\mathbf{v} = \mathbf{v}^0$ the reference (perfect) structure. The superscript $(\,\cdot\,)^0$ denotes a variable associated with the reference structure throughout this book. Consider a modified structure, being defined by *design variation vector* $\mathbf{d}_i \in \mathbb{R}^\nu$ $(i = 1, \dots, \rho)$ and the associated scaling *design parameter* ξ_i $(i = 1, \dots, \rho)$ as

$$\mathbf{v} = \mathbf{v}^0 + \sum_{i=1}^{\rho} \xi_i \mathbf{d}_i \tag{1.12}$$

Dependent on problem formulation, we often use a single scaling parameter ξ and employ a simplified expression

$$\mathbf{v} = \mathbf{v}^0 + \xi \sum_{i=1}^{\rho} \mathbf{d}_i \tag{1.13}$$

Remark 1.3.1 In stability analysis, $\mathbf{v}$ is called the *imperfection parameter* vector, $\mathbf{v}^0$ corresponds to the *perfect structure*, $\mathbf{d}_i$ is the *imperfection pattern vector* and ξ_i is the *imperfection parameter*. The use of multiple vectors $\mathbf{d}_i$ is vital generalization in the development of the *worst imperfection* in Chapter 11, and the *probabilistic variation of critical loads* in Chapters 14 and 15. □

We simply call ξ_i the *parameter* and present a unified formulation for design sensitivity and imperfection sensitivity. Our task of sensitivity analysis is to quantitatively evaluate the differential coefficients of responses with respect to ξ_i.

For simplicity, in the remainder of this chapter, we consider only a single pair of parameter ξ_1 and vector $\mathbf{d}_1$, and set $\xi_1 = \xi$, $\mathbf{d}_1 = \mathbf{d}$. Then (1.12) reduces to

$$\mathbf{v} = \mathbf{v}^0 + \xi \mathbf{d} \tag{1.14}$$

Accordingly, the total potential energy is written as $\Pi(\mathbf{U}, \Lambda, \xi)$.

In the sensitivity analysis, the *total differentiation* and *partial differentiation* with respect to ξ should be distinguished and denoted by $(\cdot)'$ and $(\cdot)_{,\xi}$, respectively. For example, in the total differentiation, $\mathbf{U}$ is to be conceived as a function of ξ, as $\mathbf{U}$ for a fixed value of Λ is obtained by solving the equilibrium equation (1.1). The total differentiation $\mathbf{U}'$ of $\mathbf{U}(\xi)$ with respect to ξ is found by total differentiation of the equilibrium equation. The argument ξ will be often suppressed in this chapter, for simplicity.

1.4 Design Sensitivity Analysis for Linear Response

Consider a linear response of a structure, and denote by $\mathbf{K}_{\mathrm{L}}(\xi) \in \mathbb{R}^{n \times n}$ the linear (infinitesimal) stiffness matrix, which is a function of ξ. The total potential energy is given as

$$\Pi(\mathbf{U}, \Lambda, \xi) = \frac{1}{2}\mathbf{U}^\top \mathbf{K}_{\mathrm{L}}(\xi)\mathbf{U} - \Lambda \mathbf{p}(\xi)^\top \mathbf{U} \tag{1.15}$$

Partial differentiation of $\Pi(\mathbf{U}, \Lambda, \xi)$ with respect to $\mathbf{U}$, with the use of symmetry of $\mathbf{K}_{\mathrm{L}}$, leads to the equilibrium equation

$$\mathbf{K}_{\mathrm{L}}(\xi)\mathbf{U} - \Lambda \mathbf{p}(\xi) = \mathbf{0} \tag{1.16}$$

Here $\mathbf{U}$ is to be obtained as a function of ξ as a solution to (1.16). Total differentiation of (1.16) with respect to ξ gives

$$\mathbf{K}_{\mathrm{L}}\mathbf{U}' + \mathbf{K}_{\mathrm{L}}'\mathbf{U} - \Lambda \mathbf{p}' = \mathbf{0} \implies \mathbf{K}_{\mathrm{L}}\mathbf{U}' = -\mathbf{K}_{\mathrm{L}}'\mathbf{U} + \Lambda \mathbf{p}' \tag{1.17}$$

In the direct differentiation method [115], the sensitivity coefficient vector $\mathbf{U}'$ is directly computed by (1.17), as $\mathbf{U} = \Lambda(\mathbf{K}_{\mathrm{L}})^{-1}\mathbf{p}$ can be computed from (1.16) and $\mathbf{p}'$ and $\mathbf{K}_\mathrm{L}'$ can be easily obtained with explicit dependence of $\mathbf{p}$ and $\mathbf{K}_\mathrm{L}$ on ξ. For example, if $\mathbf{p}$ is the self-weight of a truss and ξ corresponds to the cross-sectional area, $\mathbf{p}$ and $\mathbf{K}_\mathrm{L}$ are explicit linear functions of ξ.

Remark 1.4.1 The second equation in (1.17) has the same form as (1.16) if the right-hand-side terms are regarded as nodal loads; therefore, it is computationally efficient to factorize $\mathbf{K}_\mathrm{L}$ in the process of solving (1.16) for $\mathbf{U}$. Note that the *adjoint variable method* is more computationally efficient if we have many design parameters and few response quantities, sensitivity coefficients for which are to be evaluated [49, 162]. □

Remark 1.4.2 Upon obtaining $\mathbf{U}$ and its sensitivity coefficient $\mathbf{U}'$, we can compute the sensitivity coefficients of stresses and strains in a straightforward manner. For example, for a truss with n^m members, the relation between $\mathbf{U}$ and the axial force vector $\mathbf{N} \in \mathbb{R}^{n^\mathrm{m}}$ can be written as

$$\mathbf{N} = \mathbf{D}\mathbf{U} \tag{1.18}$$

where $\mathbf{D} \in \mathbb{R}^{n^\mathrm{m} \times n}$ is generally a function of ξ. Accordingly, the sensitivity coefficients of $\mathbf{N}$ can be obtained from

$$\mathbf{N}' = \mathbf{D}'\mathbf{U} + \mathbf{D}\mathbf{U}' \tag{1.19}$$

□

1.5 Design Sensitivity Analyses for Nonlinear Responses

Design sensitivity analyses for nonlinear responses, including a linear buckling load and a limit point load, are presented.

1.5.1 Linear buckling load

We start with the sensitivity analysis of a linear buckling load, which is weakly nonlinear. Consider a structure subjected to the proportional load $\mathbf{P} = \Lambda\mathbf{p}$ of (1.9) with sufficiently small prebuckling deformation $\mathbf{U} \simeq \mathbf{0}$. The tangent stiffness matrix $\mathbf{S}$ is expressed as the sum of the linear stiffness matrix $\mathbf{K}_\mathrm{L}$, which does not depend on deformation, and the geometrical stiffness matrix $\mathbf{K}_\mathrm{G} \in \mathbb{R}^{n \times n}$, which is a function of $\mathbf{U}$ through the internal forces or stresses at the current (reference) state.

For the small deformation $\mathbf{U} \simeq \mathbf{0}$, the internal forces or stresses can be assumed to be proportional to Λ, and $\mathbf{K}_\mathrm{G}$ is given as the product of Λ and a constant matrix $\mathbf{K}_{\mathrm{G}0} \in \mathbb{R}^{n \times n}$, which is a function of stress under the given load $\mathbf{p}$, namely,

$$\mathbf{K}_\mathrm{G} = \Lambda \mathbf{K}_{\mathrm{G}0} \tag{1.20}$$

The tangent stiffness matrix $\mathbf{S}$ becomes

$$\mathbf{S} = \mathbf{K}_\mathrm{L} + \Lambda \mathbf{K}_{\mathrm{G}0} \tag{1.21}$$

In this case, the criticality condition for eigenvalue $\lambda_r = 0$ in (1.5) is expressed as

$$[\mathbf{K}_{\mathrm{L}} + \Lambda_{\mathrm{L}r}\mathbf{K}_{\mathrm{G0}}]\mathbf{\Phi}_r = \mathbf{0}, \quad (r = 1, \dots, n) \tag{1.22}$$

where $\Lambda_{\mathrm{L}r}$ is the rth linear buckling load factor, and eigenvector $\mathbf{\Phi}_r$ is normalized by $\mathbf{K}_{\mathrm{L}}$, which is positive definite, namely,

$$\mathbf{\Phi}_r^\top \mathbf{K}_{\mathrm{L}}\mathbf{\Phi}_r = 1, \quad (r = 1, \dots, n) \tag{1.23}$$

Then the following relationship holds from (1.22) and (1.23) at a stable initial state with $\Lambda_{\mathrm{L}r} \neq 0$:

$$\mathbf{\Phi}_r^\top \mathbf{K}_{\mathrm{G0}}\mathbf{\Phi}_r = -\frac{1}{\Lambda_{\mathrm{L}r}}, \quad (r = 1, \dots, n) \tag{1.24}$$

Eq. (1.22) defines a generalized eigenvalue problem, which yields n-eigenpairs of $\Lambda_{\mathrm{L}r}$ and $\mathbf{\Phi}_r$. The smallest positive eigenvalue[3] $\Lambda_{\mathrm{L}r}$ is called the *linear buckling load factor*, and the corresponding $\mathbf{\Phi}_r$ is called the *linear buckling mode*. The critical load factor can be approximated by the linear buckling load factor with good accuracy if the assumption $\mathbf{U} \simeq \mathbf{0}$ on prebuckling deformation is satisfied.

Total differentiation of (1.22) and (1.23) with respect to ξ gives, respectively,

$$\mathbf{K}_{\mathrm{L}}'\mathbf{\Phi}_r + \mathbf{K}_{\mathrm{L}}\mathbf{\Phi}_r' + \Lambda_{\mathrm{L}r}'\mathbf{K}_{\mathrm{G0}}\mathbf{\Phi}_r + \Lambda_{\mathrm{L}r}\mathbf{K}_{\mathrm{G0}}'\mathbf{\Phi}_r + \Lambda_{\mathrm{L}r}\mathbf{K}_{\mathrm{G0}}\mathbf{\Phi}_r' = \mathbf{0} \tag{1.25}$$

$$\mathbf{\Phi}_r^\top \mathbf{K}_{\mathrm{L}}'\mathbf{\Phi}_r + 2\mathbf{\Phi}_r^\top \mathbf{K}_{\mathrm{L}}\mathbf{\Phi}_r' = 0 \tag{1.26}$$

for $r = 1, \dots, n$. Sensitivity coefficients $\Lambda_{\mathrm{L}r}'$ and $\mathbf{\Phi}_r'$ can be obtained by solving the set of $n+1$ simultaneous linear equations (1.25) and (1.26). Since $\mathbf{K}_{\mathrm{G0}}$ is a function of stresses under the load $\mathbf{p}$, the sensitivity coefficients of the stresses are required to compute $\mathbf{K}_{\mathrm{G0}}'$ in (1.25) (cf., Remark 1.5.1).

Premultiplying $\mathbf{\Phi}_r^\top$ to both sides of (1.25) and using symmetry of $\mathbf{K}_{\mathrm{L}}$ and $\mathbf{K}_{\mathrm{G0}}$, we obtain

$$\mathbf{\Phi}_r^\top [\mathbf{K}_{\mathrm{L}}' + \Lambda_{\mathrm{L}r}\mathbf{K}_{\mathrm{G0}}']\mathbf{\Phi}_r + \mathbf{\Phi}_r'^\top [\mathbf{K}_{\mathrm{L}} + \Lambda_{\mathrm{L}r}\mathbf{K}_{\mathrm{G0}}]\mathbf{\Phi}_r + \Lambda_{\mathrm{L}r}'\mathbf{\Phi}_r^\top \mathbf{K}_{\mathrm{G0}}\mathbf{\Phi}_r = 0 \tag{1.27}$$

By (1.22) and (1.24), (1.27) gives

$$\Lambda_{\mathrm{L}r}' = \Lambda_{\mathrm{L}r}\mathbf{\Phi}_r^\top [\mathbf{K}_{\mathrm{L}}' + \Lambda_{\mathrm{L}r}\mathbf{K}_{\mathrm{G0}}']\mathbf{\Phi}_r \tag{1.28}$$

Hence, $\Lambda_{\mathrm{L}r}'$ can be found from (1.28) by simple matrix computation if the sensitivity coefficient $\mathbf{\Phi}_r'$ of $\mathbf{\Phi}_r$ is not needed.

Remark 1.5.1 The evaluation of $\mathbf{K}_{\mathrm{G0}}'$ is the most difficult process in the computation of the sensitivity coefficients $\Lambda_{\mathrm{L}r}'$ in (1.28). For example, for a truss with n^{m} members with a parameter ξ defining the cross-sectional areas, $\mathbf{K}_{\mathrm{G0}}'$ is evaluated to

$$\mathbf{K}_{\mathrm{G0}}' = \sum_{j=1}^{n^{\mathrm{m}}} \frac{\partial \mathbf{K}_{\mathrm{G0}}}{\partial N_j} N_j' \tag{1.29}$$

[3] The negative eigenvalues correspond to the buckling loads against the proportional load in the opposite direction. However, in customary linear buckling analysis, the directions of the loads are fixed and Λ is assumed to be positive.

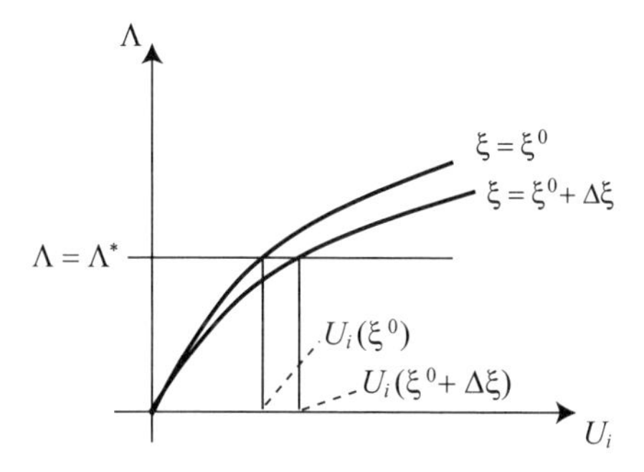

Fig. 1.3 Variation of an equilibrium path with respect to parameter modification $\xi = \xi^0 \longrightarrow \xi^0 + \Delta\xi$.

where N_j is the axial force of the jth member, and its sensitivity coefficient N_j' can be computed by (1.19) after obtaining $\mathbf{U}'$ by (1.17). □

1.5.2 Responses at a regular state

Consider a regular state of a nonlinear response. Since the critical load cannot be defined for the regular state, design sensitivity of displacements and stresses for a fixed load $\Lambda = \Lambda^*$, instead of the critical load, is investigated.

Consider an elastic structure satisfying

- the structure is in a regular state at a load level Λ^*, where the tangent stiffness matrix $\mathbf{S}$ is nonsingular, and
- the displacements and stresses are monotonically increasing functions of Λ.

Fig. 1.3 illustrates the variation of the equilibrium path with respect to the modification of the parameter ξ from ξ^0 to $\xi^0 + \Delta\xi$. In this case, sensitivity coefficients $\mathbf{U}'$ of $\mathbf{U}$ at $\xi = \xi^0$ can be obtained based only on the response at the load level $\Lambda = \Lambda^*$ without incrementally updating or accumulating $\mathbf{U}'$ along the fundamental path [45].

Since the total potential energy $\Pi(\mathbf{U}, \Lambda, \xi)$ is a function of $\mathbf{U}$, Λ and ξ, the total and partial differentiations of Π should be properly distinguished. The dependence of $\mathbf{U}$ on ξ is implicitly defined by the equilibrium equation (1.1). Therefore, the design sensitivity U_j' is to be found by total differentiation of (1.1) with respect to ξ for a fixed value of $\Lambda = \Lambda^*$ as

$$S_{,i}' \equiv \sum_{j=1}^{n} S_{,ij} U_j' + S_{,i\xi} = 0, \quad (i = 1, \dots, n) \tag{1.30}$$

where in the partial differentiation

$$S_{,i\xi} = \frac{\partial S_{,i}(\mathbf{U}, \Lambda, \xi)}{\partial \xi} \tag{1.31}$$

the implicit dependence of $\mathbf{U}$ on ξ is not considered. Note that the sensitivity coefficient $\mathbf{U}'$ cannot be obtained from (1.30) at a critical state where $\mathbf{S} = [S_{,ij}]$ is singular.

For a proportionally loaded structure satisfying (1.11), (1.30) reduces to

$$\sum_{j=1}^{n} H_{,ij} U'_j = \Lambda p'_i - H_{,i\xi}, \quad (i = 1, \ldots, n) \tag{1.32}$$

Here $H = H(\mathbf{U}(\xi), \xi)$ implicitly depends on ξ through $\mathbf{U}$.

At the course of numerical path-tracing analysis, U'_j can be computed from (1.30) at minimal additional cost as follows (cf., [213, 262] for details):

- The tangent stiffness matrix $\mathbf{S} = [S_{,ij}]$ has already been computed during the path-tracing analysis.
- Since $S_{,i}$ has also been computed to obtain residual (unbalanced) forces and the dependence of $S_{,i}$ on ξ is known, $S_{,i\xi}$ can be computed with ease.
- Since $\mathbf{S}$ has been factorized at $\Lambda = \Lambda^*$, (1.30) can be solved with minimal additional computational cost.

Notice that the sensitivity equation (1.30) or (1.32) at a regular state has a similar form as (1.17) for the geometrically linear problem in Section 1.4; i.e., the linear stiffness matrix $\mathbf{K}_{\mathrm{L}}$ is to be replaced by the tangent stiffness matrix $\mathbf{S}$.

The sensitivity coefficients of eigenvalue λ_r and eigenvector $\mathbf{\Phi}_r$ of $\mathbf{S}$ are obtained from the following equations, which are derived by total differentiation of (1.5) and (1.7):

$$\sum_{j=1}^{n} \left(\sum_{k=1}^{n} S_{,ijk} \phi_{rj} U'_k + S_{,ij\xi} \phi_{rj} + S_{,ij} \phi'_{rj} \right) = \lambda_r \phi'_{ri} + \lambda'_r \phi_{ri}, \quad (i = 1, \ldots, n) \tag{1.33}$$

$$\sum_{j=1}^{n} \phi_{rj} \phi'_{rj} = 0 \tag{1.34}$$

Note that we have $n+1$ equations for $n+1$ unknowns λ'_r and ϕ'_{ri} $(i = 1, \ldots, n)$ for each r. By multiplying ϕ_{ri} to (1.33), summing up by i and using (1.5) and (1.7), we obtain

$$\lambda'_r = \sum_{i=1}^{n} \sum_{j=1}^{n} \left(\sum_{k=1}^{n} S_{,ijk} \phi_{ri} \phi_{rj} U'_k + S_{,ij\xi} \phi_{ri} \phi_{rj} \right) \tag{1.35}$$

If ϕ'_{ri} is not needed, λ'_r can be obtained from (1.35) by arithmetical operation, after computing U'_i by (1.30).

1.5.3 Limit point load

Sensitivity coefficient $\Lambda^{\mathrm{c}\prime}$ of a limit point load can be found only from the equilibrium equations (1.1) in a similar manner as the case of the regular state presented in Section 1.5.2.

By considering Λ^{c} as a function of ξ, and by differentiating the equilibrium equations (1.1) with respect to ξ at the limit point load $\Lambda = \Lambda^{\mathrm{c}}$, we obtain

$$\sum_{j=1}^{n} S_{,ij} U_j' + S_{,i\Lambda} \Lambda^{\mathrm{c}\prime} + S_{,i\xi} = 0, \quad (i = 1, \ldots, n) \tag{1.36}$$

where $(\,\cdot\,)_{,\Lambda}$ denotes partial differentiation with respect to Λ.

Premultiplying the critical eigenmode ϕ_{1i}^{c} to the both sides of (1.36), taking summation over i, and using $\lambda_1 = 0$ for (1.5), we can obtain the sensitivity coefficient of the limit point load

$$\Lambda^{\mathrm{c}\prime} = -\sum_{i=1}^{n} S_{,i\xi} \phi_{1i}^{\mathrm{c}} \Big/ \sum_{i=1}^{n} S_{,i\Lambda} \phi_{1i}^{\mathrm{c}} \tag{1.37}$$

By (1.37), the sensitivity coefficient $\Lambda^{\mathrm{c}\prime}$ can be found directly without resort to the sensitivity coefficients $\mathbf{U}'$ of displacements and $\mathbf{\Phi}_1^{\mathrm{c}\prime}$ of the eigenmode. Since $\Lambda^{\mathrm{c}\prime}$ in (1.37) does not depend on ξ, the critical load $\Lambda^{\mathrm{c}}(\xi)$ of an imperfect system is written as a linear function of ξ as

$$\Lambda^{\mathrm{c}}(\xi) = \Lambda^{\mathrm{c}}(0) + \Lambda^{\mathrm{c}\prime} \xi \tag{1.38}$$

Remark 1.5.2 Eq. (1.37) in principle agrees with preexisting results of stability analysis of elastic conservative systems based on the derivatives with respect to the displacement in the direction of $\mathbf{\Phi}_1^{\mathrm{c}}$ (cf., [283, 285] and Section 3.5.2). However, we prefer formulations based on physical coordinates, as employed in (1.37), so as to be consistent with the conventional finite element analysis. □

Remark 1.5.3 The denominator of (1.37) vanishes at a bifurcation point (cf., (3.25) and (3.27) in Section 3.3); therefore, (1.37) is not extendable to a bifurcation point [239]. It should also be noted that the sensitivity equation presented in this section does not depend on the symmetry of the imperfection unlike the case of bifurcation loads (cf., Section 2.2). □

Remark 1.5.4 For a proportional loading, (1.36) and (1.37) are to be rewritten by using the following formulas (cf., (1.11)):

$$S_{,i\Lambda} = -p_i, \quad S_{,i\xi} = H_{,i\xi} - \Lambda p_i', \quad S_{,ij} = H_{,ij} \tag{1.39}$$

□

1.6 Historical Development

Design sensitivity analysis and shape sensitivity analysis were initiated mainly in the field of structural optimization [49, 115], and established works on design sensitivity analysis of linear elastic responses, eigenvalues of vibration, and so on, are at hand. In the 1980's, geometrically nonlinear formulations were developed for optimization against buckling of simple structures [150, 158, 249, 316].

For structures exhibiting limit point instability, the algorithm for design sensitivity analysis was developed [239, 314, 316], and the optimum design for a specified nonlinear buckling load factor was conducted [150, 158, 179, 249, 316]. Mathematical equivalence between design sensitivity and imperfection sensitivity was recognized [225]. The formulas for imperfection sensitivity coefficients were implemented into optimality conditions for problems under constraint on a limit point load, and the optimization for nonlinear buckling was studied [217, 232, 233, 254]. Thus general theory of elastic stability can effectively be used in design sensitivity analysis for structural optimization.

A semi-analytical approach for sensitivity analysis of critical loads, and an optimization algorithm implementing design modifications were presented [253]. Sensitivity analysis of critical loads based on asymptotic approach was also presented, where computation of the third-order derivatives of the total potential energy was necessitated for bifurcation loads [202]. Examples of optimum design with limit points were studied [203]. An approach considering postbuckling states was presented [97].

Several numerical approaches to design sensitivity analyses for bifurcation loads were developed in the framework of finite element analysis. For example, sensitivity analyses of bifurcation loads were conducted by an interpolation approach [227], and an approach using linear eigenvalue analysis [214, 316].

Another branch of sensitivity analysis of nonlinear response is found in elastoplastic continuum mechanics, and formulations suitable for computational implementation were presented [162, 177, 214, 262, 305]. However, there are only a few studies on design sensitivity analysis of elastoplastic critical loads [215].

1.7 Summary

In this chapter,

- general frameworks of elastic stability of conservative systems have been presented, and
- design sensitivity analyses for linear and nonlinear responses have been classified and briefly been introduced.

The major findings of this chapter are as follows.

- The design sensitivity analyses presented herein are expressed by simple formulas. Yet these formulas contain the essence of design and imperfection sensitivity analyses of buckling loads to be introduced in the following chapters.
- Design sensitivity analysis for regular states can be carried out easily based on the response quantities at the final load level. However, sensitivity coefficients at the critical states cannot be obtained similarly, because the tangent stiffness matrix is singular at the critical point.
- The sensitivity coefficient of a limit point load can be obtained simply by differentiation of the equilibrium equations.

2

Methods of Design Sensitivity Analysis

2.1 Introduction

The design sensitivity analysis for the critical state at a limit point permits simple formulation, as we have seen in Chapter 1. A straightforward application of this formulation to a bifurcation point encounters a problem in that the design (imperfection) sensitivity coefficient of the load factor generally is unbounded, as cautioned in Remark 1.5.3.

A hint to resolve this problem can be found in the fact that bifurcation takes place in general in a symmetric structure subjected to a set of symmetric loads [128, 137]. Roorda [258] classified initial imperfections into minor (second-order) imperfections preserving symmetry and major (first-order) imperfections breaking symmetry, and developed asymptotic forms of the critical loads of the imperfect or modified systems. The formulation of sensitivity for minor imperfections by Roorda [258], which utilized Liapunov–Schmidt–Koiter reduction (cf., Section 3.2.3), serves as a theoretical background of design sensitivity analysis, but is not convenient for numerical implementation.

For the sensitivity analysis of a bifurcation load, the relative importance of minor and major modifications (imperfections) changes dependent on the purposes of analyses.

- For sensitivity analysis to be employed in the gradient-based optimization algorithm, it is customary to preserve the symmetry of a structure and consider only minor modifications preserving symmetry.

- For stability analysis, major modifications breaking symmetry are of great importance, as such imperfections are usually more influential on buckling strength than minor imperfections.

The design sensitivity analysis for minor modifications is studied in this chapter compatibly with the framework of the finite element analysis, whereas the imperfection sensitivity analysis for major modifications will be studied in Chapter 3. The results in structural *stability* are applicable to structural *optimization*, and vice versa, by interpreting

$$\left\{\begin{array}{lcl}\text{perfect system} & \longleftrightarrow & \text{reference system}\\ \text{imperfect system} & \longleftrightarrow & \text{modified system}\\ \text{imperfection sensitivity} & \longleftrightarrow & \text{design sensitivity}\\ \text{initial imperfection} & \longleftrightarrow & \text{design modification}\end{array}\right.$$

Optimization of structures undergoing bifurcation often produces coincident (multiple) critical points [219, 286, 290]. For example, these points are generated by optimizing column-type structures based on linear buckling formulations [114, 196, 231]. As a foundation of the optimization of structures, the sensitivity analysis at coincident critical points is included in this chapter.

This chapter is organized as follows. Design sensitivity of a simple pedagogic example structure with respect to minor and major design modifications is studied in Section 2.2. The design sensitivity is shown to be regular for minor design modifications in Section 2.3. We introduce several numerical approaches to design sensitivity analyses, including:

- linear eigenvalue analysis approach in Section 2.4,
- interpolation approach in Section 2.5, and
- explicit diagonalization approach in Section 2.6.

Numerical examples are presented in Section 2.7.

2.2 Sensitivity of Bifurcation Load: Pedagogic Example

We investigate the design sensitivity of the bifurcation load of a pedagogic example structure, the propped cantilever of Fig. 2.1. This structure is comprised of a vertical truss member, simply supported at the rigid foundation and supported by a horizontal elastic spring at the top.

Let Λp denote the vertical load normalized with respect to EA of the truss member, where E is Young's modulus and A is the cross-sectional area. The extensional stiffness of the spring is denoted by K. The initial locations of nodes 1 and 2 are given as (x_1, y_1) and (x_2, y_2), respectively. The displaced coordinate of node 2 is denoted by (x, y). $\mathbf{U} = (U_x, U_y)^\top = (x - x_2, y - y_2)^\top$ denotes the displacement vector of node 2. The spring remains horizontal, and its strain is defined by the horizontal displacement U_x of node 2. The lengths of the truss member before and after deformation, denoted by L and $\widehat{L}$, respectively, are defined as

$$\begin{aligned} L &= [(x_2 - x_1)^2 + (y_2 - y_1)^2]^{\frac{1}{2}} \\ \widehat{L}(\mathbf{U}) &= [(U_x + x_2 - x_1)^2 + (U_y + y_2 - y_1)^2]^{\frac{1}{2}} \end{aligned} \qquad (2.1)$$

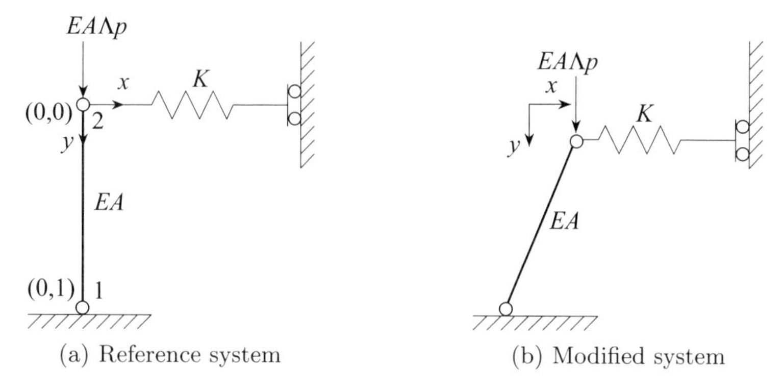

Fig. 2.1 Propped cantilever.

Remark 2.2.1 Nodal coordinates are often used to define imperfections and deformations in stability theory. Yet, it is pertinent to use nodal displacements to develop unified formulations for design sensitivity and imperfection sensitivity in the framework of finite element analysis. □

The total potential energy of this cantilever is

$$\Pi(\mathbf{U}, \Lambda) = \frac{EA}{2L}(\widehat{L} - L)^2 + \frac{1}{2}K(U_x)^2 - EA\Lambda p\, U_y \tag{2.2}$$

Stationary condition of Π with respect to $\mathbf{U}$ leads to the equilibrium equations

$$\begin{pmatrix} S_{,1} \\ S_{,2} \end{pmatrix} = EA \begin{pmatrix} (1/L - 1/\widehat{L})(U_x + x_2 - x_1) + KU_x \\ (1/L - 1/\widehat{L})(U_y + y_2 - y_1) - \Lambda p \end{pmatrix} = \begin{pmatrix} 0 \\ 0 \end{pmatrix} \tag{2.3}$$

In the following, let $p = 1$ for simplicity.

The design parameter vector in (2.2) and (2.3) is chosen as

$$\mathbf{v} = (x_1, y_1, x_2, y_2)^\top \tag{2.4}$$

We consider two design modifications, including the antisymmetric modification ξ_1 and the symmetric modification ξ_2, which turn out to be the major and minor modifications, respectively. Namely, (cf., (1.12))

$$\mathbf{v} = \mathbf{v}^0 + \xi_1 \mathbf{d}_1 + \xi_2 \mathbf{d}_2 \tag{2.5}$$

where

$$\mathbf{v}^0 = \begin{pmatrix} 0 \\ 1 \\ 0 \\ 0 \end{pmatrix}, \quad \mathbf{d}_1 = \begin{pmatrix} 1 \\ 0 \\ -1 \\ 0 \end{pmatrix}, \quad \mathbf{d}_2 = \begin{pmatrix} 0 \\ 1 \\ 0 \\ -1 \end{pmatrix} \tag{2.6}$$

and the modification patterns $\mathbf{d}_1$ and $\mathbf{d}_2$ are illustrated in Figs. 2.2(a) and (b), respectively.

In what follows, *exact* and *asymptotic* analyses for the reference and modified systems are conducted in order to investigate the parameter sensitivity of the propped cantilever.

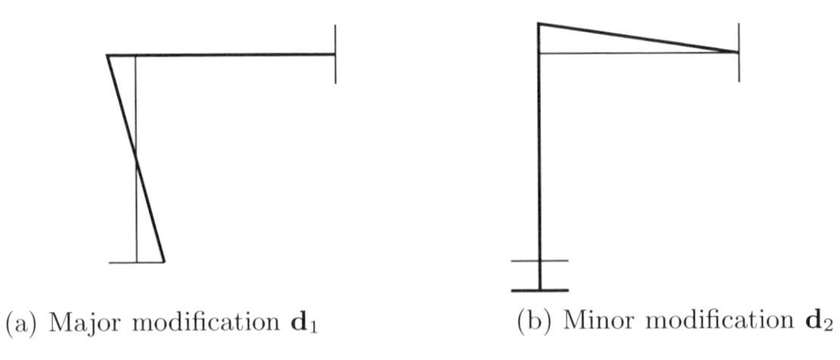

(a) Major modification $\mathbf{d}_1$ (b) Minor modification $\mathbf{d}_2$

Fig. 2.2 Design modification patterns of the propped cantilever.

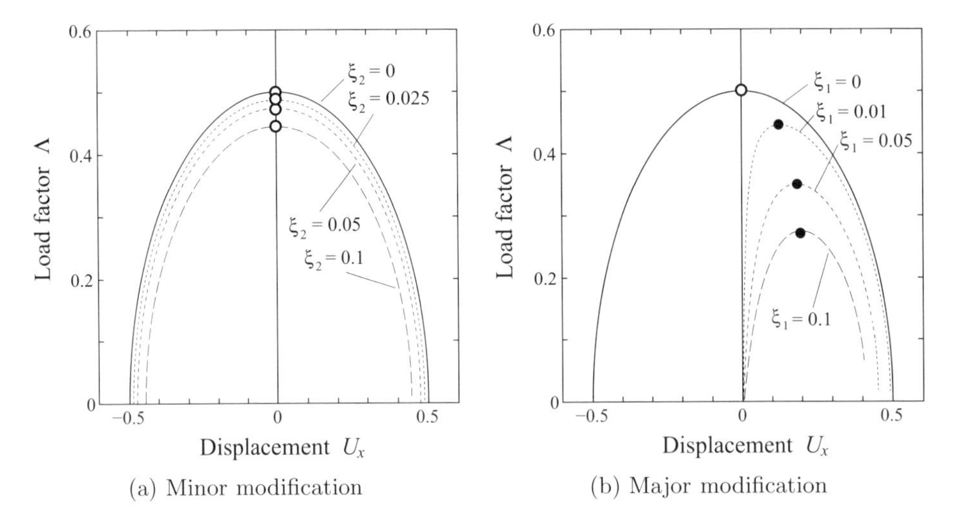

(a) Minor modification (b) Major modification

Fig. 2.3 Equilibrium paths of the propped cantilever. Solid curve: path for the reference system; dashed curve: path for a modified system; ◦: unstable bifurcation point; •: limit point.

2.2.1 Exact analysis

First, the equilibrium paths for this reference system are obtained by setting $\mathbf{v} = \mathbf{v}^0$ in (2.1) and (2.3) and are shown in Fig. 2.3 by the solid curves. The Λ-axis with $U_x = 0$ is the fundamental path, and a bifurcation path branches at the bifurcation point shown by $\circ$ at $(U_x^{c0}, U_y^{c0}, \Lambda^{c0}) = (0, 1/2, 1/2)$, with bifurcation mode $\mathbf{\Phi}_1^{c0} = (1, 0)^\top$. Since the bifurcation path is symmetric with respect to the Λ-axis and Λ decreases along the bifurcation path, this bifurcation point is an unstable-symmetric bifurcation point (cf., Section 3.3.1).

Next, modified cantilevers are considered. For the symmetric design modification $\xi_2 \neq 0$ $(\xi_1 = 0)$, equilibrium paths consist of the fundamental path $(U_x = 0)$ and a bifurcation path that intersect at a bifurcation point, as shown by the dashed curves in Fig. 2.3(a) for $\xi_2 = 0.025$, 0.05 and 0.1. The design sensitivity relationship between the design modification parameter ξ_2 and the bifurcation load Λ^c shown by $\circ$ in Fig. 2.4 displays a slow and almost linear reduction of Λ^c in association with an increase in ξ_2.

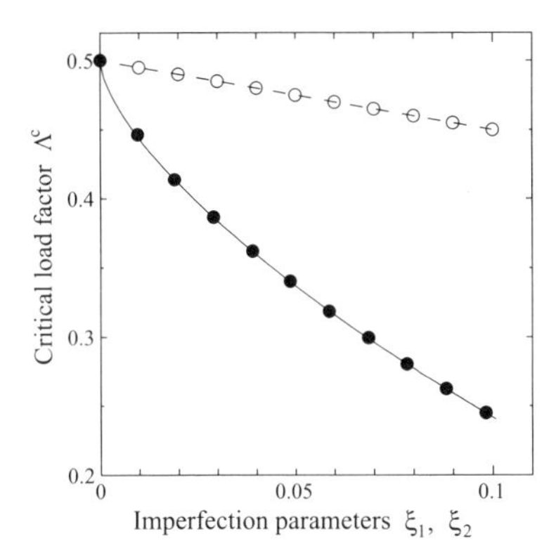

Fig. 2.4 Design sensitivity of the propped cantilever. Solid and dashed curves: sensitivity law (2.11); ∘: result of path-tracing analysis for a symmetric modification; •: that for an antisymmetric modification.

For the antisymmetric design modification $\xi_1 \neq 0$ ($\xi_2 = 0$), an equilibrium path has a limit point as shown in Fig. 2.3(b) for $\xi_1 = 0.01, 0.05$, and 0.1. The design sensitivity relationship shown by • in Fig. 2.4 displays a sharp and nonlinear reduction of $\Lambda^{\rm c}$, especially for small ξ_1.

Thus the symmetric modification ξ_2 serves as a minor (second-order) modification; by contrast, the antisymmetric modification ξ_1 serves as a major (first-order) modification. The sensitivity analyses for minor and major modifications are presented in Sections 2.4–2.6 and Chapter 3, respectively.

2.2.2 Asymptotic analysis

As we have seen, bifurcation behavior is complex even for a system with a few degrees of freedom. Such complexity is resolved by an asymptotic analysis[1] by the power series expansion method presented below to untangle the mechanism of bifurcation and parameter sensitivity (cf., Section 3.4 for details).

In order to investigate the properties in the neighborhood of the bifurcation point $(U_x^{\rm c0}, U_y^{\rm c0}, \Lambda^{\rm c0}) = (0, 1/2, 1/2)$ of the reference system, we define incremental variables

$$(\widetilde{U}_x, \widetilde{U}_y, \widetilde{\Lambda}) = (U_x, U_y, \Lambda) - (U_x^{\rm c0}, U_y^{\rm c0}, \Lambda^{\rm c0}) \tag{2.7}$$

Expanding the equilibrium equation (2.3) at $(U_x^{\rm c0}, U_y^{\rm c0}, \Lambda^{\rm c0}) = (0, 1/2, 1/2)$ and $(\xi_1, \xi_2) = (0, 0)$, we obtain a set of incremental equations

$$\begin{aligned} &EA[4(\widetilde{U}_x)^3 - 4\widetilde{U}_x\widetilde{U}_y + 2\xi_1 + 2\widetilde{U}_x\xi_2] + \text{h.o.t.} = 0 \\ &EA[\widetilde{U}_y - \widetilde{\Lambda} - 2(\widetilde{U}_x)^2] + \text{h.o.t.} = 0 \end{aligned} \tag{2.8}$$

[1] *Asymptotic* means that all results are valid only for sufficiently small values of design modification (initial imperfection) parameters in a sufficiently close neighborhood of the critical point of the reference (perfect) system.

where h.o.t. denotes higher-order terms.

The elimination of $\widetilde{U}_y$ from (2.8) leads to the so called bifurcation equation:

$$F \equiv EA[-4(\widetilde{U}_x)^3 - 4\widetilde{U}_x\widetilde{\Lambda} + 2\xi_1 + 2\widetilde{U}_x\xi_2] + \text{h.o.t.} = 0 \tag{2.9}$$

The criticality condition of the bifurcation equation (2.9) reads

$$\frac{\partial F}{\partial \widetilde{U}_x} = EA[-12(\widetilde{U}_x)^2 - 4\widetilde{\Lambda} + 2\xi_2] + \text{h.o.t.} = 0 \tag{2.10}$$

The location $(\widetilde{U}_x^{\mathrm{c}}, \widetilde{\Lambda}^{\mathrm{c}})$ of a critical point of a modified system can be obtained by simultaneously solving (2.9) and (2.10). In particular, $\widetilde{\Lambda}^{\mathrm{c}}$ gives the sensitivity law

$$\widetilde{\Lambda}^{\mathrm{c}} = -\frac{3}{4}(2\xi_1)^{\frac{2}{3}} + \frac{1}{2}\xi_2 + \text{h.o.t.} \tag{2.11}$$

which serves as a unified expression of the 2/3-power law for the antisymmetric modification ξ_1, and the linear law for the symmetric modification ξ_2 (cf., Section 3.5.2). The sensitivity law (2.11) is plotted by the solid curve in Fig. 2.4 for the antisymmetric modification ξ_1 and by the dashed curve for the symmetric modification ξ_2. These curves are in good agreement with the computed values by path-tracing analysis shown by $\bullet$ for $\xi_1 \neq 0$ and $\circ$ for $\xi_2 \neq 0$. The accuracy of the asymptotic law (2.11) has thus been assessed.

From (2.11), the sensitivity coefficients are calculated to

$$\begin{aligned} \widetilde{\Lambda}^{\mathrm{c}}_{,\xi_1} &= -(2\xi_1)^{-\frac{1}{3}} + \text{h.o.t.} \\ \widetilde{\Lambda}^{\mathrm{c}}_{,\xi_2} &= \frac{1}{2} + \text{h.o.t.} \end{aligned} \tag{2.12}$$

At $(\xi_1, \xi_2) = (0, 0)$, the value of $\widetilde{\Lambda}^{\mathrm{c}}_{,\xi_1}$ is unbounded and $\widetilde{\Lambda}^{\mathrm{c}}$ has a singular sensitivity regarding ξ_1; $\widetilde{\Lambda}^{\mathrm{c}}_{,\xi_2}$ takes a finite value 1/2 and has a regular sensitivity regarding ξ_2.

Remark 2.2.2 The bifurcation equation (2.9) has been derived from the full system of equations (2.3) by eliminating the variable $\widetilde{U}_y$, which is not associated with the critical eigenmode. This derivation is called elimination of passive coordinates, or the Liapunov–Schmidt–Koiter reduction (cf., Section 3.2.3). □

2.3 Minor and Major Design Modifications

Design modifications are classified into minor and major modifications and their design sensitivities are investigated.

2.3.1 Symmetry and classification of design modifications

Consider the initial reference system with reflection symmetry subjected to symmetric loadings as shown in Fig. 2.5, the critical load of which is governed by an

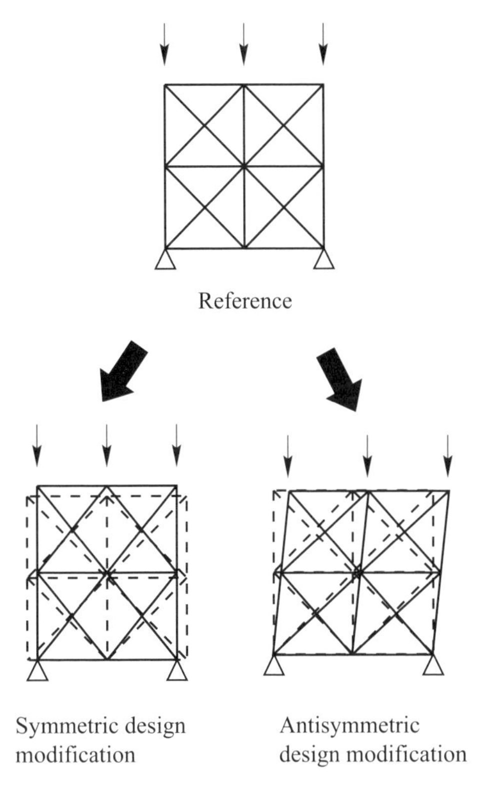

Fig. 2.5 Classification of design modifications.

unstable-symmetric bifurcation point, and undergoes symmetric and antisymmetric design modifications (cf., Section 3.3). For the symmetric modification, the deformation **U** before bifurcation remains symmetric.

For a general *asymmetric* design modification ξ, which contains both symmetric and antisymmetric modifications, the critical load factor $\Lambda^{\mathrm{c}}(\xi)$ in the vicinity of the reference system is expressed by the following asymptotic equation (cf., (2.11)):

$$\Lambda^{\mathrm{c}}(\xi) \simeq \Lambda^{\mathrm{c}0} + \widetilde{C}_1 \xi^{\frac{2}{3}} + \widetilde{C}_2 \xi \tag{2.13}$$

with $\Lambda^{\mathrm{c}0} = \Lambda^{\mathrm{c}}(0)$. Concrete forms of $\widetilde{C}_1$ and $\widetilde{C}_2$ are given in Sections 3.5. and 2.6, respectively. The sensitivity law (2.13) represents the 2/3-power law for $\widetilde{C}_1 \neq 0$ but reduces to the linear law for $\widetilde{C}_1 = 0$ and $\widetilde{C}_2 \neq 0$. Design modifications, accordingly, are classified as [258]

$$\begin{cases} \text{minor (second-order) modification:} & \widetilde{C}_1 = 0, \quad \widetilde{C}_2 \neq 0 \\ \text{major (first-order) modification:} & \widetilde{C}_1 \neq 0 \end{cases} \tag{2.14}$$

The vanishing and non-vanishing of $\widetilde{C}_1$ in general depend on the symmetry of a structure under consideration

$$\widetilde{C}_1 = \begin{cases} 0: & \text{symmetric design modification} \\ \text{nonzero}: & \text{asymmetric design modification} \end{cases} \tag{2.15}$$

as studied in detail in [142] and also treated briefly in Section 3.6.5. Then from (2.14), design modifications for a symmetric structure can be classified as (cf., Section (3.6.5)):

$$\begin{cases} \text{minor (second-order) modification:} & \text{symmetric (symmetry preserving)} \\ \text{major (first-order) modification:} & \text{asymmetric (symmetry breaking)} \end{cases} \tag{2.16}$$

2.3.2 Regular sensitivity for minor design modification

For a minor modification, the sensitivity is given by a linear law (cf., (2.13))

$$\Lambda^{\mathrm{c}}(\xi) = \Lambda^{\mathrm{c}0} + \widetilde{C}_2\xi \tag{2.17}$$

Thus Λ^{c} is differentiable with respect to ξ. The differentiation of (2.17) with respect to ξ gives the sensitivity coefficient

$$\Lambda^{\mathrm{c}\prime} = \widetilde{C}_2 \tag{2.18}$$

Thus the sensitivity coefficient $\Lambda^{\mathrm{c}\prime}$ for a minor modification takes a finite value $\widetilde{C}_2$, and, therefore, is called regular sensitivity. In the following sections, several numerical methods to evaluate the sensitivity coefficient $\widetilde{C}_2$ compatibly with a large-scale finite element model are presented focusing on minor modifications, for which $\Lambda^{\mathrm{c}\prime}$ exists.

2.3.3 Finite Difference Approach

The sensitivity coefficient of the bifurcation load $\Lambda^{\mathrm{c}}(\xi)$ at $\xi = \xi^0$ for a minor modification is numerically approximated within numerical errors by, e.g., the central *finite difference approach*

$$\Lambda^{\mathrm{c}\prime}(\xi^0) = \frac{1}{2\Delta\xi}[\Lambda^{\mathrm{c}}(\xi^0 + \Delta\xi) - \Lambda^{\mathrm{c}}(\xi^0 - \Delta\xi)] \tag{2.19}$$

However, selection of appropriate value of the increment $\Delta\xi$ is very difficult; too large value gives inaccurate sensitivity, and too small value also leads to a large error due to a truncation error.

Furthermore, the central difference approach demands large computational cost. Suppose we have 1000 design variables. Then path-tracing analysis should be carried out 2000 times to obtain all the necessary sensitivity coefficients by (2.19). On the contrary, path-tracing analysis should be carried out only once for the reference structure, if the sensitivity analysis approaches presented below are used. The sensitivity coefficients obtained by several approaches are compared with those by the central finite difference method based on path-tracing analysis to ensure their accuracy.

2.4 Linear Eigenvalue Analysis Approach

In the linear eigenvalue analysis approach, singularity of $\mathbf{S}$ at a bifurcation point can be circumvented by a linear estimation of the bifurcation load Λ^{c} from a load level Λ^* that is moderately smaller than Λ^{c} [214, 316]. Consider a linear estimation of $\mathbf{K}_{\mathrm{G}}(\Lambda)$ at $\Lambda = \Lambda^*$ as

$$\mathbf{K}_{\mathrm{G}}(\mathbf{U}(\xi, \Lambda), \xi) \simeq \frac{\Lambda}{\Lambda^*}\mathbf{K}_{\mathrm{G}}(\mathbf{U}(\xi, \Lambda^*), \xi) \tag{2.20}$$

Then (1.22) for the lowest critical load factor Λ^{c} is rewritten for the lowest positive eigenvalue $\mu^{\mathrm{c}} = \Lambda^{\mathrm{c}}/\Lambda^*$ of the following generalized eigenvalue problem:

$$[\mathbf{K}_{\mathrm{L}}(\xi) + \mu^{\mathrm{c}}\mathbf{K}_{\mathrm{G}}(\mathbf{U}(\xi, \Lambda^*), \xi)]\mathbf{\Psi} = \mathbf{0} \tag{2.21}$$

where $\mathbf{\Psi} \in \mathbb{R}^n$ is the eigenvector corresponding to μ^{c} that is normalized by

$$\mathbf{\Psi}^\top\mathbf{K}_{\mathrm{L}}\mathbf{\Psi} = 1, \quad \mathbf{\Psi}^\top\mathbf{K}_{\mathrm{G}}\mathbf{\Psi} = -\frac{1}{\mu^{\mathrm{c}}} \tag{2.22}$$

Note that $\mathbf{\Psi}$ implicitly depends on ξ and $\mathbf{U}(\xi)$ by (2.21) and (2.22).

By differentiating (2.21) with respect to ξ for fixed $\Lambda = \Lambda^*$, premultiplying $\mathbf{\Psi}^\top$ and using (2.21) and (2.22), we obtain the sensitivity coefficient of the critical load as

$$\mu^{\mathrm{c}\prime} = \mu^{\mathrm{c}}\mathbf{\Psi}^\top\mathbf{K}_{\mathrm{L}}'\mathbf{\Psi} + (\mu^{\mathrm{c}})^2\mathbf{\Psi}^\top\left[\mathbf{K}_{\mathrm{G},\xi} + \sum_{j=1}^{n}\frac{\partial\mathbf{K}_{\mathrm{G}}}{\partial U_j}U_j'\right]\mathbf{\Psi} \tag{2.23}$$

where $\mathbf{K}_{\mathrm{G},\xi}$ is the partial differentiation of $\mathbf{K}_{\mathrm{G}}$ at fixed $\mathbf{U}$, and U_j' can be computed from (1.30) at $\Lambda = \Lambda^*$. Note that $\mu^{\mathrm{c}\prime}$ computed by (2.23) gives an approximate value, as the difference of deformations between $\Lambda = \Lambda^*$ and $\Lambda = \Lambda^{\mathrm{c}}$ is neglected in (2.21). The value of Λ^* often is chosen elaborately.

Remark 2.4.1 In (2.23), the computation of the term $(\partial\mathbf{K}_{\mathrm{G}}/\partial U_j)U_j'$ is most costly. In the semi-analytical approach [61, 105], this term can be rewritten in a matrix–vector form

$$\sum_{j=1}^{n}\frac{\partial\mathbf{K}_{\mathrm{G}}}{\partial U_j}U_j^{*\prime}\mathbf{\Psi} = \sum_{j=1}^{n}\frac{\partial(\mathbf{K}_{\mathrm{G}}\mathbf{\Psi})}{\partial U_j}U_j^{*\prime} \tag{2.24}$$

for fixed $\mathbf{\Psi}$, and a finite difference method is used at each element to be assembled through the structure to compute the sensitivity of the vector $\mathbf{K}_{\mathrm{G}}\mathbf{\Psi}$ in the right-hand-side of (2.24) with respect to U_j. □

Remark 2.4.2 A sensitivity analysis procedure compatible with a cylindrical arc-length method [56] for tracing the fundamental path was developed [213]. In this procedure, the eigenvector corresponding to the vanishing eigenvalue of $\mathbf{S}$ is replaced by $\mathbf{h}^*$ computed from

$$\mathbf{h}^* = \frac{1}{\sqrt{\mathbf{h}^\top\mathbf{K}_{\mathrm{L}}\mathbf{h}}}\mathbf{h}, \quad \mathbf{h} = \mathbf{S}^{-1}\Delta\mathbf{U} \tag{2.25}$$

where $\Delta\mathbf{U}$ is the displacement increment vector to be obtained in the process of path-tracing. To circumvent the singularity of $\mathbf{S}$ at the critical point, $\mathbf{h}$ should be evaluated at an equilibrium state near the critical point. This procedure was extended to be compatible with one- and two-point approximations for detecting the critical points along the fundamental path by an incremental approach [175]. □

2.5 Interpolation Approach

In the interpolation approach, the sensitivity coefficient $\Lambda^{c\prime}$ of the bifurcation load factor Λ^{c} under a constraint of the vanishing of the lowest eigenvalue $\lambda_1 = 0$ of $\mathbf{S}$ is numerically approximated.

In a numerical analysis, it is difficult to arrive at the critical point, at which $\lambda_1 = 0$ is satisfied exactly. Instead of this, Λ^{c} is to be approximated from the two loads Λ^{I} and Λ^{II} at which $\lambda_1 \simeq 0$.

By total differentiation of (1.1), (1.5) and (1.7) for $r = 1$ with respect to ξ for a specified value λ_1^* of λ_1, and by use of $\lambda_1' = 0$ that arises from $\lambda_1 = 0$, we arrive at the following set of relations [227]

$$\sum_{j=1}^{n} S_{,ij} U_j^{*\prime} + S_{,i\xi} + S_{,i\Lambda}\Lambda^{*\prime} = 0, \quad (i = 1, \dots, n) \tag{2.26}$$

$$\sum_{j=1}^{n} \Big(\sum_{k=1}^{n} S_{,ijk}\phi_{1j} U_k^{*\prime} + S_{,ij\xi}\phi_{1j} + S_{,ij\Lambda}\phi_{1j}\Lambda^{*\prime} + S_{,ij}{\phi_{1j}^*}' \Big) = \lambda_1 {\phi_{1i}^*}', \quad (i = 1, \dots, n) \tag{2.27}$$

$$\sum_{j=1}^{n} \phi_{1j}{\phi_{1j}^*}' = 0 \tag{2.28}$$

Here $(\,\cdot\,)^{*\prime}$ denotes the sensitivity coefficient at $\lambda_1 = \lambda_1^*$.

Remark 2.5.1 It is desirable in practice to avoid the direct computation of $S_{,ijk}$ in the first term in the left-hand-side of (2.27), as $S_{,ijk}$ amounts to the third-order differentiation of the total potential energy. Note that tangent stiffness matrix $\mathbf{S} = [S_{,ij}]$, which requires the second-order differentiation of the potential, is employed in path-tracing analysis. A semi-analytical approach was proposed to avoid the computation of the third-order terms [61, 105] (cf., Remark 2.4.1). □

2.5.1 Regular state

At a regular state with nonsingular $\mathbf{S} = [S_{,ij}]$, there are several methods to compute $\Lambda^{*\prime}$ and $U_j^{*\prime}$.

The $2n+1$ sensitivity coefficients $\Lambda^{*\prime}$, $\mathbf{U}^{*\prime}$ and $\mathbf{\Phi}_1^{*\prime}$ at $\lambda_1 = \lambda_1^*$ can be obtained from a set of $2n+1$ simultaneous linear equations (2.26)–(2.28), by considering Λ^*, $\mathbf{U}^*$ and $\mathbf{\Phi}_1^*$ as functions of ξ.

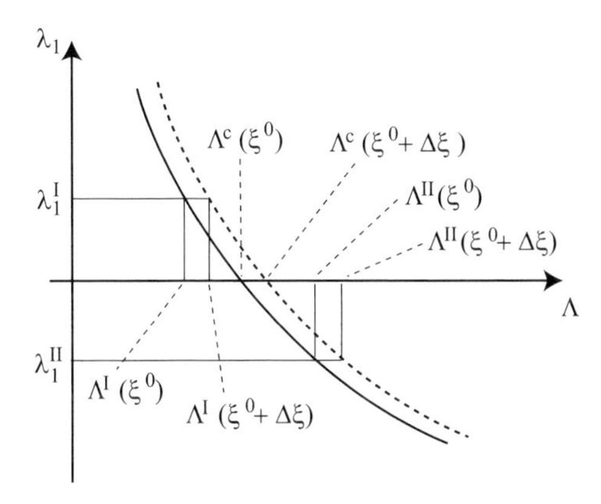

Fig. 2.6 Variation of an eigenvalue λ_1 by design modification $\xi = \xi^0 \longrightarrow \xi^0 + \Delta\xi$. Solid curve: λ_1 at $\xi = \xi^0$; dashed curve: λ_1 at $\xi = \xi^0 + \Delta\xi$.

Alternatively, (2.26) can be solved for $U_j^{*\prime}$ in terms of $\Lambda^{*\prime}$ as

$$U_j^{*\prime} = a_j + b_j \Lambda^{*\prime}, \quad (j = 1, \ldots, n) \tag{2.29}$$

with

$$(a_j) = -[S_{,ij}]^{-1}(S_{,i\xi}), \quad (b_j) = -[S_{,ij}]^{-1}(S_{,i\Lambda}) \tag{2.30}$$

By multiplying ϕ_{1i} to (2.27), taking summation over i, and incorporating (1.5) and (1.7), we obtain

$$\Lambda^{*\prime} = -\frac{1}{\sum_{i=1}^{n}\sum_{j=1}^{n} S_{,ij\Lambda}\phi_{1i}\phi_{1j}} \sum_{i=1}^{n}\sum_{j=1}^{n}\left(\sum_{k=1}^{n} S_{,ijk}\phi_{1i}\phi_{1j}U_k^{*\prime} + S_{,ij\xi}\phi_{1i}\phi_{1j}\right) \tag{2.31}$$

The relation (2.29) is incorporated to (2.31) to obtain $\Lambda^{*\prime}$. Thus, the $2n+1$ linear equations need not be solved simultaneously, if the sensitivity coefficient $\mathbf{\Phi}_1^{*\prime}$ is not needed. Note that (2.31) is not applicable to a proportional loading (cf., Section 1.2.3), for which $S_{,ij\Lambda}$ vanishes.

2.5.2 Bifurcation state

At a bifurcation point, the set of equations (2.26)–(2.28) encounters the singularity of $\mathbf{S} = [S_{,ij}]$ [228]. Among several methods for overcoming this singularity, we employ an interpolation technique to find the sensitivity coefficients of the bifurcation load factor Λ^c [227]. The solid and dashed curves in Fig. 2.6 illustrate the variations of λ_1 of two designs defined by $\xi = \xi^0$ and $\xi = \xi^0 + \Delta\xi$, respectively. Let $\Lambda^{\rm I}$ and $\Lambda^{\rm II}$ denote two load factors near Λ^c satisfying $\lambda_1(\Lambda^{\rm I}) > 0$ and $\lambda_1(\Lambda^{\rm II}) < 0$ for $\xi = \xi^0$, respectively. Then the value of Λ^c is approximated by the interpolation

$$\Lambda^c \simeq \frac{\lambda_1^{\rm I}\Lambda^{\rm II} - \lambda_1^{\rm II}\Lambda^{\rm I}}{\lambda_1^{\rm I} - \lambda_1^{\rm II}} \tag{2.32}$$

where $\lambda_1^{\rm I} = \lambda_1(\Lambda^{\rm I})$, $\lambda_1^{\rm II} = \lambda_1(\Lambda^{\rm II})$.

By differentiating (2.32) with respect to ξ for fixed values of $\lambda^{\rm I}$ and $\lambda^{\rm II}$, we obtain

$$\Lambda^{\rm c\prime} \simeq \frac{\lambda_1^{\rm I}\Lambda^{\rm II\prime} - \lambda_1^{\rm II}\Lambda^{\rm I\prime}}{\lambda_1^{\rm I} - \lambda_1^{\rm II}} \tag{2.33}$$

Note that $\Lambda^{\rm I\prime}$ can be computed from (2.26)–(2.28) for $\lambda_1^* = \lambda_1^{\rm I}$, and $\Lambda^{\rm II\prime}$ for $\lambda_1^* = \lambda_1^{\rm II}$. The accuracy of (2.33) is investigated in the numerical examples in Section 2.7.

As an alternative way to evaluate the sensitivity coefficient $\Lambda^{\rm c\prime}$ of $\Lambda^{\rm c}$, (2.32) is differentiated with respect to ξ for fixed $\Lambda^{\rm I}$ and $\Lambda^{\rm II}$, instead of fixed $\lambda^{\rm I}$ and $\lambda^{\rm II}$, to arrive at

$$\Lambda^{\rm c\prime} \simeq \frac{\lambda_1^{\rm I}{}'\Lambda^{\rm II} - \lambda_1^{\rm II}{}'\Lambda^{\rm I}}{\lambda_1^{\rm I} - \lambda_1^{\rm II}} - \Lambda^{\rm c}\frac{\lambda_1^{\rm I}{}' - \lambda_1^{\rm II}{}'}{\lambda_1^{\rm I} - \lambda_1^{\rm II}} \tag{2.34}$$

Here $\lambda_1^{\rm I}{}'$ and $\lambda_2^{\rm I}{}'$ can be computed by (1.35) for $r = 1$.

2.6 Explicit Diagonalization Approach

An explicit form of the coefficient $\widetilde{C}_2$ in (2.17) is derived by the explicit diagonalization approach for a simple unstable-symmetric bifurcation point [227] and a coincident bifurcation point [218] of a symmetric system (cf., Section 3.3 for the classification of critical points), while mathematical details are worked out in Appendix A.3. The formulas presented in this section, although they are computationally costly, conform with the finite element analysis and lead to the exact sensitivity coefficients. In the following, all quantities such as $U_i^{\rm c}$, $\Lambda^{\rm c}$, $\lambda_r^{\rm c}$, $\mathbf{\Phi}_r^{\rm c}$ and the differential coefficients $S_{,ijk}$, $S_{,i\Lambda}$, $S_{,i\xi}$, and so on, are evaluated at the bifurcation point of the reference system with $\xi = 0$.

2.6.1 Simple unstable-symmetric bifurcation point

For a simple unstable-symmetric bifurcation point, we find the sensitivity $\Lambda^{\rm c\prime}$ of Λ under constraint of $\lambda_1 = 0$. Differentiation of (1.1) and (1.5) for $r = 1$ with respect to ξ with the use of $\lambda_1 = \lambda_1{}' = 0$ leads, respectively, to

$$\sum_{j=1}^{n} S_{,ij}U_j^{\rm c\prime} + S_{,i\xi} + S_{,i\Lambda}\Lambda^{\rm c\prime} = 0, \quad (i = 1, \dots, n) \tag{2.35}$$

$$\sum_{j=1}^{n}\left(\sum_{k=1}^{n} S_{,ijk}\phi_{1j}^{\rm c}U_k^{\rm c\prime} + S_{,ij\xi}\phi_{1j}^{\rm c} + S_{,ij\Lambda}\phi_{1j}^{\rm c}\Lambda^{\rm c\prime} + S_{,ij}\phi_{1j}^{\rm c}{}'\right) = 0, \quad (i = 1, \dots, n) \tag{2.36}$$

From (2.35) and (2.36), as derived in Appendix A.3.1, the sensitivity coefficient is obtained as

$$\Lambda^{c\prime} = \frac{1}{\mu}\Big[\sum_{i=1}^{n}\sum_{j=1}^{n} S_{,ij\xi}\phi_{1i}^{c}\phi_{1j}^{c} - \sum_{r=2}^{n}\frac{1}{\lambda_r}\Big(\sum_{i=1}^{n}\sum_{j=1}^{n}\sum_{k=1}^{n} S_{,ijk}\phi_{1i}^{c}\phi_{1j}^{c}\phi_{rk}^{c}\sum_{i=1}^{n} S_{,i\xi}\phi_{ri}^{c}\Big)\Big] \tag{2.37}$$

where μ is independent of ξ and is given as

$$\mu = -\sum_{i=1}^{n}\sum_{j=1}^{n} S_{,ij\Lambda}\phi_{1i}^{c}\phi_{1j}^{c} + \sum_{r=2}^{n}\frac{1}{\lambda_r}\Big(\sum_{i=1}^{n}\sum_{j=1}^{n}\sum_{k=1}^{n} S_{,ijk}\phi_{1i}^{c}\phi_{1j}^{c}\phi_{rk}^{c}\sum_{i=1}^{n} S_{,i\Lambda}\phi_{ri}^{c}\Big) \tag{2.38}$$

A possible and simpler alternative of this derivation is given in Section 3.4. The formula of sensitivity coefficient (2.37) with (2.38) is applicable to numerical implementation, but is computationally costly, since higher derivatives of the total potential energy with respect to displacements as well as all eigenvalues and eigenvectors are necessitated.

2.6.2 Coincident bifurcation point of a symmetric system

The explicit diagonalization approach for a simple unstable-symmetric bifurcation point is extended to be applicable to a coincident bifurcation point of a symmetric system. The validity of the proposed formula is assessed through an application to a simple bar–spring model in Section 2.7.3.

Consider a *symmetric* structure with a reflection and/or rotation symmetry, and assume that the deformation of this structure along the fundamental path is symmetric. We further consider a minor modification at an m-fold coincident symmetric bifurcation point in the sense that

$$\sum_{i=1}^{n} S_{,i\xi}\phi_{\alpha i}^{c} = 0, \quad (\alpha = 1,\ldots,m) \tag{2.39}$$

Such modification serves as a minor modification for all m critical eigenvectors $\mathbf{\Phi}_{\alpha}^{c} = (\phi_{\alpha i}^{c})$, and is called *completely minor modification.*

A critical eigenvector $\mathbf{\Psi}^{c} = (\psi_{i}^{c})$ that satisfies $\mathbf{S}\mathbf{\Psi}^{c} = \mathbf{0}$ is expressed as a linear combination of $\mathbf{\Phi}_{\alpha}^{c}$ $(\alpha = 1,\ldots,m)$ as

$$\mathbf{\Psi}^{c} = \sum_{\alpha=1}^{m} a_{\alpha}\mathbf{\Phi}_{\alpha}^{c} \tag{2.40}$$

where coefficients a_{α} $(\alpha = 1,\ldots,m)$ are normalized by

$$\sum_{\alpha=1}^{m}(a_{\alpha})^2 = 1 \tag{2.41}$$

Fig. 2.7(a) illustrates the relation between the design parameter ξ and the two critical load factors for a double bifurcation point ($m = 2$). As can be observed, the lowest critical load factor Λ_1^{c} that is plotted by the solid curve has a kink; therefore, the sensitivity coefficient $\Lambda_1^{c\prime}$ is discontinuous at the coincident

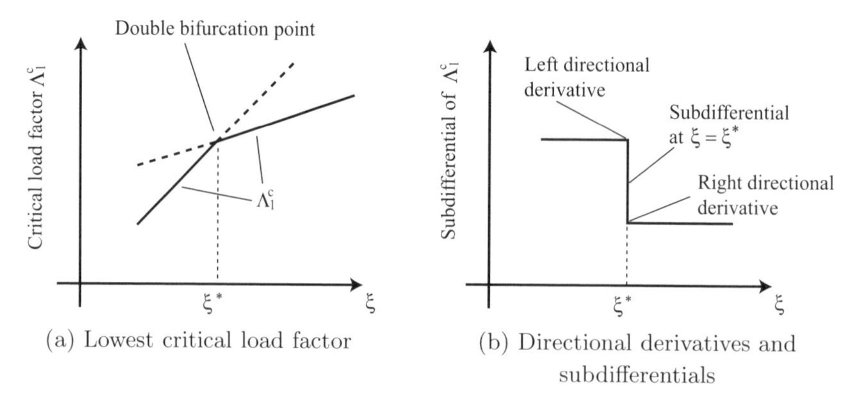

(a) Lowest critical load factor

(b) Directional derivatives and subdifferentials

Fig. 2.7 Definition of directional derivatives and subdifferentials of the lowest critical load factor.

bifurcation point, and the left and right directional derivatives can be defined as illustrated in Fig. 2.7(b). The directional derivatives are computed as the minimum and maximum values of the subdifferential [201] to be defined below[2].

As worked out in Appendix A.3.2, the subdifferential of Λ^c is formulated as

$$\Lambda^{c\prime}_{\text{sub}} = \frac{\sum_{\alpha=1}^{m} \sum_{\beta=1}^{m} a_\alpha a_\beta \Xi_{\alpha\beta}}{\sum_{\alpha=1}^{m} \sum_{\beta=1}^{m} a_\alpha a_\beta \Theta_{\alpha\beta}} \tag{2.42}$$

where

$$\Xi_{\alpha\beta} = \sum_{i=1}^{n}\sum_{j=1}^{n} S_{,ij\xi}\phi^c_{\alpha i}\phi^c_{\beta j} - \sum_{r=m+1}^{n} \frac{1}{\lambda^c_r}\Big(\sum_{i=1}^{n}\sum_{j=1}^{n}\sum_{k=1}^{n} S_{,ijk}\phi^c_{\alpha i}\phi^c_{\beta j}\phi^c_{rk} \sum_{i=1}^{n} S_{,i\xi}\phi^c_{ri}\Big) \tag{2.43}$$

depends on ξ via $S_{,ij\xi}$ and $S_{,i\xi}$; and

$$\Theta_{\alpha\beta} = -\sum_{i=1}^{n}\sum_{j=1}^{n} S_{,ij\Lambda}\phi^c_{\alpha i}\phi^c_{\beta j} + \sum_{r=m+1}^{n} \frac{1}{\lambda^c_r}\Big(\sum_{i=1}^{n}\sum_{j=1}^{n}\sum_{k=1}^{n} S_{,ijk}\phi^c_{\alpha i}\phi^c_{\beta j}\phi^c_{rk} \sum_{i=1}^{n} S_{,i\Lambda}\phi^c_{ri}\Big) \tag{2.44}$$

is independent of ξ. The formula (2.42) defines the subdifferential $\Lambda^{c\prime}_{\text{sub}}$ of Λ^c, and its value changes according to the choice of the vector (a_α) as shown in the bar–spring model in Section 2.7.3. The left and right directional derivatives of Λ^c_1 are obtained as the maximum and minimum value of $\Lambda^{c\prime}_{\text{sub}}$.

Eq. (2.42) is rewritten as

$$\sum_{\alpha=1}^{m}\sum_{\beta=1}^{m} a_\alpha a_\beta [\Xi_{\alpha\beta} - \Lambda^{c\prime}_{\text{sub}}\Theta_{\alpha\beta}] = 0 \tag{2.45}$$

which is satisfied by a nontrivial vector (a_α) when the symmetric matrix $[\,\Xi_{\alpha\beta} - \Lambda^{c\prime}_{\text{sub}}\Theta_{\alpha\beta}] \in \mathbb{R}^{m\times m}$ is singular, as is also the case for the multiple eigenvalues of

[2]Subdifferential is originally defined for a convex function, and the extended form for a non-convex function is called Clarke subdifferential [55].

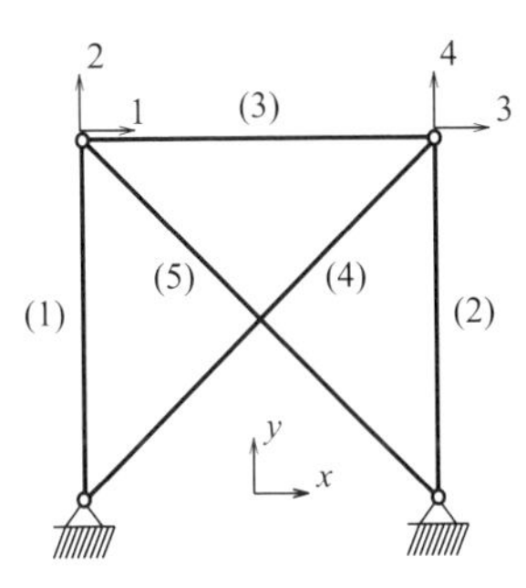

Fig. 2.8 Five-bar truss. $1, \ldots, 4$ denote degrees of freedom; $(1), \ldots, (5)$ denote member numbers.

the linear buckling problem [49, 115, 271]. Hence, the eigenvalue problem

$$\sum_{\beta=1}^{m} \Xi_{\alpha\beta}\eta_{\gamma\beta} = \tau_\gamma \sum_{\beta=1}^{m} \Theta_{\alpha\beta}\eta_{\gamma\beta}, \quad (\alpha, \gamma = 1, \ldots, m) \tag{2.46}$$

is to be solved to find the left and right directional derivatives of Λ_1^{c} as the maximum and minimum eigenvalues τ_γ.

2.7 Numerical Examples for Design Sensitivity

Design sensitivity analysis of structures is conducted by the interpolation approach, the explicit diagonalization approach, and the block diagonalization approach (cf., Sections 2.5, 2.6 and Appendix A.4). The results by these approaches are compared with those by the central finite difference method with $\xi = \xi^0 = 0$ in (2.19) based on path-tracing analysis.

2.7.1 Five-bar truss

Consider the five-bar truss as shown in Fig. 2.8 subjected to the proportional vertical loads $\Lambda \mathbf{p}$, where $\mathbf{p} = (0, -1, 0, -1)^\top$. The units of force N and of length m are suppressed in the following. The extensional stiffness of the ith member is denoted by K_i, for which the reference structure is given such that $K_1^0 = K_2^0 = K_3^0 = 1.0$ and $K_4^0 = K_5^0 = 0.70711$. The bifurcation load factor is $\Lambda^{\mathrm{c}0} = 0.19146$. Note that the deformation along the fundamental path is symmetric and the buckling mode $\mathbf{\Phi}_1^{\mathrm{c}}$ is antisymmetric with respect to the y-axis.

Consider a symmetric modification of K_1 and K_2 as

$$K_1 = K_1^0 + \xi, \quad K_2 = K_2^0 + \xi \tag{2.47}$$

i.e., $\mathbf{v} = (K_1, K_2)^\top$, $\mathbf{v}^0 = (K_1^0, K_2^0)^\top$, $\mathbf{d} = (1, 1)^\top$ in (1.14).

Fig. 2.9 shows the variation of the sensitivity coefficient $\Lambda^{*\prime}$ for the fixed lowest eigenvalue $\lambda_1 = \lambda_1^*$ of $\mathbf{S}$ plotted against Λ. As can be seen, $\Lambda^{*\prime}$ is a smooth function of Λ; hence, the use of interpolation approach for the computation of $\Lambda^{\mathrm{c}\prime}$ is valid. The sensitivity coefficient found by the interpolation approach with

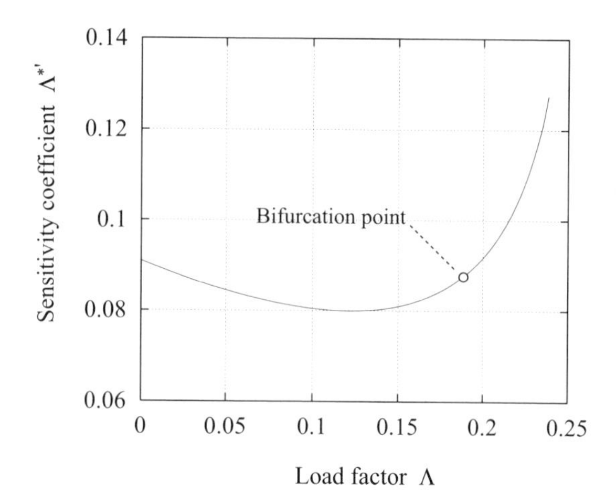

Fig. 2.9 Variation of sensitivity coefficient $\Lambda^{*\prime}$ for a fixed value of $\lambda_1 = \lambda_1^*$ plotted against load factor Λ for the five-bar truss.

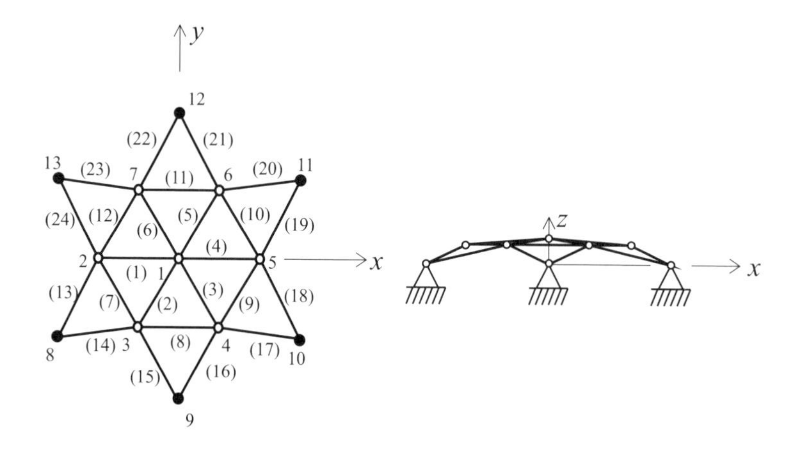

Fig. 2.10 Symmetric shallow truss dome. $1, \ldots, 13$ denote node numbers; $(1), \ldots, (24)$ denote member numbers.

$\lambda_1^{\mathrm{I}} = 1.1741 \times 10^{-3}$ and $\lambda_1^{\mathrm{II}} = -1.0131 \times 10^{-3}$ is equal to $\Lambda^{c\prime} = 0.093286$, and agrees up to five digits with the results by the explicit diagonalization approach, the block diagonalization approach, and the central finite difference approach ($\Delta\xi = 1.0290 \times 10^{-4}$), which is supposed to give the exact value within a numerical error. The accuracy of the interpolation approach has thus been ensured.

2.7.2 Symmetric shallow truss dome

Consider the symmetric shallow truss dome as shown in Fig. 2.10 subjected to proportional vertical loads at nodes 1–7. The z-directional load at the ith node

Table 2.1 Nodal coordinates of the truss dome.

Node number	x	y	z
1	0.0	0.0	821.6
2	−2500.0	0.0	621.6
3	−1250.0	−2165.0	621.6
4	1250.0	−2165.0	621.6
5	2500.0	0.0	621.6
6	1250.0	2165.0	621.6
7	−1250.0	2165.0	621.6
8	−4330.0	−2500.0	0.0
9	0.0	−5000.0	0.0
10	4330.0	−2500.0	0.0
11	4330.0	2500.0	0.0
12	0.0	5000.0	0.0
13	−4330.0	2500.0	0.0

Table 2.2 Sensitivity coefficients of displacements of the truss dome.

Node	Direction	U_i^{c}	$U_i^{\mathrm{I}\prime}$	$U_i^{\mathrm{B}\prime}$	γ_i
1	z	−17.920	0.19827	0.19559	1.0000
2	x	5.8858	0.012045	0.012033	0.99999
2	z	−63.122	−0.11351	−0.11389	1.0000
3	x	2.9429	0.0060224	0.0060167	0.99999
3	y	−5.0972	−0.010431	−0.010421	1.0000
3	z	−63.122	−0.11351	−0.11389	1.0000

is denoted by Λp_i, where

$$p_i = \begin{cases} -4.9: & \text{for} \quad i = 1 \\ -9.8: & \text{for} \quad i = 2, \dots, 7 \end{cases} \tag{2.48}$$

The nodal coordinates are listed in Table 2.1. The units kN of force and mm of length are suppressed in the following. Young's modulus of each member is $E = 205.8$. Let A_i denote the cross-sectional area of the ith member. The reference structure is defined by $A_i^0 = 100$ for all members. The bifurcation load factor of the reference structure is $\Lambda^{\mathrm{c}} = 1815.5$ and is associated with a simple unstable-symmetric bifurcation point.

Consider a minor design modification preserving symmetry specified such that $A_i = A_i^0 + \xi$ for the members (1)–(6) connected to the center node, and $A_i = A_i^0 = 100$ for other members (7)–(24). The sensitivity coefficient computed by the explicit diagonalization approach is $\Lambda^{\mathrm{c}\prime} = 2.4786$, which agrees up to 13 digits with the result by the block diagonalization approach, and seven digits with the result by the central finite difference method ($\Delta\xi = 0.001$).

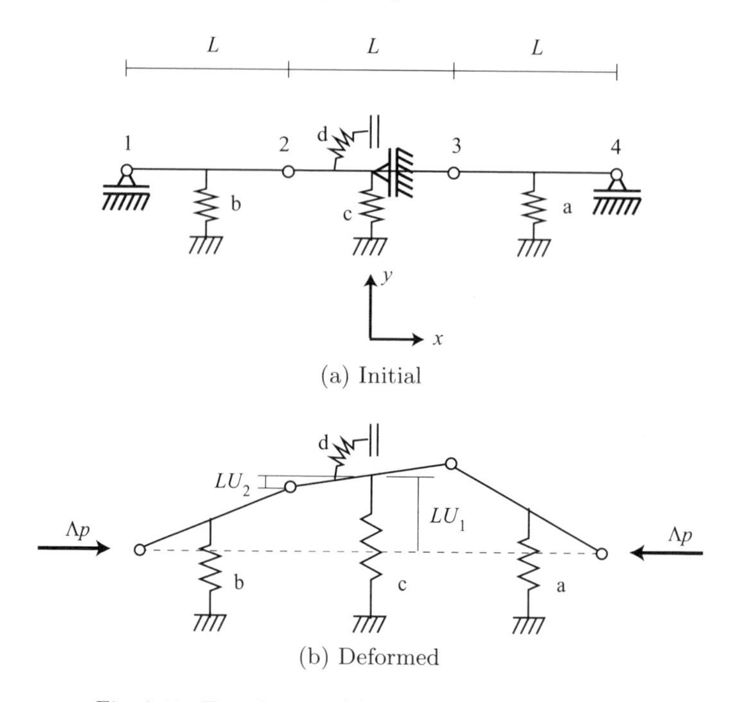

Fig. 2.11 Two-degree-of-freedom bar–spring system.

Table 2.2 lists the values of displacements U_i^{c} at the bifurcation point, the sensitivity coefficients of displacements $U_i^{\mathrm{I}\prime}$ computed by the interpolation approach, and $U_i^{\mathrm{B}\prime}$ by the block diagonalization approach.

Denote by γ_i the ratio of $U_i^{\mathrm{B}\prime}$ to the coefficient calculated by the central finite difference method. A good agreement ($\gamma_i \simeq 1$) is observed between $U_i^{\mathrm{B}\prime}$ and those by the finite difference method. Slight differences are present between $U_i^{\mathrm{I}\prime}$ and $U_i^{\mathrm{B}\prime}$ to show that the interpolation approach is slightly less accurate. Yet the interpolation approach may be successfully used in engineering application despite such slight inaccuracy, because it demands less computational effort than the other two methods and is compatible with the finite element analysis.

2.7.3 Two-degree-of-freedom bar–spring system

The explicit diagonalization approach for coincident bifurcation points presented in Section 2.6.2 is applied to the two-degree-of-freedom bar–spring system as shown in Fig. 2.11(a) [48, 129]. This corresponds to a special case with $m = n$ (= 2), where the second terms in (2.43) and (2.44) are non-existent.

The bar–spring system is supported at nodes 1 and 4 in the y-direction and at the center in the x-direction. The horizontal axial load Λp is applied to the system. This system is symmetric with respect to the y-axis. The generalized displacements U_1 and U_2 are defined as shown in Fig. 2.11(b) and the extensions $e_{\mathrm{a}}, e_{\mathrm{b}}, e_{\mathrm{c}}$ of the springs 'a', 'b', 'c', and the rotation θ_{d} of the rotational spring 'd' are defined as

$$\begin{aligned} e_{\mathrm{a}} &= \frac{1}{2}L(U_1 + U_2), \quad e_{\mathrm{b}} = \frac{1}{2}L(U_1 - U_2), \quad e_{\mathrm{c}} = LU_1 \\ \theta_{\mathrm{d}} &= \sin^{-1}(2U_2) \end{aligned} \tag{2.49}$$

The total potential energy Π is written as

$$\Pi = E_{\mathrm{a}} + E_{\mathrm{b}} + E_{\mathrm{c}} + E_{\mathrm{d}} - \Lambda p v \tag{2.50}$$

Here

$$v = L\{(U_1)^2 + 3(U_2)^2 + \frac{1}{4}[(U_1)^4 + 6(U_1)^2(U_2)^2 + 9(U_2)^4]\} \tag{2.51}$$

is the x-directional relative displacement between nodes 1 and 4, and the strain energies of the springs are defined as

$$\begin{aligned} E_{\mathrm{a}} &= \frac{2A_1}{L^2}(e_{\mathrm{a}})^2, \quad E_{\mathrm{b}} = \frac{2A_1}{L^2}(e_{\mathrm{b}})^2, \quad E_{\mathrm{c}} = \frac{A_2}{2L^2}(e_{\mathrm{c}})^2 \\ E_{\mathrm{d}} &= \frac{\overline{A}_3}{2}(\theta_{\mathrm{d}})^2 \simeq \frac{1}{2}A_3(U_2)^2 \end{aligned} \tag{2.52}$$

where A_1, A_2, A_3 and $\overline{A}_3$ are constants, and $\overline{A}_3 \simeq \frac{1}{4}A_3$, $\theta_{\mathrm{d}} \simeq 2U_2$.

In the following, the load is normalized so that $pL = 1$. By successively differentiating Π with respect to U_1 and U_2, and by substituting the trivial solution $U_1 = U_2 = 0$ of the fundamental path, we obtain

$$\begin{aligned} \mathbf{S} &= \begin{bmatrix} S_{,11} & S_{,12} \\ S_{,21} & S_{,22} \end{bmatrix} = \begin{bmatrix} 2A_1 + A_2 - 2\Lambda & 0 \\ 0 & 2A_1 + A_3 - 6\Lambda \end{bmatrix} \\ S_{,111} &= S_{,112} = S_{,122} = S_{,222} = 0 \end{aligned} \tag{2.53}$$

From (2.53), the two critical load factors are

$$\Lambda_1^{\mathrm{c}} = \frac{1}{2}(2A_1 + A_2), \quad \Lambda_2^{\mathrm{c}} = \frac{1}{6}(2A_1 + A_3) \tag{2.54}$$

and the associated eigenvectors are $\mathbf{\Phi}_1^{\mathrm{c}} = (1, 0)^\top$, $\mathbf{\Phi}_2^{\mathrm{c}} = (0, 1)^\top$.

The design parameter is chosen to be A_1. From (2.54), the sensitivity coefficients of Λ_1^{c} and Λ_2^{c} are written as

$$\Lambda_1^{\mathrm{c}\prime} = 1, \quad \Lambda_2^{\mathrm{c}\prime} = \frac{1}{3} \tag{2.55}$$

where $\Lambda_1^{\mathrm{c}\prime}$ and $\Lambda_2^{\mathrm{c}\prime}$ correspond to left and right directional derivatives, respectively.

In order to realize a coincident bifurcation point with $\Lambda_1^{\mathrm{c}} = \Lambda_2^{\mathrm{c}}$, we set $A_3 = 4A_1 + 3A_2$. Then the null space of $\mathbf{S}$ is two-dimensional at the coincident bifurcation point, and the critical eigenvector, which is not unique, may be written as

$$\mathbf{\Psi}^{\mathrm{c}} = a_1 \begin{pmatrix} 1 \\ 0 \end{pmatrix} + a_2 \begin{pmatrix} 0 \\ 1 \end{pmatrix}, \quad (a_1)^2 + (a_2)^2 = 1 \tag{2.56}$$

The eigenvalue problem (2.46) for this example structure becomes

$$\begin{bmatrix} 2 & 0 \\ 0 & 2 \end{bmatrix} \begin{pmatrix} \eta_{\gamma 1} \\ \eta_{\gamma 2} \end{pmatrix} = \tau_\gamma \begin{bmatrix} 2 & 0 \\ 0 & 6 \end{bmatrix} \begin{pmatrix} \eta_{\gamma 1} \\ \eta_{\gamma 2} \end{pmatrix}, \quad (\gamma = 1, 2) \tag{2.57}$$

and the directional derivatives are computed as eigenvalues $\tau_r = 1,\ 1/3$, which give the values of $\Lambda_1^{\mathrm{c}\prime}$ and $\Lambda_2^{\mathrm{c}\prime}$ in (2.55), respectively.

Let $a_1 = \sin\nu$ and $a_2 = \cos\nu$ $(0 \leq \nu < 2\pi)$. The subdifferential $\Lambda_{\mathrm{sub}}^{\mathrm{c}\prime}$ in (2.42) is given as

$$\Lambda_{\mathrm{sub}}^{\mathrm{c}\prime} = \frac{\sum_{\alpha=1}^{2} \sum_{\beta=1}^{2} \Xi_{\alpha\beta} a_\alpha a_\beta}{\sum_{\alpha=1}^{2} \sum_{\beta=1}^{2} \Theta_{\alpha\beta} a_\alpha a_\beta} = \frac{1}{1 + 2\cos^2\nu} \tag{2.58}$$

The maximum 1 and the minimum 1/3 of the subdifferential $\Lambda_{\mathrm{sub}}^{\mathrm{c}\prime}$ in (2.58) agree with the sensitivity coefficients in (2.55).

2.8 Summary

In this chapter,

- regular design sensitivity of a bifurcation load for a minor modification has been studied, and
- numerical approaches for sensitivity analysis have been proposed and applied to example structures.

The major findings of this chapter are as follows.

- Design modifications can be classified into minor and major design modifications. The sensitivity coefficient of the bifurcation load is bounded for minor modifications, and unbounded for major modifications.
- The numerical approaches for design sensitivity analyses for minor imperfections introduced herein may look complicated at a glance. Yet, their physical and theoretical backgrounds will be elucidated in the light of modern stability theory in Chapter 3, and their practical importance will be demonstrated in Part II.
- The sensitivity coefficients computed by the explicit diagonalization and the block diagonalization approaches are very accurate. Those by the interpolation approach are less accurate. Yet this approach is suggested for use in engineering application, because its computational effort is smaller than those of the other two methods and can be readily implemented into existing finite element analysis programs.
- The formulation for sensitivity analysis of a simple bifurcation load factor with respect to a minor modification has been extended to coincident bifurcation. The directional derivatives of the coincident bifurcation load factor can be computed, in a similar manner as those of multiple linear buckling load factor, by solving an eigenvalue problem to obtain the proper choice of the coefficients for the eigenmodes.

3
Imperfection Sensitivity Analysis

3.1 Introduction

Structures undergoing buckling can be optimized using a gradient-based optimization algorithm with design sensitivity analysis, as we will see in Part II. Nevertheless, optimized structures often become highly imperfection-sensitive and their buckling loads erode sharply by initial imperfections that arise from errors in manufacturing process, material defects, and other causes. This problem was cautioned as the "danger of naive optimization [287]." It is certainly ironic that dangerous structures are produced in the pursuit of optimized performance. Safety of the optimized structures against imperfections must be ensured in their designs.

In this chapter, theoretical backgrounds to evaluate the performance of imperfect structures are presented on the basis of stability theory. The imperfection sensitivity law is highlighted as a systematic means to describe the dependence of buckling strength on initial imperfections. Since this law varies according to the type of critical point, the classification of critical points is conducted on the basis of the differential coefficients of the total potential energy. These theoretical backgrounds are presented in a manner that is readily accessible for engineers by minimizing mathematical details. Readers who are interested in further studies are encouraged to proceed to standard textbooks [95, 129, 142, 209, 285, 287].

Optimization for nonlinear buckling often produces coincident critical points, which can potentially increase imperfection sensitivity dramatically [288]. Complex bifurcation behaviors are encountered at these points, which are the center of singularity. Important coincident critical points, for example, are

- a hilltop branching point at which bifurcation point(s) and a limit point coincide, and

- a semi-symmetric double bifurcation point at which a symmetric bifurcation point and an asymmetric bifurcation point coincide.

Static perturbation of the total potential energy was performed to investigate perfect and imperfect behaviors of these points [277, 285].

A hilltop branching point was observed in mechanical instability of stressed atomic crystal lattices [147, 289], steel specimens [210], and structural models [216, 222]. A hilltop point for most cases has locally piecewise linear imperfection sensitivity and is not imperfection-sensitive [289], as actually demonstrated for an example structure in this chapter. Readers, accordingly, are encouraged to carry out optimization-based design with confidence to produce such a safe structure.

A semi-symmetric point occurs often via optimization for a cylindrical shell exhibiting diamond-pattern and elephant-foot-type bifurcation modes. The semi-symmetric point encounters the 1/2-power law of imperfection sensitivity that erodes the strength severely, and complex imperfect bifurcation behaviors owing to mode interaction [126, 129].

In the design of imperfection-sensitive structures, it is desirable but costly to consider many imperfection patterns. An amazing feature about imperfection modes is that some modes are very influential, while other modes are not at all. In this regard, it is useful to recall the concept of minor and major design modifications presented in Section 2.3. In this chapter, the design modification in Chapter 2 is conceived as an initial imperfection, and the reference structure as the perfect structure. Then we can focus only on major imperfections, which are generally more influential.

This chapter is organized as follows. Generalized coordinates and power series expressions of the total potential energy with respect to these expressions are presented as mathematical preliminaries in Section 3.2. Classification of critical points is described in Section 3.3. Methodologies to derive imperfection sensitivity laws are presented in Section 3.4. Imperfection sensitivity laws for major imperfections at simple and coincident critical points are presented in Sections 3.5 and 3.6, respectively. Numerical examples are provided in Section 3.7. The historical development of imperfection sensitivity analysis is presented in Section 3.8.

3.2 Mathematical Preliminaries

Generalized coordinates and expanded forms of the total potential energy are presented as mathematical preliminaries for the classification of critical points. These are the heaviest theoretical ingredients of this book, and readers who are not interested in mathematical details may march on to Section 3.3.

3.2.1 Generalized coordinates

We consider a critical point $(\mathbf{U}^{\mathrm{c}}, \Lambda^{\mathrm{c}})$, at which m eigenvalues of the tangent stiffness matrix $\mathbf{S}$ vanish, namely,

$$\lambda_1 = \cdots = \lambda_m = 0, \quad 0 < \lambda_{m+1} \leq \cdots \leq \lambda_n \tag{3.1}$$

and investigate the asymptotic behavior in the vicinity of this critical point. Recall that $\mathbf{U}$ is the displacement vector, Λ is the loading parameter, and $(\,\cdot\,)^{\mathrm{c}}$ denotes a variable at the critical point.

In the description of the critical state, it is pertinent to use generalized coordinates $\mathbf{Q} = (Q_i) \in \mathbb{R}^n$ in the direction of the eigenvectors $\mathbf{\Phi}_i^{\mathrm{c}} = (\phi_{ij}^{\mathrm{c}})$ of $\mathbf{S}$ at the critical point. These coordinates are defined as

$$U_j = \sum_{i=1}^{n} \phi_{ij}^{\mathrm{c}} Q_i = \sum_{i=1}^{n} T_{ji}^{\mathrm{c}} Q_i, \quad (j = 1, \ldots, n) \tag{3.2}$$

where $\mathbf{T}^{\mathrm{c}} = [T_{ji}^{\mathrm{c}}]$ is a local transformation at the critical point, which is defined such that the ith column vector is equal to $\mathbf{\Phi}_i^{\mathrm{c}} = (\phi_{ij}^{\mathrm{c}})$, i.e.,

$$T_{ji}^{\mathrm{c}} = \phi_{ij}^{\mathrm{c}}, \quad (i, j = 1, \ldots, n) \tag{3.3}$$

Define incremental variable vector $\mathbf{q} = (q_i) \in \mathbb{R}^n$ of $\mathbf{Q}$ from $\mathbf{Q}^{\mathrm{c}}$ by

$$\mathbf{q} = \mathbf{Q} - \mathbf{Q}^{\mathrm{c}} \tag{3.4}$$

The total potential energy is rewritten as

$$\Pi^{\mathrm{D}}(\mathbf{q}, \widetilde{\Lambda}, \xi) = \Pi(\mathbf{T}^{\mathrm{c}}(\mathbf{Q}^{\mathrm{c}} + \mathbf{q}), \Lambda^{\mathrm{c}0} + \widetilde{\Lambda}, \mathbf{v}^0 + \xi \mathbf{d}) \tag{3.5}$$

where $\widetilde{\Lambda} = \Lambda - \Lambda^{\mathrm{c}0}$ is the increment of Λ from the critical load factor $\Lambda^{\mathrm{c}0}$ of the perfect system, and the imperfections are defined by $\mathbf{v} = \mathbf{v}^0 + \xi \mathbf{d}$ (cf., (1.14) in Section 1.5.2).

Partial derivative of Π^{D} with respect to q_i is

$$D_{,i} \equiv \frac{\partial \Pi^{\mathrm{D}}}{\partial q_i} = \sum_{j=1}^{n} S_{,j} \phi_{ij}^{\mathrm{c}}, \quad (i = 1, \ldots, n) \tag{3.6}$$

where (3.2)–(3.4) and $\partial \Pi / \partial U_i = S_{,i}$ in (1.2) have been used. Higher derivatives such as $D_{,ijk}$ are evaluated by successive use of (3.6).

Remark 3.2.1 It is possible to compute those higher derivatives in a customary case, e.g., where the finite element method with polynomial displacement interpolation functions is used and geometrical nonlinearity is incorporated by the total Lagrangian formulation [17, 18]. Then Π in the right-hand-side of (3.5) can be expressed in a polynomial form of $\mathbf{U}$. A symbolic computation program Maple 9 [193] is used for simple structures in the following chapters, and the differential coefficients can be computed in each element to be assembled for the whole structure. □

Remark 3.2.2 A finite difference approximation (2.24) in Section 2.5 serves as a computationally efficient alternative to compute the derivatives. For example, using (3.6), the differential coefficients with respect to Λ and ξ, respectively, are written as

$$D_{,r\Lambda} = \frac{\partial}{\partial \Lambda} \sum_{i=1}^{n} S_{,i} \phi_{ri}^{c} = \frac{\partial}{\partial \Lambda} ((S_{,i})^{\top} \mathbf{\Phi}_{r}^{c}) \tag{3.7a}$$

$$D_{,rr\xi} = \frac{\partial}{\partial \xi} \sum_{i=1}^{n} \sum_{j=1}^{n} S_{,ij} \phi_{ri}^{c} \phi_{rj}^{c} = \frac{\partial}{\partial \xi} (\mathbf{\Phi}_{r}^{c\top} \mathbf{S} \mathbf{\Phi}_{r}^{c}) \tag{3.7b}$$

and can be computed numerically by a finite difference approach. □

We introduce below two formulations based on the generalized coordinates.

3.2.2 *D*-formulation

A formulation that employs the expanded or perturbed form of $\Pi^{\mathrm{D}}(\mathbf{q}, \widetilde{\Lambda}, \xi)$ in (3.5) is called *D*-formulation in this book.

The total potential energy in (3.5) is expanded as

$$\begin{aligned} \Pi^{\mathrm{D}}(\mathbf{q}, \widetilde{\Lambda}, \xi) = {} & \Pi^{\mathrm{D}}(\mathbf{0}, 0, 0) + \sum_{i=1}^{n} D_{,i} q_i + D_{,\Lambda} \widetilde{\Lambda} + D_{,\xi} \xi \\ & + \frac{1}{2} \sum_{i=1}^{n} \sum_{j=1}^{n} D_{,ij} q_i q_j + \sum_{i=1}^{n} (D_{,i\Lambda} q_i \widetilde{\Lambda} + D_{,i\xi} q_i \xi) \\ & + \frac{1}{6} \sum_{i=1}^{n} \sum_{j=1}^{n} \sum_{k=1}^{n} D_{,ijk} q_i q_j q_k + \frac{1}{2} \sum_{i=1}^{n} \sum_{j=1}^{n} (D_{,ij\Lambda} q_i q_j \widetilde{\Lambda} + D_{,ij\xi} q_i q_j \xi) \\ & + \frac{1}{24} \sum_{i=1}^{n} \sum_{j=1}^{n} \sum_{k=1}^{n} \sum_{l=1}^{n} D_{,ijkl} q_i q_j q_k q_l + \text{h.o.t.} \end{aligned} \tag{3.8}$$

where h.o.t. denotes higher-order terms, and all derivatives are evaluated at the critical point of the perfect system.

The conditions of equilibrium read

$$D_{,i} = 0, \quad (i = 1, \ldots, n) \tag{3.9}$$

From the standard eigenvalue problem (1.5) with (1.7) and transformation (3.6), we have

$$D_{,ij} = \lambda_i \delta_{ij} = \begin{cases} \lambda_i, & (i = j = 1, \ldots, n) \\ 0, & (i \neq j;\ i, j = 1, \ldots, n) \end{cases} \tag{3.10}$$

where δ_{ij} is the Kronecker delta. Note that so-called stability coefficients $D_{,ii}$ $(i = 1, \ldots, n)$ are nonzero at regular points, but some of them become zero at a critical point.

Incremental equilibrium equations are obtained by the differentiation of $\Pi^{\mathrm{D}}(\mathbf{q}, \widetilde{\Lambda}, \xi)$ in (3.8) with respect to q_i as

$$
\begin{aligned}
D_{,i}(\mathbf{q}, \widetilde{\Lambda}, \xi) = & D_{,ii} q_i + D_{,i\Lambda}\widetilde{\Lambda} + D_{,i\xi}\xi + \frac{1}{2}\sum_{j=1}^{n}\sum_{k=1}^{n} D_{,ijk} q_j q_k \\
& + \sum_{j=1}^{n}(D_{,ij\Lambda} q_j \widetilde{\Lambda} + D_{,ij\xi} q_j \xi) \\
& + \frac{1}{6}\sum_{j=1}^{n}\sum_{k=1}^{n}\sum_{l=1}^{n} D_{,ijkl} q_j q_k q_l + \text{h.o.t.} = 0, \quad (i = 1, \ldots, n)
\end{aligned}
\tag{3.11}
$$

3.2.3 V-formulation

A nonlinear governing equation of a structure subjected to buckling in general involves a large number of independent variables and nonlinear terms and, hence, is highly complex. In the *general nonlinear theory of stability* [166], to obtain asymptotic general forms of imperfection sensitivity laws, the nonlinear governing equation was simplified on the basis of the following two steps:

Step 1: The governing equation is reduced to the *bifurcation equation* with only a few active independent variables by the *Liapunov–Schmidt–Koiter reduction* [50, 99, 266]. For a potential system, the elimination of passive coordinates was employed for the same purpose [285].

Step 2: Higher-order terms of the bifurcation equation are truncated by an asymptotic assumption.

A formulation based on this reduction is called V-formulation and is employed throughout this book [123, 148, 183, 265, 287, 291]. The formalism of V-formulation is suitable for the classification of critical points and the derivation of imperfection sensitivity laws.

The generalized coordinates $\mathbf{q}$ are decomposed into

- active coordinates $\mathbf{q}^{\mathrm{a}} = (q_1, \ldots, q_m)^\top$ associated with zero eigenvalues, and
- passive coordinates $\mathbf{q}^{\mathrm{p}} = (q_{m+1}, \ldots, q_n)^\top$ associated with nonzero ones,

namely, $\mathbf{q} = (\mathbf{q}^{\mathrm{a}\top}, \mathbf{q}^{\mathrm{p}\top})^\top$. Compatibly with this decomposition, equilibrium equations (3.11) can be decomposed into two parts: m equilibrium equations associated with m zero eigenvalues $D_{,ii} = 0$ $(i = 1, \ldots, m)$:

$$
\begin{aligned}
D_{,i}(\mathbf{q}, \widetilde{\Lambda}, \xi) = & \, 0 \times q_i + D_{,i\Lambda}\widetilde{\Lambda} + D_{,i\xi}\xi + \frac{1}{2}\sum_{j=1}^{n}\sum_{k=1}^{n} D_{,ijk} q_j q_k \\
& + \sum_{j=1}^{n} D_{,ij\Lambda} q_j \widetilde{\Lambda} + \text{h.o.t.} = 0, \quad (i = 1, \ldots, m)
\end{aligned}
\tag{3.12}
$$

and $n - m$ equilibrium equations associated with $n - m$ nonzero eigenvalues $D_{,ii} = \lambda_i \neq 0$ $(i = m + 1, \ldots, n)$

$$D_{,i}(\mathbf{q}, \widetilde{\Lambda}, \xi) = \lambda_i q_i + D_{,i\Lambda}\widetilde{\Lambda} + D_{,i\xi}\xi + \frac{1}{2}\sum_{j=1}^{n}\sum_{k=1}^{n} D_{,ijk}q_j q_k + \sum_{j=1}^{n} D_{,ij\Lambda}q_j\widetilde{\Lambda} + \text{h.o.t.} = 0, \quad (i = m+1, \ldots, n) \tag{3.13}$$

By the implicit function theorem, $\mathbf{q}^{\mathrm{p}}$ can be locally expressed as a function of $\mathbf{q}^{\mathrm{a}}$ by solving (3.13) for $\mathbf{q}^{\mathrm{p}}$ as

$$q_i = -\frac{1}{\lambda_i}\Big(D_{,i\Lambda}\widetilde{\Lambda} + D_{,i\xi}\xi + \frac{1}{2}\sum_{j=1}^{m}\sum_{k=1}^{m} D_{,ijk}q_j q_k + \text{h.o.t.}\Big), \quad (i = m+1, \ldots, n) \tag{3.14}$$

namely, we have the implicit expression

$$\mathbf{q}^{\mathrm{p}} = \mathbf{q}^{\mathrm{p}}(\mathbf{q}^{\mathrm{a}}, \widetilde{\Lambda}, \xi) \quad \text{or} \quad \mathbf{q} = \mathbf{q}(\mathbf{q}^{\mathrm{a}}, \widetilde{\Lambda}, \xi) \tag{3.15}$$

In the *V-formulation* [285], passive coordinates $\mathbf{q}^{\mathrm{p}}$ are eliminated from the total potential energy $\Pi^{\mathrm{D}}(\mathbf{q}, \widetilde{\Lambda}, \xi)$ of *D*-formulation in (3.8) by using (3.15) as

$$\Pi^{\mathrm{V}}(\mathbf{q}^{\mathrm{a}}, \widetilde{\Lambda}, \xi) = \Pi^{\mathrm{D}}(\mathbf{q}(\mathbf{q}^{\mathrm{a}}, \widetilde{\Lambda}, \xi), \widetilde{\Lambda}, \xi) \tag{3.16}$$

Then Π^{V} is expanded as

$$\begin{aligned}\Pi^{\mathrm{V}}(\mathbf{q}^{\mathrm{a}}, \widetilde{\Lambda}, \xi) = {} & \Pi^{\mathrm{V}}(\mathbf{0}, 0, 0) + \sum_{i=1}^{m} V_{,i}q_i + V_{,\Lambda}\widetilde{\Lambda} + V_{,\xi}\xi \\ & + \frac{1}{2}\sum_{i=1}^{m}\sum_{j=1}^{m} V_{,ij}q_i q_j + \sum_{i=1}^{m}(V_{,i\Lambda}q_i\widetilde{\Lambda} + V_{,i\xi}q_i\xi) \\ & + \frac{1}{6}\sum_{i=1}^{m}\sum_{j=1}^{m}\sum_{k=1}^{m} V_{,ijk}q_i q_j q_k + \frac{1}{2}\sum_{i=1}^{m}\sum_{j=1}^{m}(V_{,ij\Lambda}q_i q_j\widetilde{\Lambda} + V_{,ij\xi}q_i q_j \xi) \\ & + \frac{1}{24}\sum_{i=1}^{m}\sum_{j=1}^{m}\sum_{k=1}^{m}\sum_{l=1}^{m} V_{,ijkl}q_i q_j q_k q_l + \text{h.o.t.}\end{aligned} \tag{3.17}$$

and all derivatives are evaluated at the critical point of the perfect system. Differential coefficients $V_{,ij}$, $V_{,ijk}, \ldots$ $(i, j, k = 1, \ldots, m)$ with respect to active coordinates q_i $(i = 1, \ldots, m)$ are employed in Section 3.3 to classify critical points.

The stationary condition of Π^{V} with respect to q_i^{a}, which is called *bifurcation equation*, is expressed as

$$\frac{\partial \Pi^{\mathrm{V}}}{\partial q_i^{\mathrm{a}}} = 0, \quad (i = 1, \ldots, m) \tag{3.18}$$

3.2.4 Correspondence between D-formulation and V-formulation

By substituting (3.14) into (3.8) and comparing with (3.17), we have obvious correspondence between the coefficients of D-formulation and V-formulation:

$$V_{,i} = D_{,i}, \quad V_{,i\Lambda} = D_{,i\Lambda}, \quad V_{,i\xi} = D_{,i\xi}, \quad (i = 1, \ldots, m) \tag{3.19a}$$

$$V_{,ij} = D_{,ij}, \quad (i, j = 1, \ldots, m) \tag{3.19b}$$

$$V_{ijk} = D_{ijk}, \quad (i, j, k = 1, \ldots, m) \tag{3.19c}$$

and non-obvious correspondence suffering from the *contamination by passive coordinates*

$$V_{,ij\Lambda} = D_{,ij\Lambda} - \sum_{k=m+1}^{n} \frac{D_{,ijk} D_{,k\Lambda}}{\lambda_k}, \quad (i, j = 1, \ldots, m) \tag{3.20a}$$

$$V_{,ij\xi} = D_{,ij\xi} - \sum_{k=m+1}^{n} \frac{D_{,ijk} D_{,k\xi}}{\lambda_k}, \quad (i, j = 1, \ldots, m) \tag{3.20b}$$

$$V_{,ijrs} = D_{,ijrs} - 3 \sum_{k=m+1}^{n} \frac{D_{,ijk} D_{,rsk}}{\lambda_k}, \quad (i, j, r, s = 1, \ldots, m) \tag{3.20c}$$

which gives

$$V_{,iiii} = D_{,iiii} - 3 \sum_{k=m+1}^{n} \frac{(D_{,iik})^2}{\lambda_k}, \quad (i = 1, \ldots, m) \tag{3.21}$$

For large-scale finite-dimensional systems, it is cumbersome to explicitly solve (3.13) for $\mathbf{q}^{\mathrm{p}}$; even computation of the derivative of $\mathbf{q}^{\mathrm{p}}$ with respect to $\mathbf{q}^{\mathrm{a}}$ is not practically feasible. Therefore, in this book, the formulations by the physical coordinates $\mathbf{U}$ are derived, wherever possible, by incorporating the effect of contamination by passive coordinates via (3.20).

Remark 3.2.3 Since the direct analytical evaluation of $V_{,ii\Lambda}$ by (3.20a) is costly, $V_{,ii\Lambda}$ can be numerically approximated by the backward finite difference

$$V_{,ii\Lambda} \simeq \frac{1}{\Delta\Lambda} [\lambda_i(\Lambda) - \lambda_i(\Lambda - \Delta\Lambda)] \tag{3.22}$$

where $\Delta\Lambda$ denotes the increment of Λ along the fundamental path. Note that the forward or central finite difference approach can also be used. □

3.3 Classification of Critical Points

Critical points that were classified to simple and coincident points in Section 1.2 are further classified herein. Classification of critical points is important in the study of imperfection sensitivity laws, the explicit forms of which change according to the type of critical point.

3.3.1 Simple critical points

Equilibrium paths of simple critical points are illustrated in Fig. 3.1. Focus on the *first* simple critical point on the fundamental path that satisfies

$$\lambda_1 = 0, \quad 0 < \lambda_2 \leq \cdots \leq \lambda_n \tag{3.23}$$

$$V_{,1} = 0, \quad V_{,11} = 0 \tag{3.24}$$

The simple critical point is classified to

$$\begin{cases} \text{limit point :} & V_{,1\Lambda} \neq 0 \\ \text{bifurcation point :} & V_{,1\Lambda} = 0 \end{cases} \tag{3.25}$$

where (cf., (3.6) and (3.19a))

$$V_{,1\Lambda} = D_{,1\Lambda} = \sum_{i=1}^{n} S_{,i\Lambda}\phi_{1i}^{\mathrm{c}} \tag{3.26}$$

Eq. (3.25) with (3.26) means that the buckling mode $\boldsymbol{\Phi}_1^{\mathrm{c}}$ is orthogonal to the instantaneous load pattern vector $(S_{,i\Lambda})$ at a bifurcation point, but is not orthogonal at a limit point. The classification in (3.25) is readily applicable to numerical analysis, as the vector $(S_{,i\Lambda})$ and the eigenvector $\boldsymbol{\Phi}_1^{\mathrm{c}} = (\phi_{1i}^{\mathrm{c}})$ in (3.26) are obtained during the path-tracing analysis.

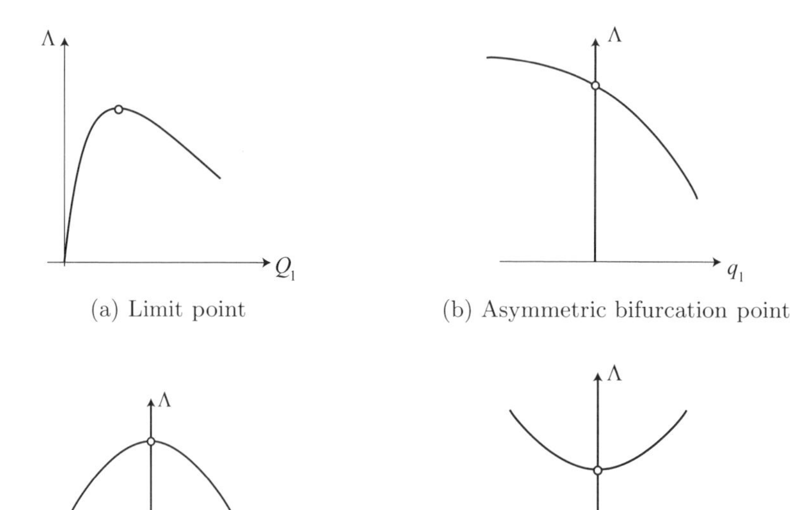

(a) Limit point

(b) Asymmetric bifurcation point

(c) Unstable-symmetric bifurcation point

(d) Stable-symmetric bifurcation point

Fig. 3.1 Equilibrium paths of simple critical points. Q_1, q_1: displacements in the direction of the critical modes; ○: critical point.

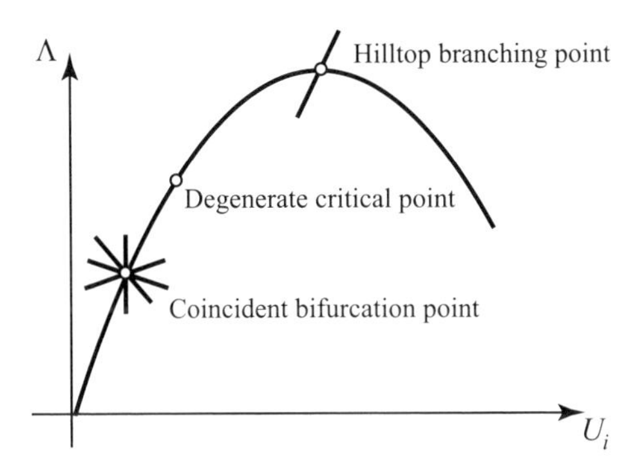

Fig. 3.2 Equilibrium paths of coincident critical points. U_i: representative displacement; ∘: critical point.

For the proportional loading (1.11), the instantaneous load pattern vector $(S_{,i\Lambda})$ becomes a constant vector $(-p_i)$ and

$$V_{,1\Lambda} = \sum_{i=1}^{n} S_{,i\Lambda}\phi^{\rm c}_{1i} = -\sum_{i=1}^{n} p_i\phi^{\rm c}_{1i} \tag{3.27}$$

A simple bifurcation point is classified to

$$\begin{cases} \text{asymmetric bifurcation point}: & V_{,111} \neq 0 \\ \text{symmetric bifurcation point}: & V_{,111} = 0 \end{cases} \tag{3.28}$$

The bifurcation path has a zero slope at a symmetric bifurcation point, but has a nonzero slope at an asymmetric bifurcation point as shown in Fig. 3.1.

A symmetric bifurcation point is further classified to

$$\begin{cases} \text{unstable-symmetric bifurcation point}: & V_{,1111} < 0 \\ \text{stable-symmetric bifurcation point}: & V_{,1111} > 0 \end{cases} \tag{3.29}$$

The bifurcation path branches downward for the unstable-symmetric point, and branches upward for the stable-symmetric point as shown in Figs. 3.1(c) and (d).

A simple *degenerate*[1] (isolated) critical point has a special feature that it is not either a limit point or a bifurcation point as shown in Fig. 3.2.

A limit point will play a pivotal role in the optimization of structures in Chapters 4–6. An asymmetric bifurcation point will appear below in defining coincident critical points. Examples of an unstable-symmetric bifurcation point and a stable-symmetric one will be provided in Section 11.5.1 and Chapter 13, respectively.

3.3.2 Coincident critical points

Coincident critical points appear generically through optimization (cf., Part II), and also owing to geometrical symmetry. Consider a coincident critical point with

---

[1] A coincident degenerate critical point possibly has a bifurcation path (cf., Appendices A.8 and A.9).

m (≥ 2) zero eigenvalues

$$\lambda_1 = \cdots = \lambda_m = 0, \quad 0 < \lambda_{m+1} \leq \cdots \leq \lambda_n \tag{3.30}$$

In this case, an arbitrary linear combination of critical eigenvectors $\mathbf{\Phi}_i^{\mathrm{c}}$ ($i = 1, \ldots, m$) serves as a critical eigenvector, because

$$\sum_{j=1}^{n} \sum_{k=1}^{m} S_{,ij} a_k \phi_{kj}^{\mathrm{c}} = 0, \quad (i = 1, \ldots, n) \tag{3.31}$$

is satisfied for any set of coefficients a_k ($k = 1, \ldots, m$).

Equilibrium paths of coincident critical points are illustrated in Fig. 3.2, and coincident critical points consist of:

- **Hilltop branching point**[2] at which a limit point and bifurcation point(s) coincide.
- **Coincident bifurcation point** (compound branching point) at which bifurcation points coincide.
- **Group-theoretic double point** which arises from the symmetry of a structure.

Remark 3.3.1 Structures with dihedral-group symmetry appear for shells of revolution and reticulated regular-polygonal truss domes. These structures have particular double bifurcation points due to symmetry, in addition to limit points and simple-symmetric bifurcation points. These double points have thoroughly been studied by group-theoretic bifurcation theory [91, 100, 117, 142, 144, 266]. □

A hilltop branching point with multiplicity $m = 2$, at which two eigenvalues of $\mathbf{S}$ vanish simultaneously as shown in Fig. 3.3, is classified to

$$\begin{cases} \text{hilltop point with symmetric bifurcation:} & V_{,111} = 0, V_{,112} \neq 0 \\ \text{hilltop point with asymmetric bifurcation:} & V_{,111} \neq 0, V_{,112} \neq 0 \\ \text{degenerate hilltop point:} & V_{,112} = 0 \end{cases} \tag{3.32}$$

where q_1 and q_2 are associated with the bifurcation mode and the limit point mode, respectively; therefore,

$$V_{,1\Lambda} = 0, \quad V_{,2\Lambda} \neq 0 \tag{3.33}$$

are satisfied. The variation of eigenvalues at a degenerate hilltop point is illustrated in Fig. 3.3(b). A bifurcation path(s) branches at a hilltop point. Optimal designs with hilltop branching will be highlighted in Chapters 8 and 10; examples of hilltop points will be provided also in Sections 3.7, 4.5 and 11.4.

A coincident bifurcation point with $m = 2$ can be classified in view of the symmetries as

[2] A hilltop branching point is often called a hilltop point.

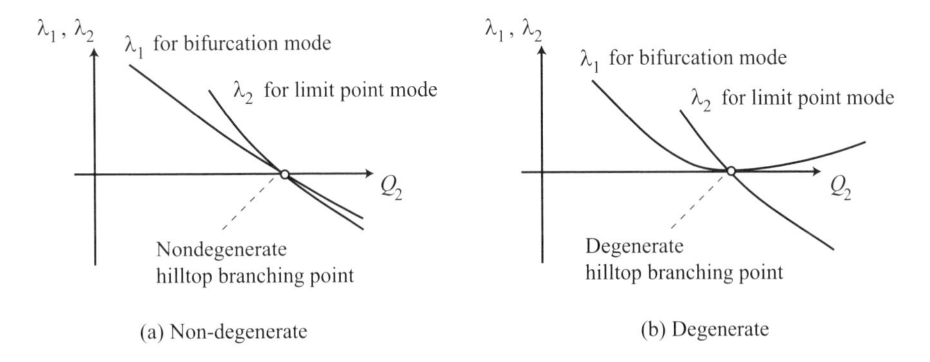

Fig. 3.3 Variation of eigenvalues at a hilltop branching point with multiplicity $m = 2$.

- A *semi-symmetric double bifurcation point* emerges as a coincidence of a simple symmetric bifurcation point and a simple asymmetric one. Suppose q_1 and q_2 correspond to symmetric and asymmetric bifurcation points, respectively. Since Π^{V} at this point is symmetric with respect to q_1 and asymmetric to q_2, the third-order differential coefficients satisfy

$$V_{,111} = 0, \quad V_{,112} \neq 0, \quad V_{,122} = 0, \quad V_{,222} \neq 0 \tag{3.34}$$

  An example of this point will be provided in Section 7.4.

- A *completely-symmetric double bifurcation point*[3] is a point where Π^{V} is symmetric both for q_1 and q_2, corresponding to two bifurcation modes. Then all third-order terms vanish, namely,

$$V_{,111} = V_{,112} = V_{,122} = V_{,222} = 0 \tag{3.35}$$

  This critical point emerges generically as a result of optimization of symmetric structures that has two planes of reflection symmetry, such as rectangular plates and thin-walled columns with symmetric sections.

3.4 Derivation of Imperfection Sensitivity Laws

We introduce two methodologies to derive imperfection sensitivity laws through the application to a simple unstable-symmetric bifurcation point:

- *Power series expansion method* that is insightful and rigorous is employed throughout this book [137, 142, 205].
- *Static perturbation method* that is robust and standard [285].

---

[3]Masur [195] defined completely-symmetric system such that the cubic terms of all the possible modes vanish. Thompson [287] used the term *fully-asymmetric system* when no cubic term of the active coordinates vanishes.

3.4.1 Power series expansion method

In the power series expansion method, the expanded form $\Pi^{\rm V}$ in (3.17) is employed. For a simple unstable-symmetric bifurcation point, we have (cf., (3.24), (3.25) and (3.28))

$$V_{,1} = V_{,11} = V_{,111} = 0, \quad V_{,1\Lambda} = 0 \tag{3.36}$$

Then $\Pi^{\rm V}$ becomes [142]

$$\begin{aligned}\Pi^{\rm V}(q_1, \widetilde{\Lambda}, \xi) = &\Pi^{\rm V}(0,0,0) + V_{,1\xi} q_1 \xi + \frac{1}{2} V_{,11\Lambda}(q_1)^2 \widetilde{\Lambda} + \frac{1}{2} V_{,1\Lambda\Lambda} q_1 \widetilde{\Lambda}^2 \\ &+ \frac{1}{2} V_{,11\xi}(q_1)^2 \xi + \frac{1}{24} V_{,1111}(q_1)^4 + \text{h.o.t.}\end{aligned} \tag{3.37}$$

The bifurcation equation for this point is expressed as:

$$\frac{\partial \Pi^{\rm V}}{\partial q_1} = V_{,1\xi}\xi + V_{,11\Lambda} q_1 \widetilde{\Lambda} + \frac{1}{2} V_{,1\Lambda\Lambda} \widetilde{\Lambda}^2 + V_{,11\xi} q_1 \xi + \frac{1}{6} V_{,1111}(q_1)^3 = 0 \tag{3.38}$$

and its criticality condition is

$$\frac{\partial^2 \Pi^{\rm V}}{\partial {q_1}^2} = V_{,11\Lambda}\widetilde{\Lambda} + V_{,11\xi}\xi + \frac{1}{2} V_{,1111}(q_1)^2 = 0 \tag{3.39}$$

When ξ is small, the critical point $(q_1^{\rm c}, \widetilde{\Lambda}^{\rm c})$ of an imperfect system exists in the neighborhood of the bifurcation point $(q_1, \widetilde{\Lambda}) = (0,0)$ of the perfect system. The critical load $\widetilde{\Lambda}^{\rm c}$ can be determined by simultaneously solving (3.38) and (3.39) to arrive at the sensitivity law

$$\Lambda^{\rm c}(\xi) = \Lambda^{\rm c0} - \frac{3^{\frac{2}{3}}(V_{,1111})^{\frac{1}{3}}}{2V_{,11\Lambda}}(V_{,1\xi}\xi)^{\frac{2}{3}} - \frac{V_{,11\xi}}{V_{,11\Lambda}}\xi \tag{3.40}$$

3.4.2 Static perturbation method

In the static perturbation method, the critical coordinate $q_1^{\rm c}$ can be chosen as the independent parameter for a solution curve for the unstable-symmetric bifurcation point. Differentiating the equilibrium equation and the criticality condition

$$V_{,1} = 0, \quad V_{,11} = 0 \tag{3.41}$$

with respect to $q_1^{\rm c}$ at the critical point, and using (3.41) and $V_{,111} = V_{,1\Lambda} = 0$ in (3.36), we obtain

$$V_{,11} + V_{,1\Lambda}\Lambda^{\rm c}_{,1} + V_{,1\xi}\xi_{,1} = 0 \quad \Longrightarrow \quad V_{,1\xi}\xi_{,1} = 0 \tag{3.42a}$$

$$V_{,111} + V_{,11\Lambda}\Lambda^{\rm c}_{,1} + V_{,11\xi}\xi_{,1} = 0 \quad \Longrightarrow \quad V_{,11\Lambda}\Lambda^{\rm c}_{,1} + V_{,11\xi}\xi_{,1} = 0 \tag{3.42b}$$

Here $\Lambda^{\rm c}_{,1}$ and $\xi_{,1}$ denote total differentiation of $\Lambda^{\rm c}$ and ξ with respect to $q_1^{\rm c}$, respectively.

For the minor imperfection with $V_{,1\xi} = 0$, (3.42a) is always satisfied and (3.42b) gives the sensitivity law:

$$\Lambda^{\rm c}_{,1} = -\frac{V_{,11\xi}}{V_{,11\Lambda}}\xi_{,1} \Longrightarrow \Lambda^{\rm c} = \Lambda^{\rm c0} - \frac{V_{,11\xi}}{V_{,11\Lambda}}\xi \tag{3.43}$$

For the major imperfection with $V_{,1\xi} \neq 0$ and $V_{,11\xi} = 0$, (3.42) gives

$$\Lambda^{\rm c}_{,1} = 0, \quad \xi_{,1} = 0 \tag{3.44}$$

Differentiation of (3.41) with respect to $q_1^{\rm c}$ twice and use of $V_{,111} = 0$ of (3.36) result in

$$\Lambda^{\rm c}_{,11} = -\frac{V_{,1111}}{V_{,11\Lambda}}, \quad \xi_{,11} = 0, \quad \xi_{,111} = \frac{2V_{,1111}}{V_{,1\xi}} \tag{3.45}$$

From (3.45), $\Lambda^{\rm c}$ and ξ are expanded with respect to q_1 as

$$\Lambda^{\rm c} = \Lambda^{\rm c0} - \frac{V_{,1111}}{2V_{,11\Lambda}}(q_1^{\rm c})^2, \quad \xi = \frac{V_{,1111}}{3V_{,1\xi}}(q_1^{\rm c})^3 \tag{3.46}$$

The elimination of $q_1^{\rm c}$ from these equations leads to

$$\Lambda^{\rm c}(\xi) = \Lambda^{\rm c0} - \frac{3^{\frac{2}{3}}(V_{,1111})^{\frac{1}{3}}}{2V_{,11\Lambda}}(V_{,1\xi}\xi)^{\frac{2}{3}} \tag{3.47}$$

The combination of (3.43) and (3.47) gives the law (3.40) obtained previously by the power series expansion method.

3.5 Imperfection Sensitivity for Simple Critical Points

Imperfection sensitivity for simple critical points is investigated for general asymmetric imperfections, which contain both minor and major imperfections. This serves as a generalization of the study in Section 2.2 for the pedagogic example, the propped cantilever. Recall that the sensitivity law (2.11) for an unstable-symmetric bifurcation point is governed by the 2/3-power of a major imperfection, followed by 1-power of a minor imperfection.

3.5.1 Imperfect behaviors

Perfect and imperfect equilibrium paths for simple critical points are illustrated in Figs. 3.4(a)–(d). The imperfect fundamental equilibrium paths for general asymmetric imperfection behave as follows:

- **Limit point:** Another limit point exists on an imperfect equilibrium path. The limit point load is reduced for $\xi > 0$ and is increased for $\xi < 0$, or vice versa.
- **Asymmetric bifurcation point:** A limit point exists on an imperfect equilibrium path for $\xi > 0$, but is absent on the path for $\xi < 0$, or vice versa. The critical load is governed by this limit point.
- **Unstable-symmetric bifurcation point:** Imperfect equilibrium paths for $\pm\xi$ are symmetric with respect to the Λ-axis and have limit points, which yield the identical critical load.
- **Stable-symmetric bifurcation point:** Imperfect equilibrium paths for $\pm\xi$ are symmetric with respect to the Λ-axis and the load Λ increases stably

beyond the bifurcation load Λ^{c0} for the perfect system. The critical load Λ^{c} for an imperfect system is non-existent because there is no critical point along these imperfect paths.

3.5.2 Imperfection sensitivity laws

For an imperfect system with an asymmetric imperfection, the critical load factor $\Lambda^{c}(\xi)$ in the vicinity of the perfect system is expressed asymptotically for sufficiently small ξ by imperfection sensitivity laws

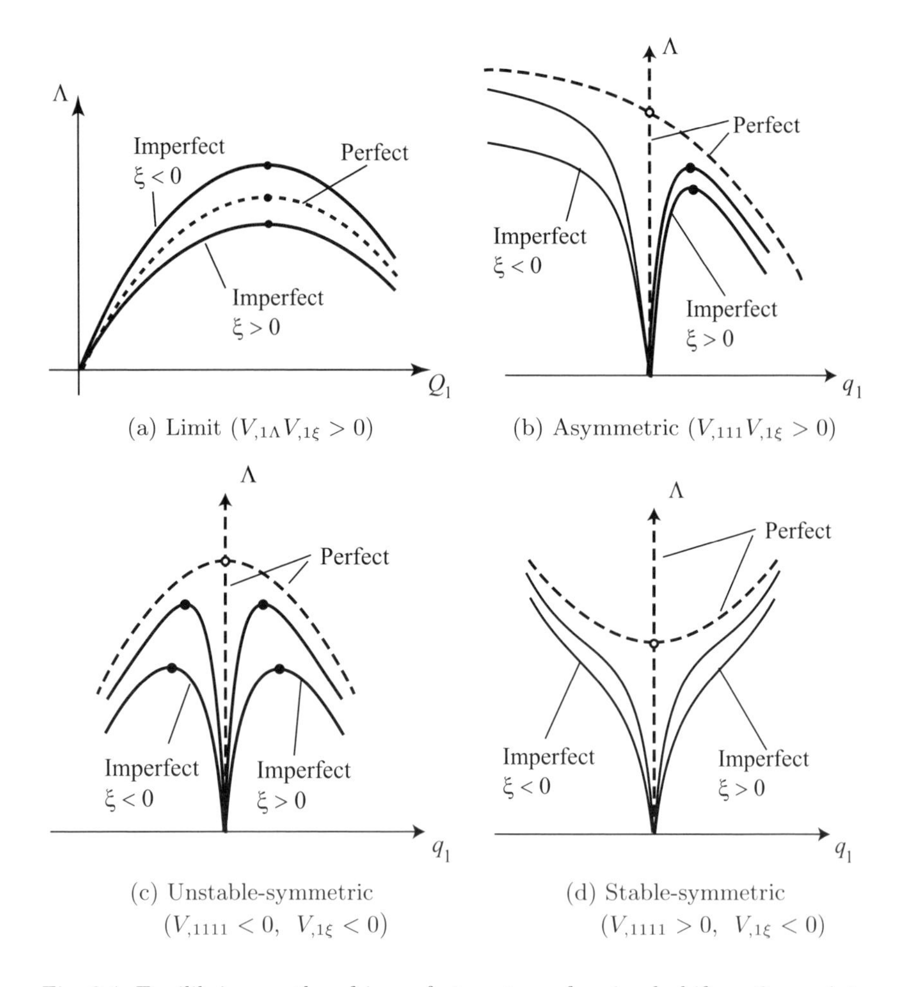

Fig. 3.4 Equilibrium paths of imperfect systems for simple bifurcation points. Q_1, q_1: displacements in the direction of critical mode $\boldsymbol{\Phi}_1^{c}$; dashed curve: perfect equilibrium path; solid curve: imperfect equilibrium path; ○: bifurcation point; •: limit point.

$$\Lambda^{\rm c}(\xi) = \begin{cases} \Lambda^{\rm c0} + C_0 V_{,1\xi}\xi : & \text{for limit point} \\ \Lambda^{\rm c0} + C_1 |V_{,1\xi}\xi|^{\frac{1}{2}} + C_2 V_{,11\xi}\xi : & \\ & \text{for asymmetric bifurcation point} \\ & (V_{,111}V_{,1\xi}\xi > 0) \\ \Lambda^{\rm c0} + C_3 (V_{,1\xi}\xi)^{\frac{2}{3}} + C_2 V_{,11\xi}\xi : & \\ & \text{for unstable-symmetric bifurcation point} \\ & (V_{,1111} < 0) \\ \text{non-existent} : & \text{for stable-symmetric bifurcation point} \end{cases} \tag{3.48}$$

where

$$C_0 = -\frac{1}{V_{,1\Lambda}}, \quad C_1 = \frac{|V_{,111}|^{\frac{1}{2}}}{V_{,11\Lambda}}, \quad C_2 = -\frac{1}{V_{,11\Lambda}}, \quad C_3 = -\frac{3^{\frac{2}{3}}(V_{,1111})^{\frac{1}{3}}}{2V_{,11\Lambda}} \tag{3.49}$$

For an asymmetric bifurcation point, all the results in this section are defined conditionally given that a limit point exists on the fundamental path of the imperfect system ($V_{,111}V_{,1\xi}\xi > 0$), and the expressions of C_1 and C_2 for an asymmetric point corresponds to a case where the fundamental path has the trivial solution $q_1 = 0$. For an unstable-symmetric bifurcation point, the sensitivity law (3.48) and the explicit forms of C_2 and C_3 were derived in (3.40) in Section 3.4.

For the nondegenerate first bifurcation point along the fundamental path, $V_{,11} = \lambda_1$ changes its sign from positive to negative as Λ is increased along the fundamental path; hence,

$$V_{,11\Lambda} = \lambda_{1,\Lambda} < 0 \tag{3.50}$$

holds. Therefore, $C_1 < 0$ and $C_3 < 0$ are satisfied. The critical loads $\Lambda^{\rm c}$ of imperfect systems for the asymmetric bifurcation and the unstable-symmetric bifurcation ($V_{,1111} < 0$) are sharply reduced, and, therefore, are *imperfection-sensitive*.

3.5.3 Sensitivity coefficients

The sensitivity laws in (3.48) capture the asymptotic influence of minor and major imperfections. Since the leading order terms of ξ in (3.48) change according to the vanishing and non-vanishing of $V_{,1\xi}$, the minor and major imperfections are classified by [258, 283]:

$$\begin{cases} \text{minor (second-order) imperfection:} & V_{,1\xi} = 0 \\ \text{major (first-order) imperfection:} & V_{,1\xi} \neq 0 \end{cases} \tag{3.51}$$

For the minor imperfection with $V_{,1\xi} = 0$, (3.48) gives the sensitivity coefficient

$$\Lambda^{\rm c\prime}(0) = \begin{cases} 0 : & \text{for limit point} \\ C_2 V_{,11\xi} : & \text{for bifurcation point} \end{cases} \tag{3.52}$$

Since this sensitivity does not have any singularity, the sensitivity related to a minor imperfection is called *regular sensitivity* [202].

For the major imperfection with $V_{,1\xi} \neq 0$, the sensitivity coefficient of $\Lambda^{\rm c}(\xi)$ is computed as

$$\Lambda^{\rm c\prime}(\xi) = \begin{cases} \frac{1}{2}C_1 \,\mathrm{sign}\,\xi \,|V_{,1\xi}|^{\frac{1}{2}}|\,|\xi|^{-\frac{1}{2}} + C_2 V_{,11\xi} : & \text{for asymmetric point} \\ \frac{2}{3}C_3 (V_{,1\xi})^{\frac{2}{3}}\xi^{-\frac{1}{3}} + C_2 V_{,11\xi} : & \text{for unstable-symmetric point} \end{cases} \tag{3.53}$$

where $\mathrm{sign}\,\xi$ denotes the sign of ξ. This equation indicates that the value of $\Lambda^{\rm c\prime} = \Lambda^{\rm c\prime}(0)$ for the perfect system is unbounded. For this reason, the sensitivity for a major imperfection is called *singular sensitivity* [202, 283].

Remark 3.5.1 The sensitivity law (3.48) with (3.49) for a simple bifurcation point for a minor imperfection with $V_{,1\xi} = 0$ reduces to

$$\Lambda^{\rm c}(\xi) = \Lambda^{\rm c0} - \frac{V_{,11\xi}}{V_{,11\Lambda}}\xi \tag{3.54}$$

and $\Lambda^{\rm c\prime} = \Lambda^{\rm c\prime}(0)$ is evaluated to

$$\Lambda^{\rm c\prime} = -\frac{V_{,11\xi}}{V_{,11\Lambda}} = -\Big(D_{,11\xi} - \sum_{r=2}^{n} \frac{D_{,11r}D_{,r\xi}}{D_{,rr}}\Big)\Big/\Big(D_{,11\Lambda} - \sum_{r=2}^{n} \frac{D_{,11r}D_{,r\Lambda}}{D_{,rr}}\Big) \tag{3.55}$$

with the use of the expansions (3.20a) and (3.20b). The transformation (3.6) is used in (3.55) to obtain the explicit form (2.37) with (2.38). □

3.6 Imperfection Sensitivity for Coincident Critical Points

It is a characteristic of a coincident critical point that its bifurcation behavior may become completely different from the behaviors of individual simple critical points that compose the coincident point. The coincident point, in turn, often has completely different imperfection sensitivity from those of the simple points. Such difference arising from the presence of mode interaction terms, such as $V_{,112}$, will be studied in Chapters 7–10. In this section, as its prologue, imperfection sensitivity laws for coincident critical points are briefly introduced, and are shown to vary according to the type of a coincident critical point.

3.6.1 Hilltop branching point

Imperfection sensitivity laws of hilltop branching points with simple bifurcation ($m = 2$) are presented briefly, while details are given in Section 8.2 and Appendices A.8 and A.9. Recall that this point is classified by (3.32) according to whether the bifurcation point at the limit point is symmetric or asymmetric, and whether the point is nondegenerate ($V_{,112} \neq 0$) or degenerate ($V_{,112} = 0$).

The nondegenerate hilltop points have piecewise linear laws, which are not that imperfection-sensitive as the 2/3- and 1/2-power laws for simple bifurcation points [145, 146, 284]. This is a major claim of this book to be revived repeatedly.

When critical eigenmodes $\boldsymbol{\Phi}_1^{\rm c}$ and $\boldsymbol{\Phi}_2^{\rm c}$ are assumed to be related to a bifurcation point and a limit point, respectively, these laws are given as follows:

- For a nondegenerate hilltop point with simple symmetric bifurcation, the imperfection sensitivity law[4] for an asymmetric imperfection ξ becomes

$$\Lambda^{\rm c}(\xi) = \Lambda^{\rm c0} + C_4 |V_{,1\xi}\xi| + C_5 V_{,2\xi}\xi \tag{3.56}$$

  Here $C_4|V_{,1\xi}\xi|$ corresponds to a major (antisymmetric) imperfection, and $C_5 V_{,2\xi}\xi$ corresponds to a minor (symmetric) imperfection[5]. The curve of $\Lambda^{\rm c}(\xi)$ has a kink at $\xi = 0$. The sensitivity coefficient $\Lambda^{\rm c\prime}(\xi)$ is bounded but discontinuous at $\xi = 0$. For an antisymmetric imperfection with $V_{,1\xi} \neq 0$ and $V_{,2\xi} = 0$, the law (3.56) is symmetric.

- For a nondegenerate hilltop point with simple asymmetric bifurcation, the imperfection sensitivity law is given by

$$\Lambda^{\rm c}(\xi) = \begin{cases} \Lambda^{\rm c0} + C_6 |V_{,1\xi}\xi| + C_5 V_{,2\xi}\xi : & \text{for } \xi > 0 \\ \Lambda^{\rm c0} + C_7 |V_{,1\xi}\xi| + C_5 V_{,2\xi}\xi : & \text{for } \xi < 0 \end{cases} \tag{3.57}$$

- For a degenerate hilltop point, the imperfection sensitivity laws have many variants in accordance with the vanishing and non-vanishing of the differential coefficients $V_{,111}, V_{,1111}, \ldots$, as worked out in Appendices A.8 and A.9. Its imperfection sensitivity is possibly as severe as those of simple bifurcation points.

3.6.2 Semi-symmetric double bifurcation point

We investigate imperfect behaviors of a semi-symmetric double bifurcation point, which emerges as a coincidence of a simple symmetric bifurcation point and a simple asymmetric one (cf., (3.34) in Section 3.3.2).

Suppose the critical eigenmodes $\boldsymbol{\Phi}_1^{\rm c}$ and $\boldsymbol{\Phi}_2^{\rm c}$ are related to symmetric and asymmetric bifurcation points, respectively. Since imperfect behaviors at a semi-symmetric point in general are very complicated, we focus only on the imperfection mode in the direction of $\boldsymbol{\Phi}_2^{\rm c}$.

The imperfection sensitivity for the major imperfection under consideration is given by the 1/2-power law [129]

$$\Lambda^{\rm c}(\xi) = \Lambda^{\rm c0} + C_8(\kappa) |V_{,2\xi}\xi|^{\frac{1}{2}} \tag{3.58}$$

where κ is a structural parameter, which is independent of ξ. The concrete form of $C_8(\kappa)$ is derived in Appendix A.7 and more discussion on this issue can be found in Section 7.3.

[4] The law developed in [289] corresponds to $V_{,2\xi} = 0$.

[5] Although these imperfections exert the same order of influence on the law (3.56), the words major and minor are used to be consistent with other critical points.

3.6.3 Completely-symmetric double bifurcation point

Consider a completely-symmetric double bifurcation point defined in (3.35) in Section 3.3.2. If $V_{,1111} < 0$ and $V_{,1122} = 0$, the sensitivity law for the imperfection in the direction of $\boldsymbol{\Phi}_1^{\mathrm{c}}$ is expressed by the 2/3-power law

$$\Lambda^{\mathrm{c}}(\xi) = \Lambda^{\mathrm{c}0} + C_9 (V_{,1\xi}\xi)^{\frac{2}{3}} \tag{3.59}$$

This is of the same form as the sensitivity law (3.48) for a major imperfection for a simple unstable-symmetric bifurcation point.

Mode interaction is influential on imperfect behaviors. By virtue of the complete symmetry, the third-order terms vanish (cf., (3.35)). Therefore, mode interaction in the third-order terms is absent, and *fourth-order interaction*[6] becomes predominant [284].

Consider, for simplicity, a two-degree-of-freedom system with the completely-symmetric double bifurcation point, and suppose that the two bifurcation paths are both stable and symmetric. If the structure is slightly deviated by preserving symmetry such that two bifurcation points do not coincide, we have $V_{,1111} > 0$, $V_{,2222} > 0$. If a fourth-order cross-term $V_{,1122}$ is negative, there may be an unstable secondary bifurcation point along the bifurcation path near the first bifurcation point. In this case, coincidence of two *stable* bifurcation points may lead to emergence of additional two pairs of *unstable* bifurcation paths that is highlighted as the "danger of naive optimization [285]."

Remark 3.6.1 A thin-walled shell and a frame structure are well known to be subjected to a coincidence of local buckling mode $\boldsymbol{\Phi}^{\mathrm{L}}$ and global buckling mode $\boldsymbol{\Phi}^{\mathrm{G}}$. The local mode $\boldsymbol{\Phi}^{\mathrm{L}}$, e.g., represents member buckling of frames or buckling of ribs of a stiffened plate. An imperfection in the direction of $\boldsymbol{\Phi}^{\mathrm{L}}$ accelerates deformation in the direction of $\boldsymbol{\Phi}^{\mathrm{L}}$, but, at the same time, contributes to an imperfection for $\boldsymbol{\Phi}^{\mathrm{G}}$ and accelerates deformation in the direction of $\boldsymbol{\Phi}^{\mathrm{G}}$. By the recurrence of this interacting process, the maximum load factor is drastically reduced. This interaction of local and global modes is the third- or fourth-order according to cases. □

3.6.4 Group-theoretic double bifurcation point

The imperfection sensitivity laws of a group-theoretic double bifurcation point of a structure with dihedral-group symmetry vary with the number $\widehat{n}$ of bifurcation paths, and are given as [142]

$$\Lambda^{\mathrm{c}}(\xi) = \begin{cases} \Lambda^{\mathrm{c}0} + C_{10}(\psi)\, |\{(V_{,1\xi})^2 + (V_{,2\xi})^2\}^{\frac{1}{2}}\xi|^{\frac{1}{2}} : & \text{for } \widehat{n} = 3 \\ \Lambda^{\mathrm{c}0} + C_{11}(\psi)\, \{[(V_{,1\xi})^2 + (V_{,2\xi})^2]^{\frac{1}{2}}\xi\}^{\frac{2}{3}} : & \text{for } \widehat{n} = 4 \\ \Lambda^{\mathrm{c}0} + C_{12}\, \{[(V_{,1\xi})^2 + (V_{,2\xi})^2]^{\frac{1}{2}}\xi\}^{\frac{2}{3}} : & \text{for } \widehat{n} \geq 5 \end{cases} \tag{3.60}$$

where $\psi = \tan^{-1}(V_{,2\xi}/V_{,1\xi})$, and $\tan^{-1}(\,\cdot\,)$ takes the principal value in the range $[-\pi/2, \pi/2]$.

---

[6] The fourth-order interaction is observed for the Augusti model [287].

3.6.5 Symmetry of a structure

Symmetry has a close relation with bifurcation and the associated imperfection sensitivity, as bifurcation usually takes place in association with the loss of symmetry of the system in question. Consider the initial perfect system with reflection symmetry subjected to a symmetric load that encounters a critical point with the multiplicity m, and q_1 is associated with an antisymmetric bifurcation mode. The results below are applicable to simple bifurcation point by setting $m = 1$.

For the perfect system, the total potential energy Π^{V} is invariant with respect to the reflection $q_1 \mapsto -q_1$, namely,

$$\Pi^{\mathrm{V}}(q_1, q_2, \ldots, q_m, \widetilde{\Lambda}, 0) = \Pi^{\mathrm{V}}(-q_1, q_2, \ldots, q_m, \widetilde{\Lambda}, 0) \tag{3.61}$$

which means Π^{V} is an even function of q_1. Then from the expanded form (3.17) of Π^{V}, we have

$$\begin{aligned} &V_{,1\Lambda} = 0, \quad V_{,111} = 0 \\ &V_{,1i\Lambda} = 0, \quad V_{,1ij} = 0, \quad (i, j = 2, \ldots, m) \end{aligned} \tag{3.62}$$

By $V_{,111} = 0$, this point is symmetric for q_1.

From a standpoint of symmetry, initial imperfections are classified as follows:

- **Symmetric imperfection** that preserves symmetry of the imperfect structure satisfies the symmetry condition [142]

$$\Pi^{\mathrm{V}}(q_1, q_2, \ldots, q_m, \widetilde{\Lambda}, \xi) = \Pi^{\mathrm{V}}(-q_1, q_2, \ldots, q_m, \widetilde{\Lambda}, \xi) \tag{3.63}$$

  which yields

$$\begin{aligned} &V_{,1\xi} = 0, \quad V_{,1i\xi} = 0, \quad (i = 2, \ldots, m) \\ &V_{,11\xi}, \ V_{,i\xi}, \ (i = 2, \ldots, m): \ \text{possibly nonzero} \end{aligned} \tag{3.64}$$

- **Antisymmetric imperfection** that breaks symmetry satisfies the symmetry condition [142]

$$\Pi^{\mathrm{V}}(q_1, q_2, \ldots, q_m, \widetilde{\Lambda}, \xi) = \Pi^{\mathrm{V}}(-q_1, q_2, \ldots, q_m, \widetilde{\Lambda}, -\xi) \tag{3.65}$$

  which yields

$$\begin{aligned} &V_{,11\xi} = 0, \quad V_{,i\xi} = 0, \quad (i = 2, \ldots, m) \\ &V_{,1\xi}, \ V_{,1i\xi}, \ (i = 2, \ldots, m): \ \text{possibly nonzero} \end{aligned} \tag{3.66}$$

- **Asymmetric imperfection** that is the mixture of symmetric and antisymmetric imperfections has no geometrical conditions. Hence all terms associated with imperfections, such as $V_{,1\xi}$, $V_{,11\xi}$, $V_{,i\xi}$ and $V_{,1i\xi}$ $(i = 2, \ldots, m)$, in general are nonzero.

For a general case where occasional vanishing of coefficients does not take place, we have the correspondence

$$\left\{ \begin{array}{ll} \text{symmetric imperfection } (V_{,1\xi} = 0, \ V_{,11\xi} \neq 0): & \text{minor imperfection} \\ \text{antisymmetric imperfection } (V_{,1\xi} \neq 0, \ V_{,11\xi} = 0): & \text{major imperfection} \\ \text{asymmetric imperfection } (V_{,1\xi} \neq 0, \ V_{,11\xi} \neq 0): & \text{major imperfection} \end{array} \right. \tag{3.67}$$

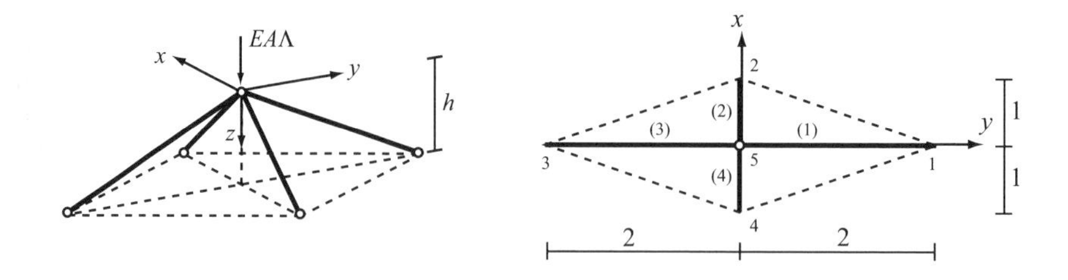

Fig. 3.5 Four-bar truss tent. $1, \ldots, 5$ denote node numbers; $(1), \ldots, (4)$ denote member numbers.

3.7 Imperfection Sensitivity of Four-Bar Truss Tent

The imperfection sensitivity of the four-bar truss tent in Fig. 3.5 with a hilltop point with simple symmetric bifurcation is investigated. The height $h = h^0 = 1.8852$ was chosen elaborately to realize this hilltop branching point. For slightly different values of h than $h^0 = 1.8852$, the hilltop point splits into a pair of simple critical points: a limit point and a symmetric bifurcation point. All truss members have the same cross-sectional area A and the same Young's modulus E.

The set of equilibrium equations of the four-bar truss under the vertical load $EA\Lambda$ is described by

$$EA \sum_{i=1}^{4} \left(\frac{1}{L_i} - \frac{1}{\widehat{L}_i} \right) \begin{pmatrix} U_x + x_5 - x_i \\ U_y + y_5 - y_i \\ U_z + z_5 - z_i \end{pmatrix} - EA\Lambda \begin{pmatrix} 0 \\ 0 \\ 1 \end{pmatrix} = \begin{pmatrix} 0 \\ 0 \\ 0 \end{pmatrix} \tag{3.68}$$

where (x_i, y_i, z_i) is the initial location of node i $(i = 1, \ldots, 5)$, $\mathbf{U} = (U_x, U_y, U_z)^\top$ is the displacement vector of node 5 and

$$\begin{aligned} L_i &= [(x_5 - x_i)^2 + (y_5 - y_i)^2 + (z_5 - z_i)^2]^{\frac{1}{2}} \\ \widehat{L}_i &= [(U_x + x_5 - x_i)^2 + (U_y + y_5 - y_i)^2 + (U_z + z_5 - z_i)^2]^{\frac{1}{2}} \end{aligned} \tag{3.69}$$

for $i = 1, \ldots, 4$ are member lengths before and after deformation, respectively.

We are interested in the change $\widetilde{\Lambda}^{\rm c} = \Lambda^{\rm c} - \Lambda^{\rm c0}$ of the critical load $\Lambda^{\rm c}$ by small imperfections[7] of member lengths L_i $(i = 1, \ldots, 4)$. In $\mathbf{v} = \mathbf{v}^0 + \xi \sum_{i=1}^{\rho} \mathbf{d}_i$ in (1.13) for defining imperfections, the vector $\mathbf{v}$ for representing the mechanical properties of the structure is chosen to be

$$\mathbf{v} = (L_1, \ldots, L_4)^\top \tag{3.70}$$

which, for the perfect system $(h = h^0 = 1.8852)$, is equal to

$$\mathbf{v}^0 = (\sqrt{4 + (h^0)^2}, \sqrt{1 + (h^0)^2}, \sqrt{4 + (h^0)^2}, \sqrt{1 + (h^0)^2}\,)^\top \tag{3.71}$$

[7]For given imperfections of the member lengths, the initial location of node 5 for $\Lambda = 0$ is computed from the equilibrium equations (3.68), while the locations of nodes 1–4 are fixed. Thus, member forces exist at the initial state.

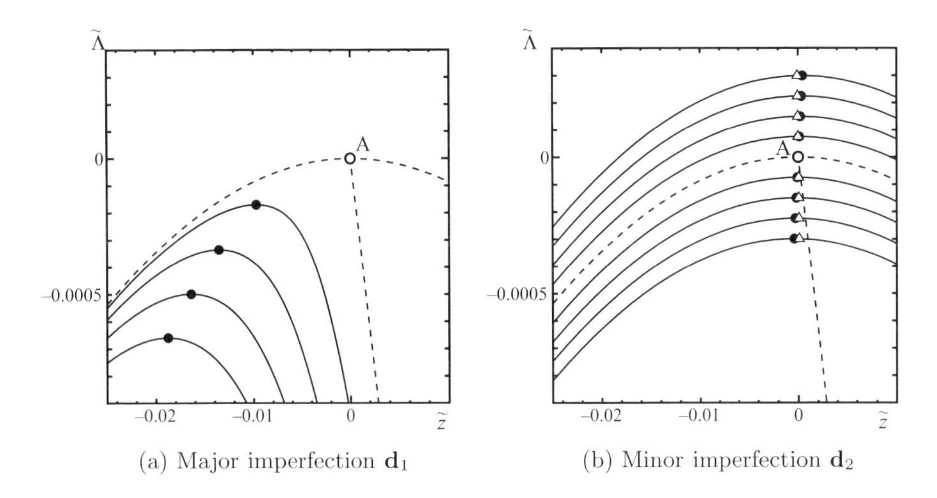

(a) Major imperfection $\mathbf{d}_1$ (b) Minor imperfection $\mathbf{d}_2$

Fig. 3.6 Equilibrium paths of the four-bar truss. $\widetilde{\Lambda} = \Lambda - \Lambda^{c0}; \widetilde{z} = U_z + z_5 - z_5^{c0}$; solid curve: imperfect equilibrium path; dashed curve: perfect equilibrium path; ○: hilltop branching point; •: limit point; △: unstable-symmetric bifurcation point.

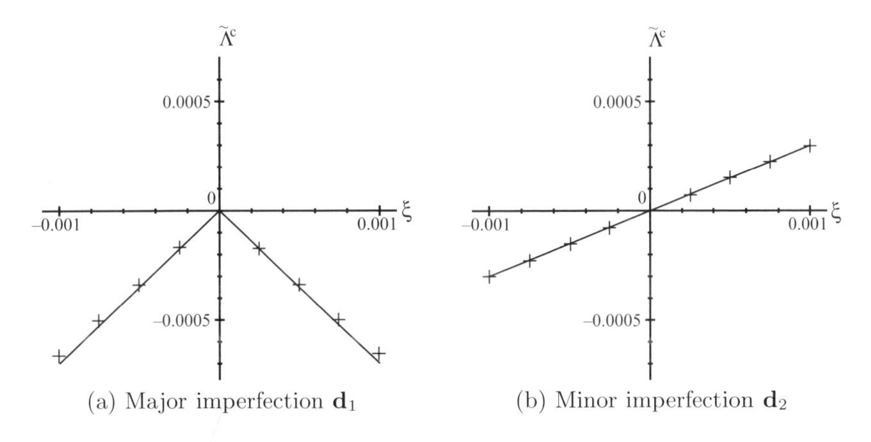

(a) Major imperfection $\mathbf{d}_1$ (b) Minor imperfection $\mathbf{d}_2$

Fig. 3.7 Imperfection sensitivity of $\widetilde{\Lambda}^c = \Lambda^c - \Lambda^{c0}$. +: result of path-tracing analysis; solid line: piecewise linear law or linear law.

Consider two imperfection patterns

$$\begin{cases} \text{major imperfection}: \mathbf{d}_1 = (0, 1/\sqrt{2}, 0, -1/\sqrt{2})^\top \\ \text{minor imperfection}: \mathbf{d}_2 = (1/2, 1/2, 1/2, 1/2)^\top \end{cases} \tag{3.72}$$

Equilibrium paths of the perfect system are shown in Fig. 3.6 by the dashed curves. The hilltop branching point 'A' is located at $(U_x^{c0}, U_y^{c0}, U_z^{c0}, \Lambda^{c0}) = (0, 0, 1.0356, 0.6624)$ on the fundamental path, and has two critical eigenvectors $\mathbf{\Phi}_1^{c0} = (1, 0, 0)^\top$ and $\mathbf{\Phi}_2^{c0} = (0, 0, 1)^\top$, which correspond to a bifurcation mode and a limit point mode, respectively.

Major imperfection $\mathbf{d}_1$ breaks the symmetry of the system and accelerates a x-directional sway. Minor imperfection $\mathbf{d}_2$ preserves the symmetry and increases the height of the truss.

The imperfect equilibrium paths shown by the solid curves in Fig. 3.6 are computed for the two initial imperfection patterns in (3.72) for several imperfection magnitudes ξ. The limit points shown by $\bullet$ exist on the equilibrium paths for major imperfections $\xi\mathbf{d}_1$. For a minor imperfection $\xi\mathbf{d}_2$, the hilltop branching point shown by $\circ$ breaks into a pair of simple critical points: a limit point $\bullet$ and an unstable-symmetric bifurcation point $\triangle$.

In order to investigate imperfection sensitivity at the hilltop branching point 'A', critical load $\widetilde{\Lambda}^{\rm c} = \Lambda^{\rm c} - \Lambda^{\rm c0}$ is plotted against ξ in Fig. 3.7, where '+' denotes the result by a path-tracing analysis and the solid line denotes asymptotic formula (3.56). The relationship between ξ and $\widetilde{\Lambda}^{\rm c}$ for the major imperfection $\mathbf{d}_1$, for which $\pm\xi$ reduce critical load $\widetilde{\Lambda}^{\rm c}$ at the same amount, agrees well with the piecewise linear law (3.56) for $C_5 = 0$ with a kink at $\xi = 0$. The relationship for the minor imperfection $\mathbf{d}_2$ follows well the linear law ($C_4 = 0$ in (3.56)).

3.8 Historical Development

The experimental buckling loads of long thin-walled cylinders fell markedly below the critical load computed by the classical linearized theory of bifurcation [83]. In order to resolve such inadequacy, *geometrical nonlinearity* and *initial imperfections* were implemented into the theory of elastic stability of shells [67, 87, 307]. Combined experimental, analytical and computational studies of imperfection-sensitive structures were motivated.

Koiter's *general nonlinear theory of stability* [166] emerged as a pertinent asymptotic tool to analyze *initial postbuckling behavior*. Critical points were classified in view of the expanded form of the total potential energy, and imperfection sensitivity laws were invented to elucidate the mechanism of the sharp reduction of the strength owing to small initial imperfections. The importance of this theory, however, was not recognized until the early 1960's when research on initial postbuckling behaviors sprung up worldwide. An overview of early development of postbuckling theory is found in [133]

The perturbation technique was applied to the total potential energy of a finite-dimensional system to derive asymptotic information on bifurcation buckling. The general theory of elastic stability played a pivotal role in dealing with initial postbuckling behavior, especially from an analytical standpoint. Although most of the perturbation approaches are the extension of this traditional method, there are several different branches: singular perturbation method to bifurcation analysis and sensitivity analysis [197], a higher-order perturbation method for sensitivity analysis [96, 128, 219], and critical imperfection magnitude method [236].

The principle of *simultaneous mode design* states, "A given form will be optimum if all failure modes which can possibly intersect occur simultaneously [276]." Thompson cautioned the danger of naive optimization without due regard to imperfection sensitivity and the erosion of optimization by compound branching [286, 290]. Imperfection sensitivity is enhanced by the interaction of buckling modes corresponding to coincident critical points [128, 287]. The basic framework to deal with a coincident critical point can be found in [127, 277, 285, 287]. Various kinds of structures were found highly imperfection-sensitive when two or more bifurcation points are nearly or strictly coincident [44, 132, 169, 288, 296]. Interaction between local and global modes are important for frame structures [171, 172] and composite structures [178].

3.9 Summary

In this chapter,

- stability theory of structures has been introduced,
- simple and coincident critical points have further been classified, and
- imperfection sensitivity laws for various types of critical points have been presented.

The major findings of this chapter are as follows.

- The V-formulation with its formalism is suitable for the classification of critical points and derivation of imperfection sensitivity laws based on the power series expansion. It is pertinent to transform the results obtained by the V-formulation to the D-formulation and, further, to the physical coordinates (nodal displacements) to make them readily accessible to practical analyses.
- The power series expansion method has turned out to be an insightful and rigorous way to deal with bifurcation problems of finite-dimensional systems, whereas the static perturbation method is a robust and standard way to deal with a nonlinear problem.
- It is widely recognized that the coincidence of critical points often leads to an extremely imperfection-sensitive structure. Nonetheless, imperfection sensitivity law varies according to the type of a critical point, and coincident critical points are not always imperfection-sensitive. The sensitivity is severe for semi-symmetric and completely-symmetric double bifurcation points, but it is less severe for a hilltop branching point.

Part II:
Optimization Methods for Stability Design

4

Optimization Under Stability Constraints

4.1 Introduction

In Part I, the theoretical and numerical foundations of the design and imperfection sensitivity analyses have been constructed. In Part II, we build on these foundations computer-aided methodologies to realize efficient and systematic optimization methods for stability designs.

An optimization problem is usually formulated to maximize structural performance under a specified cost or structural volume. Optimization procedures proposed in Part II combine the efficiency of mathematical programming and completeness of stability theory. Characteristics of various kinds of critical points of optimized structures are studied. Special cares to be taken for these critical points are made transparent. Two major claims are:

- Snapthrough is welcomed in the state-of-the-art application of optimization, shape design (topology optimization) of compliant mechanisms.
- Optimal structures with hilltop branching are in most cases safe as their imperfection sensitivity is reduced via optimization.

In connection with the second claim, recall that structural optimization against nonlinear buckling often has negative influence on imperfection sensitivity and may produce an extremely imperfection-sensitive structure owing to mode interaction [170, 286, 290]. Nonetheless, a hilltop branching point, which emerges as a result of optimization, is not sensitive to imperfections [146, 222, 284]. Investigation of imperfection sensitivity of an optimized structure is the major objective of Part II.

Structural optimization is conducted with the aid of mathematical programming and operations research developed in applied mathematics and

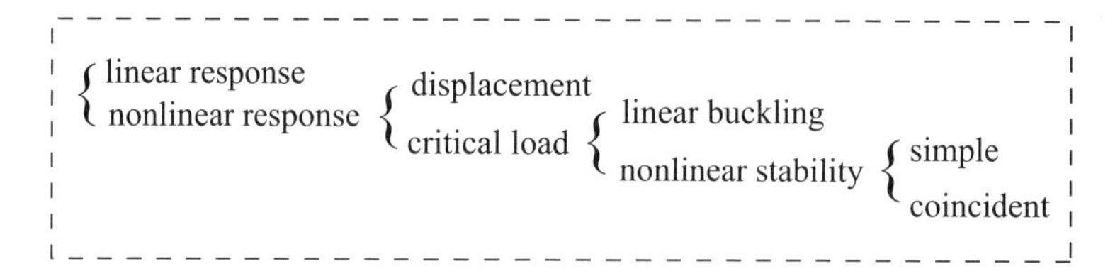

Fig. 4.1 Branches of structural optimization.

economics [13, 106]. A structural optimization problem is formally stated as either of the following forms:

Minimize *structural cost*
subject to *constraints on structural performances*

Maximize *structural performance*
subject to *constraints on structural cost*

The structural cost for a steel structure is usually defined by the total structural (material) volume. Responses to static or dynamic loads are usually taken as structural performances. The eigenvalues of vibration and linear buckling loads are also used as performance measures. In this book, we mainly use the nonlinear critical load at the first critical point for formulating optimization problems.

In a similar manner to the classification of parameter sensitivity analysis as shown in Fig. 1.1 in Section 1.1, structural optimization problems can also be classified as shown in Fig. 4.1 depending on the structural properties incorporated in the objective function and/or constraints. A gradient-based approach can be used for a NonLinear Programming (NLP) problem with continuous design variables.

This chapter is organized as follows. In Section 4.2, a short introduction to nonlinear programming problem is provided. In Section 4.3, general formulation of an optimization problem and the outline of a gradient-based approach are presented, and are extended to optimization under stability constraints in Section 4.4. Numerical examples are presented in Section 4.5. Systematic occurrence of a degenerate hilltop point via optimization and associated imperfection sensitivity are investigated for a bar–spring model in Section 4.6. The historical review of optimization of geometrically nonlinear structures is presented in Section 4.7.

4.2 Introduction to Nonlinear Programming Problem

We start with a nonlinear programming problem to present the general formulation and terminologies of the optimization problem.

Let $\mathbf{x} \in \mathbb{R}^{\nu}$ denote the vector of variables, and the objective function to be minimized is denoted by $C(\mathbf{x})$. An optimization problem under equality and inequality constraints is formulated as

$$\begin{aligned} \text{OPT1}: \text{ minimize } \quad & C(\mathbf{x}) & (4.1a) \\ \text{subject to } \quad & H_j(\mathbf{x}) \leq 0, \quad (j = 1, \ldots, s) & (4.1b) \\ & G_j(\mathbf{x}) = 0, \quad (j = 1, \ldots, e) & (4.1c) \\ & x_i^{\mathrm{L}} \leq x_i \leq x_i^{\mathrm{U}}, \quad (i = 1, \ldots, \nu) & (4.1d) \end{aligned}$$

where x_i^{U} and x_i^{L} are the upper and lower bounds for x_i, and s and e are the numbers of inequality and equality constraints, respectively.

Terminologies for optimization problem are summarized as

- Problem OPT1 is called Linear Programming (LP) problem if $C(\mathbf{x})$ and all the constraint functions $H_j(\mathbf{x})$ and $G_j(\mathbf{x})$ are linear functions of $\mathbf{x}$. It is called NonLinear Programming (NLP) problem, if some of them are nonlinear.
- A solution $\mathbf{x}$ is called *feasible* if it satisfies all the constraints. The set of feasible solutions forms a *feasible region.*
- The inequalities (4.1d) are called *side constraints*, or bound constraints, that are separately treated from general inequality constraints (4.1b) in the optimization algorithm.
- The inequality constraint $H_j(\mathbf{x}) \leq 0$ is called *active* if it is satisfied in equality as $H_j(\mathbf{x}) = 0$; otherwise, it is called *inactive*.
- The feasible solution that takes the smallest value of $C(\mathbf{x})$ is called *global optimal solution* or simply *optimal solution.* If $\mathbf{x}$ has the local minimum value in the neighboring feasible solutions, it is called *local optimal solution.*
- Lagrangian $L(\mathbf{x}, \boldsymbol{\mu}^{\mathrm{I}}, \boldsymbol{\mu}^{\mathrm{E}})$ for OPT1 is defined with Lagrange multipliers $\boldsymbol{\mu}^{\mathrm{I}} = (\mu_j^{\mathrm{I}})$ $(\mu_j^{\mathrm{I}} > 0)$ and $\boldsymbol{\mu}^{\mathrm{E}} = (\mu_j^{\mathrm{E}})$ as

$$L(\mathbf{x}, \boldsymbol{\mu}^{\mathrm{I}}, \boldsymbol{\mu}^{\mathrm{E}}) = C(\mathbf{x}) + \sum_{j=1}^{s} \mu_j^{\mathrm{I}} H_j(\mathbf{x}) + \sum_{j=1}^{e} \mu_j^{\mathrm{E}} G_j(\mathbf{x}) \tag{4.2}$$

4.3 Structural Optimization Problem and Gradient-Based Optimization Algorithm

General formulation of structural optimization problem is presented, and gradient-based optimization algorithm is introduced as a standard tool for solving the general NLP problem in Section 4.2.

Interrelations among the gradient-based optimization algorithm, design sensitivity analysis and structural analysis are illustrated in Fig. 4.2. The structural analysis and design sensitivity analysis programs are called by the optimization program, as indicated by the arrows, and the structural analysis is conducted to compute the sensitivity coefficients.

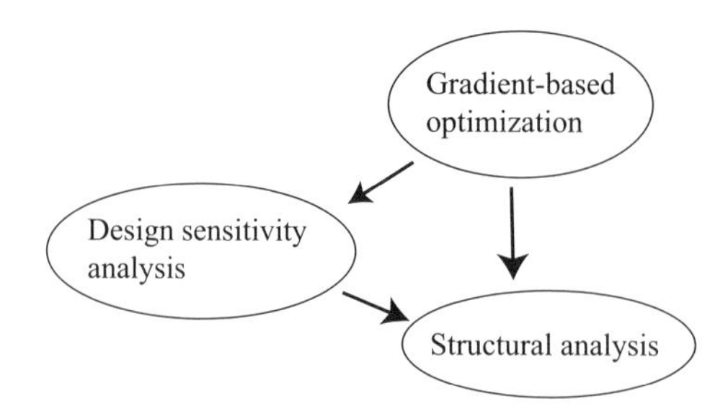

Fig. 4.2 Interrelation among optimization, sensitivity analysis and structural analysis.

4.3.1 General formulation of structural optimization problem

Denote by $\mathbf{A} = (A_i) \in \mathbb{R}^\nu$ the vector of design variables (design parameters) that represents, e.g., cross-sectional areas of trusses and thicknesses of plates discretized by finite elements (cf., Section 1.3 for more design parameters). Note that A_i takes the role of ξ_i in the general formulation of parameter sensitivity analysis in Part I.

Optimization of a structure is conducted as follows:

- The objective function $C(\mathbf{A})$ to be minimized, such as total structural volume, is given as a function of $\mathbf{A}$.
- Constraints on structural responses are formulated.
- Specify the upper bound A_i^{U} and the lower bound A_i^{L} of A_i.

The responses, such as displacements and stresses under static loads, buckling loads and eigenfrequencies, are denoted by a vector $\mathbf{Z} = (Z_i) \in \mathbb{R}^q$. Requirements on the responses are usually formulated as inequalities

$$H_j(\mathbf{Z}(\mathbf{A}), \mathbf{A}) \leq 0, \quad (j = 1, \dots, s) \tag{4.3}$$

Note that $H_j(\mathbf{Z}(\mathbf{A}), \mathbf{A})$ is generally a function of $\mathbf{Z}$ and is explicitly dependent also on $\mathbf{A}$.

Remark 4.3.1 For example, for a truss, the axial force N_i is defined as $N_i = E\varepsilon_i A_i$ using Young's modulus E, the strain ε_i and the cross-sectional area A_i. Then N_i depends on A_i explicitly and also implicitly through ε_i that is a function of A_i. □

A structural optimization problem is formulated as

$$\text{OPT2}: \quad \text{minimize} \quad C(\mathbf{A}) \tag{4.4a}$$

$$\text{subject to} \quad H_j(\mathbf{Z}(\mathbf{A}), \mathbf{A}) \leq 0, \quad (j = 1, \dots, s) \tag{4.4b}$$

$$A_i^{\mathrm{L}} \leq A_i \leq A_i^{\mathrm{U}}, \quad (i = 1, \dots, \nu) \tag{4.4c}$$

Problem OPT2 is generally categorized as an NLP problem, for which many approaches have been developed [13, 106, 191]. Although an optimization problem generally has equality and inequality constraints as formulated in OPT1 in Section 4.2, most of the design requirements in structural optimizations are assigned by inequality constraints.

4.3.2 Gradient-based optimization approach

There are many approaches to solve optimization problems.

- Accurate solutions for problems with continuous functions and variables can be obtained by gradient-based approaches, such as NLP.
- Nearly optimal solutions can be obtained by heuristics such as genetic algorithms [98], simulated annealing [1], and tabu search [94], which can be effectively used when variables take only integer values, and the problem is formulated as a combinatorial optimization problem.

In this book, we deal with optimization problems with continuous variables.

Most of NLP approaches utilize gradient (sensitivity) information for a search direction of solutions or for the transformation of the problem. Define

$$H_j^*(\mathbf{A}) = H_j(\mathbf{Z}(\mathbf{A}), \mathbf{A}) \tag{4.5}$$

and reformulate OPT2 as follows:

$$\begin{aligned} \text{OPT3}: \ \text{minimize} \quad & C(\mathbf{A}) && (4.6a) \\ \text{subject to} \quad & H_j^*(\mathbf{A}) \leq 0, \quad (j = 1, \dots, s) && (4.6b) \\ & A_i^{\mathrm{L}} \leq A_i \leq A_i^{\mathrm{U}}, \quad (i = 1, \dots, \nu) && (4.6c) \end{aligned}$$

The gradient of $C(\mathbf{A})$ and $H_j^*(\mathbf{A})$ are defined, respectively, as

$$\nabla C(\mathbf{A}) = \left(\frac{\partial C}{\partial A_1}, \dots, \frac{\partial C}{\partial A_\nu} \right)^\top \tag{4.7a}$$

$$\nabla H_j^*(\mathbf{A}) = \left(\frac{\partial H_j^*}{\partial A_1}, \dots, \frac{\partial H_j^*}{\partial A_\nu} \right)^\top \tag{4.7b}$$

Since $H_j^*(\mathbf{A})$ is an implicit function of $\mathbf{A}$ through $\mathbf{Z}(\mathbf{A})$ as defined in (4.5), its design sensitivity is computed by

$$\frac{\partial H_j^*}{\partial A_i} = \sum_{k=1}^{q} \frac{\partial H_j}{\partial Z_k} \frac{\partial Z_k}{\partial A_i} + H_{j,A_i} \tag{4.8}$$

where H_{j,A_i} is the partial differentiation of explicit terms of A_i in H_j, and the design sensitivity $\partial Z_k / \partial A_i$ of response Z_k can be computed by numerical approaches presented in Chapter 2.

One of the most popular approaches for iterative update of solutions is the Sequential Quadratic Programming (SQP), which employs the following quadratic programming subproblem to find a search direction $\Delta\mathbf{A}^{(k)}$ from the current solution $\mathbf{A}^{(k)}$ at the kth step of NLP:

$$\text{SQP}: \text{minimize} \quad \nabla C(\mathbf{A}^{(k)})^\top \Delta\mathbf{A}^{(k)} + \frac{1}{2}\Delta\mathbf{A}^{(k)\top}\mathbf{D}\Delta\mathbf{A}^{(k)} \tag{4.9a}$$

$$\text{subject to} \quad H_j^*(\mathbf{A}^{(k)}) + \nabla H_j^*(\mathbf{A}^{(k)})^\top \Delta\mathbf{A}^{(k)} \leq 0, \quad (j = 1, \dots, s) \tag{4.9b}$$

$$A_i^{\mathrm{L}} \leq A_i^{(k)} + \Delta A_i^{(k)} \leq A_i^{\mathrm{U}}, \quad (i = 1, \dots, \nu) \tag{4.9c}$$

The inequality constraints (4.6b) are linearly approximated by (4.9b). The second term in (4.9a) is the penalty term for preventing drastic update of the design variables. In principle, $\mathbf{D}$ is an arbitrary positive-definite matrix, but the Hessian of the Lagrangian (cf., (4.2) in Section 4.2) of OPT3 at $\mathbf{A} = \mathbf{A}^{(k)}$ is usually used so that the solution by SQP satisfies the optimality conditions of the original problem OPT3. Since a straightforward evaluation of this Hessian is costly, the Hessian or its inverse is approximated by the quasi-Newton method. The SQP subproblem is an approximation of the original NLP problem, and is solved only to find the direction of design modification. An incremental and iterative search, which is called *line search*, is usually used to obtain an appropriate value of norm $\|\Delta\mathbf{A}^{(k)}\|$ of the direction vector [191].

In Modified Feasible Direction Method (MFDM), the search direction $\Delta\mathbf{A}^{(k)}$ is found by solving the subproblem [301]:

$$\text{MFDM}: \text{minimize} \quad \nabla C(\mathbf{A}^{(k)})^\top \Delta\mathbf{A}^{(k)} \tag{4.10a}$$

$$\text{subject to} \quad \nabla H_j^*(\mathbf{A}^{(k)})^\top \Delta\mathbf{A}^{(k)} \leq 0, \quad (j \in J) \tag{4.10b}$$

$$\Delta\mathbf{A}^{(k)\top}\Delta\mathbf{A}^{(k)} \leq 1 \tag{4.10c}$$

where J is the set of *active constraints* satisfying $H_j^*(\mathbf{A}^{(k)}) = 0$, and the side constraints (4.6c) are included, for simplicity, in the general inequality constraints (4.6b). Note from (4.10b) that $H_j^*(\mathbf{A}^{(k)})$ for currently active constraints are allowed to be negative (inactive) at the next step if the objective value is decreased.

In other methods, such as sequential linear programming [13], augmented Lagrangian method [245], penalty function approach [191], and so on, linear or quadratic subproblems are formulated based on the gradient information, or the solutions are updated in the direction of the gradient of a modified objective function.

As we have seen, a solution to a structural optimization problem is found based on the gradients of the objective functions and constraints. Design sensitivity analysis, therefore, is indispensable for structural optimization by means of NLP approach.

Remark 4.3.2 Only a local optimal solution is obtained by an NLP approach, especially for highly nonlinear problems; such is the case for structural optimization under stability constraints. However, a local optimal solution is called an *optimal solution* for simplicity in the sequel. □

4.4 Optimization Under Stability Constraints

The optimization problem OPT3 in (4.6) is revised to implement constraint(s) on a single or multiple critical load(s) [216].

4.4.1 Direct formulation

If the stability of a structure is governed by a simple critical point, the optimization problem OPT3 in (4.6) simplifies to

$$\begin{aligned} \mathrm{P1}: \ \text{minimize} \quad & C(\mathbf{A}) && (4.11\mathrm{a}) \\ \text{subject to} \quad & \Lambda^{\mathrm{c}}(\mathbf{A}) \geq \overline{\Lambda}^{\mathrm{c}} && (4.11\mathrm{b}) \\ & A_i^{\mathrm{L}} \leq A_i \leq A_i^{\mathrm{U}}, \quad (i = 1, \dots, \nu) && (4.11\mathrm{c}) \end{aligned}$$

with a single general inequality constraint (4.11b). Here $\overline{\Lambda}^{\mathrm{c}}$ is the specified lower bound of the critical load factor Λ^{c}. Optimal solutions for problem P1 may be easily found by an optimization algorithm using the interpolation approach in Section 2.5 for design sensitivity analysis.

If the stability is governed by a coincident critical point, the optimization problem is formulated as

$$\begin{aligned} \mathrm{P2}: \ \text{minimize} \quad & C(\mathbf{A}) && (4.12\mathrm{a}) \\ \text{subject to} \quad & \Lambda_j^{\mathrm{c}}(\mathbf{A}) \geq \overline{\Lambda}^{\mathrm{c}}, \quad (j = 1, \dots, h) && (4.12\mathrm{b}) \\ & A_i^{\mathrm{L}} \leq A_i \leq A_i^{\mathrm{U}}, \quad (i = 1, \dots, \nu) && (4.12\mathrm{c}) \end{aligned}$$

where Λ_j^{c} is the jth critical load factor along the fundamental equilibrium path, and h is the largest possible multiplicity of the critical load factor.

The formulation of P2, despite its simple appearance, suffers from the following difficulties in application:

- The second and higher critical load factors are often far above the lowest factor during iterations. Then a substantial computational effort is demanded for finding all necessary critical load factors and their design sensitivity coefficients, if a gradient-based method is used.

- For a degenerate coincident critical point, discontinuity in Λ_j^{c} makes its sensitivity coefficient discontinuous or unbounded, as actually encountered in the bar–spring model in Fig. 4.7 in Section 4.6 and in the truss dome in Section 10.3. Variation of the two lowest eigenvalues of the tangent stiffness matrix $\mathbf{S}$ against Λ at a nearly degenerate coincident critical point is illustrated in Fig. 4.3. Suppose that the two eigenvalues vanish simultaneously at $\Lambda = \Lambda_1^{\mathrm{c}} = \Lambda_2^{\mathrm{c}}$ for the design $\mathbf{A} = \mathbf{A}^0$ shown by the solid curves, and the mode 'a' corresponds to a nearly degenerate critical point. For a slightly modified design $\mathbf{A} = \mathbf{A}^0 + \Delta\mathbf{A}$ shown by the dashed curves, the eigenvalue corresponding to the mode 'a' remains positive at $\Lambda = \Lambda_1^{\mathrm{c}}$. Therefore, the second critical point disappears on the fundamental path of the design $\mathbf{A} = \mathbf{A}^0 + \Delta\mathbf{A}$. In this case, Λ_2^{c} is discontinuous, and its design

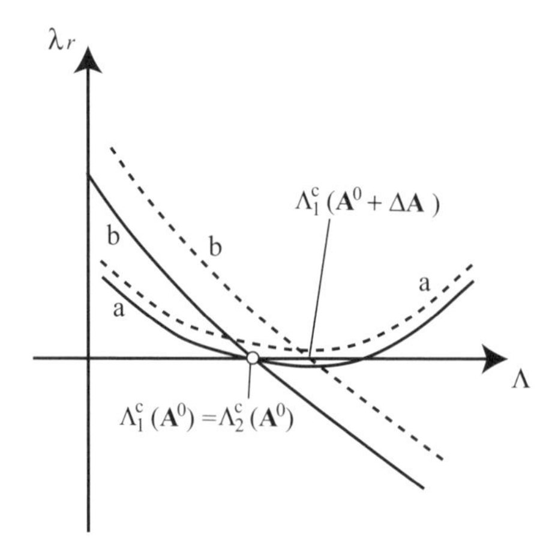

Fig. 4.3 Variations of the two lowest eigenvalues of a structure at a nearly degenerate coincident critical point denoted by ∘.

sensitivity coefficients cannot be defined. It causes a difficulty in application of nonlinear programming to problem P2.

4.4.2 Formulation with eigenvalue constraints

To address the aforementioned difficulties, the optimization problem P2 in (4.12) is revised as

$$\text{P3}: \ \text{minimize} \quad C(\mathbf{A}) \tag{4.13a}$$

$$\text{subject to} \quad \Lambda_j^{\mathrm{c}}(\mathbf{A}) \geq \overline{\Lambda}^{\mathrm{c}}, \quad (j = 1, \dots, m) \tag{4.13b}$$

$$\lambda_r^{\mathrm{c}}(\mathbf{A}) \geq 0, \quad (r = m+1, \dots, h) \tag{4.13c}$$

$$A_i^{\mathrm{L}} \leq A_i \leq A_i^{\mathrm{U}}, \quad (i = 1, \dots, \nu) \tag{4.13d}$$

where m is the multiplicity of the critical point, and λ_r^{c} $(r = m+1, \dots, h)$ are eigenvalues of $\mathbf{S}$ at the critical point that may possibly become zero at the optimal solution. In problem P3, it is possible to terminate path-tracing analysis at the first critical point, and to detect nearly coincident critical points as shown in Fig. 4.3 by monitoring the eigenvalues in the constraint (4.13c).

Remark 4.4.1 The multiplicity m of the critical point is not known a priori, but is to be found during optimization. If λ_{m+1} turns out to vanish, the value of m is increased by 1 and the constraints are reformulated. However, at the final stage of optimization, the optimal solution is searched in a small domain of the design space, and the value of m tends to converge. Furthermore, the total value h of the largest possible multiplicity can be specified a priori to be moderately large, and is kept unchanged during optimization. □

Remark 4.4.2 Although the difficulty due to discontinuity in the critical load factor of a degenerate critical point can be resolved by the problem formulation

P3 in (4.13), discontinuity in the sensitivity coefficients remains as a possible problem. However, an optimization algorithm utilizing line search as described in Section 4.3.2 is robust against such discontinuity. □

The optimization algorithm is summarized as follows:

Step 1: Initialize $\mathbf{A} = (A_i)$ and set $A_i^{\rm L}, A_i^{\rm U}, \overline{\Lambda}^{\rm c}$ and h.

Step 2: Carry out path-tracing analysis by increasing the path parameter[1] t by Δt, and find $t^{\rm II}$ where $\lambda_1 < 0$ is first satisfied. Let $t^{\rm I} = t^{\rm II} - \Delta t$. Compute $\Lambda_1^{\rm c}$ by the interpolation approach of design sensitivity analysis in Appendix A.2 that is applicable to coincident critical points.

Step 3: Find the multiplicity m of the critical load factor as the number of λ_r satisfying $\lambda_1(t^{\rm II}) \leq \lambda_r(t) \leq \lambda_1(t^{\rm I})$ at $t = t^{\rm I}$ or $t^{\rm II}$. Calculate sensitivity coefficients of $\Lambda_j^{\rm c}$ $(j = 1, \dots, m)$ and $\lambda_r^{\rm c}$ $(r = m+1, \dots, h)$ at the critical point using (A.4) and (A.6) in Appendix A.2.

Step 5: Update $\mathbf{A} = (A_i)$ by an optimization algorithm, such as SQP or MFDM presented in Section 4.3.2.

Step 6: Return to Step 2 if the iteration does not converge.

In the optimization problem P3, optimal solutions correspond in many cases to coincident critical states. For this reason, knowledge on elastic stability is vital in solving the optimization problems under stability constraints and ensuring the safety of optimal structures. The reduction of the critical load factor $\Lambda^{\rm c}$ by initial imperfections is not considered in the formulation of a structural optimization problem in this chapter. The effect of imperfections will be formulated in terms of the worst mode of imperfection in Section 11.3 [123, 137].

4.5 Optimization of a Symmetric Shallow Truss Dome

The symmetric shallow truss dome as shown in Fig. 4.4 is optimized by solving the optimization problem P3 in Section 4.4 with the use of the interpolation approach in Appendix A.2. An optimal solution with a simple bifurcation point is found by a simple application of the optimization algorithm. Optimal solutions with hilltop branching points will be presented in Sections 10.4 and 10.5.

The lower-bound cross-sectional area is $A_i^{\rm L} = 200\ {\rm mm}^2$ for all the members, and $A_i^{\rm U}$ is not specified. Young's modulus is $E = 205.8\ {\rm kN/mm}^2$. The units of force kN and of length mm are suppressed in the following. Let Λp_i denote the z-directional load applied to the ith node, and p_i is given as

$$p_i = \begin{cases} 0: & \text{for} \quad i = 1 \\ -1.31: & \text{for} \quad i = 2, \dots, 7 \end{cases} \tag{4.14}$$

[1] A general path parameter t is used instead of Λ in the path-tracing of a limit point and a hilltop branching point.

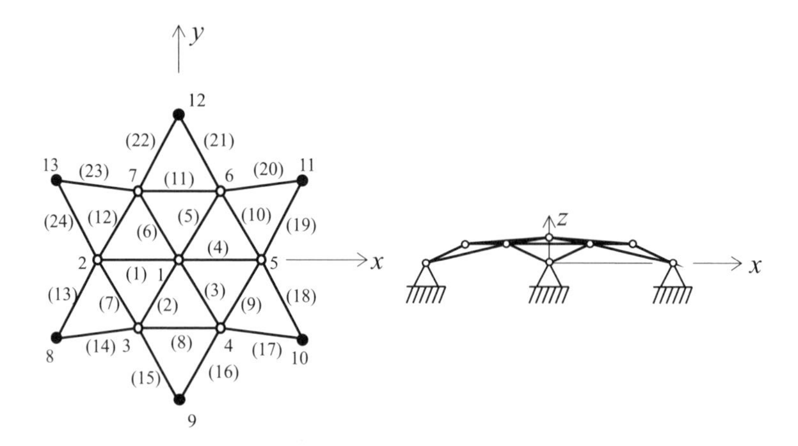

Fig. 4.4 Symmetric shallow truss dome. Nodal coordinates are listed in Table 2.1 in Section 2.7.2. $1, \ldots, 13$ denote node numbers; $(1), \ldots, (24)$ denote member numbers.

The members are divided into three groups in view of geometrical symmetry, and members in the same group have the same cross-sectional area, being denoted as

$$\begin{cases} A_1^* : & \text{for members } 1, \ldots, 6 \\ A_2^* : & \text{for members } 7, \ldots, 12 \\ A_3^* : & \text{for members } 13, \ldots, 24 \end{cases} \tag{4.15}$$

The total structural volume V is chosen as the objective function, and the largest possible multiplicity is $h = 10$. The specified critical load factor is $\overline{\Lambda}^{\mathrm{c}} = 10000$. Accordingly, the optimum design problem P3 in (4.13) is written down as

$$\text{minimize} \quad A_1^* \sum_{i=1}^{6} L_i + A_2^* \sum_{i=7}^{12} L_i + A_3^* \sum_{i=13}^{24} L_i \tag{4.16a}$$

$$\text{subject to} \quad \Lambda_j^{\mathrm{c}}(\mathbf{A}^*) \geq 10000, \quad (j = 1, \ldots, m) \tag{4.16b}$$

$$\lambda_r^{\mathrm{c}}(\mathbf{A}^*) \geq 0, \quad (r = m+1, \ldots, 10) \tag{4.16c}$$

$$A_i^* \geq 200, \quad (i = 1, \ldots, 3) \tag{4.16d}$$

where $\mathbf{A}^* = (A_1^*, A_2^*, A_3^*)^\top$ and L_i is the length of member i. Optimal solutions are found by the optimization library DOT Ver. 5.0 [308], which employs the MFDM in (4.10a)–(4.10c).

The cross-sectional areas and total structural volume of the optimal solution are $A_1^* = 688.33$, $A_2^* = 756.21$, $A_3^* = 705.38$ and $V = 4.8448 \times 10^7$. The multiplicity of the critical point is equal to 1, since $\lambda_1^{\mathrm{c}} = 0.0$, $\lambda_2^{\mathrm{c}} = \lambda_3^{\mathrm{c}} = 28.863$. Recall that the critical points are classified by (3.25) in Section 3.3.1, namely,

$$\begin{cases} \text{limit point}: & V_{,r\Lambda} \neq 0 \\ \text{bifurcation point}: & V_{,r\Lambda} = 0 \end{cases} \tag{4.17}$$

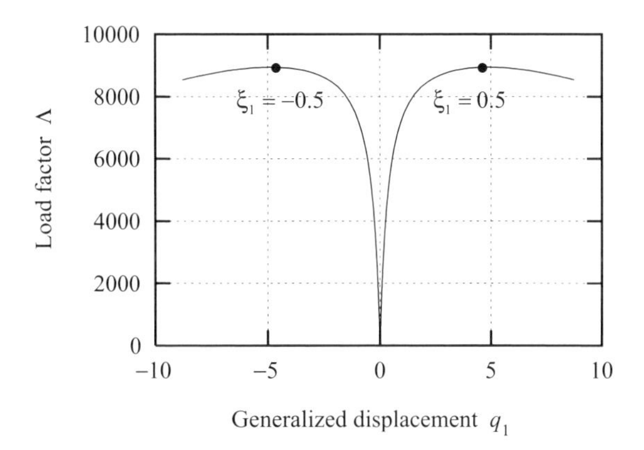

Fig. 4.5 Equilibrium paths for imperfect systems with nodal imperfection $\xi_1 \boldsymbol{\Phi}_1^{c0}$ $((\xi_1, \xi_2) = (\pm 0.5, 0))$. •: limit point.

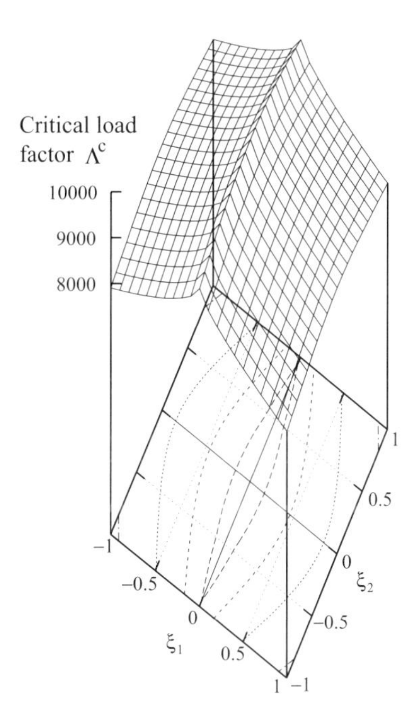

Fig. 4.6 Variation of critical load factors Λ^{c} of imperfect systems plotted against ξ_1 and ξ_2.

This critical point is a simple bifurcation point, because $V_{,1\Lambda} = 0$. It has been observed from the mode shape $\boldsymbol{\Phi}_1^{c0}$ that this bifurcation point is symmetric.

Nodal imperfections are given for the optimal solution in the direction of $\xi_i \boldsymbol{\Phi}_i^{c0}$ $(i = 1, 2)$. Fig. 4.5 shows equilibrium paths for the imperfect systems for $(\xi_1, \xi_2) = (\pm 0.5, 0)$. The critical load factors corresponding to limit points for major imperfections $\xi_1 = 0.5$ and -0.5 are identical and are equal to $\Lambda^{\mathrm{c}} = 8798.7$, which amounts to 12.013% reduction from $\Lambda_1^{c0} = 10000$ of the perfect system. Thus major imperfections are influential on critical loads.

The critical load factor Λ^c is plotted in Fig. 4.6 against the two imperfection parameters ξ_1 and ξ_2. The critical points for the imperfect systems along the line of $\xi_1 = 0$ are bifurcation points; therefore, the sensitivity is unbounded in the direction of $\mathbf{\Phi}_1^{c0}$, and follows the 2/3-power law (cf., (3.48) in Section 3.5.2) for major imperfections at a simple symmetric bifurcation point. Note that $\mathbf{\Phi}_2^{c0}$ corresponds to a minor imperfection because $\mathbf{\Phi}_1^{c0}$ and $\mathbf{\Phi}_2^{c0}$ have different planes of symmetry; therefore, imperfection sensitivity coefficient is bounded in the direction of $\mathbf{\Phi}_2^{c0}$.

4.6 Bar–Spring Model

Consider the bar–spring model as shown in Fig. 4.7(a), which consists of rigid bars and elastic springs. The length of the diagonal bars is given as $L = L^*/\cos\overline{\theta} = 1$ ($\overline{\theta} = (2/9)\pi$); we set $H = 1.5$ and $p = 1$. In the following, the units of length and force are omitted for brevity. The extensional stiffnesses of Spring 1 and 2 are denoted by k_{h1} and k_t, respectively. The two rotational springs have a nonlinear relation $M = k_{\mathrm{r1}} r + \frac{1}{2} k_{\mathrm{r2}} r^2$ between the rotation r and the moment M.

Define the displacement vector by $\mathbf{U} = (\eta, \theta)^\top$ as shown in Fig. 4.7(b), where the thin and thick lines are the states before and after deformation, respectively.

The total potential energy Π is then given as

$$
\begin{aligned}
\Pi(\mathbf{U}, \Lambda, \mathbf{v}) \;=\; & \frac{1}{2} k_{h1} H^2 \sin^2 \eta + \frac{1}{2} k_{\mathrm{t}} [2L(\cos(\overline{\theta} - \theta) - \cos\overline{\theta})]^2 \\
& + k_{\mathrm{r1}}(\theta^2 + \eta^2) + \frac{1}{3} k_{\mathrm{r2}} \theta(\theta^2 + 3\eta^2) \\
& - \Lambda p L[\sin\overline{\theta} - \sin(\overline{\theta} - \theta)] - \Lambda p H(1 - \cos\eta)
\end{aligned}
\qquad (4.18)
$$

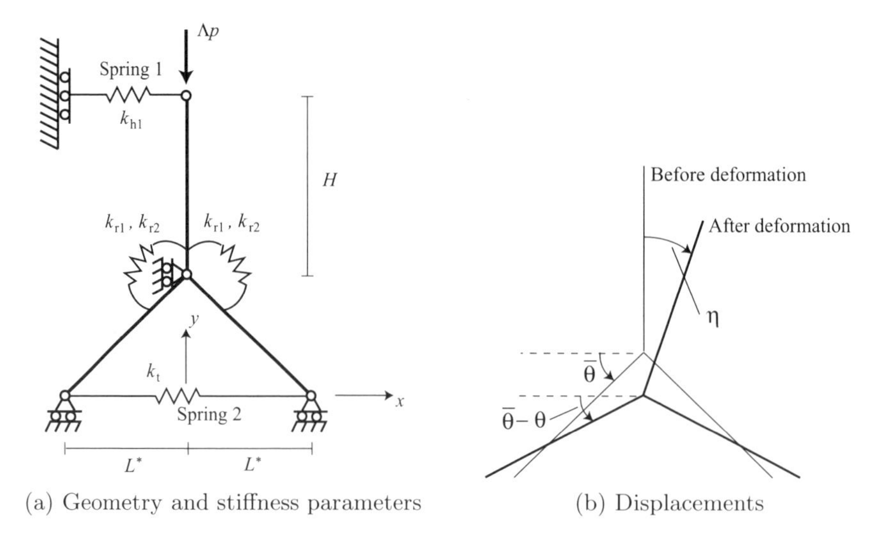

(a) Geometry and stiffness parameters (b) Displacements

Fig. 4.7 Bar–spring model.

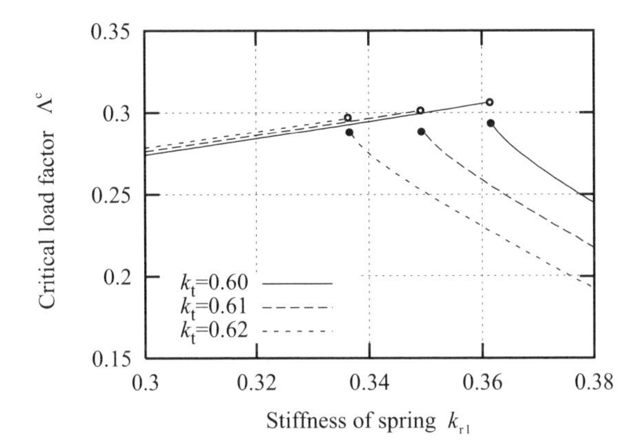

Fig. 4.8 Variations of critical load factor $\Lambda^{\rm c}$ plotted against spring stiffness $k_{\rm r1}$ ($k_{\rm r2} = -k_{\rm r1} \neq 0$).

Note that Π is symmetric with respect to η if $k_{r2} = 0$. The derivatives of Π with respect to η and θ are

$$S_{,1} = k_{h1} H^2 \sin\eta \cos\eta + 2k_{\rm r1}\eta + 2k_{\rm r2}\theta\eta - \Lambda p H \sin\eta \quad (4.19a)$$

$$S_{,2} = 4k_{\rm t} L^2 [\cos(\overline{\theta} - \theta) - \cos\overline{\theta}] \sin(\overline{\theta} - \theta) + 2k_{\rm r1}\theta + k_{\rm r2}(\theta^2 + \eta^2) - \Lambda p L \cos(\overline{\theta} - \theta) \quad (4.19b)$$

Higher derivatives are derived in Appendix A.6.

The eigenmode $\mathbf{\Phi}_r$ and eigenvalue λ_r are defined by the eigenvalue problem:

$$\begin{bmatrix} S_{,11} & 0 \\ 0 & S_{,22} \end{bmatrix} \mathbf{\Phi}_r = \lambda_r \mathbf{\Phi}_r, \quad (r = 1, 2) \quad (4.20)$$

Note that $S_{,12} = S_{,21} = 0$ is satisfied from symmetry on the fundamental path $\eta = 0$.

4.6.1 Simple degenerate

Let $\mathbf{\Phi}_1 = (1, 0)^\top$ and $\mathbf{\Phi}_2 = (0, 1)^\top$ correspond to bifurcation mode and limit point mode, respectively. Then the generalized coordinates q_1 and q_2 correspond to the direction of U_1 $(= \eta)$ and of U_2 $(= \theta)$, respectively.

An optimization problem for maximizing the critical load factor $\Lambda^{\rm c}$ is solved. We choose $k_{\rm t}$ as a parameter and the remaining three stiffness parameters $k_{\rm h1}$, $k_{\rm r1}$, and $k_{\rm r2}$ as design variables. The variables are linked, for simplicity, as $k_{\rm r2} = -k_{\rm r1}$ and a constraint is given as

$$k_{\rm h1} + k_{\rm t} + \frac{k_{\rm r1}}{L^2} = 1 \quad (4.21)$$

Hence we have the single independent design variable $k_{\rm r1}$, and the system is not symmetric with respect to η because $k_{r2} \neq 0$.

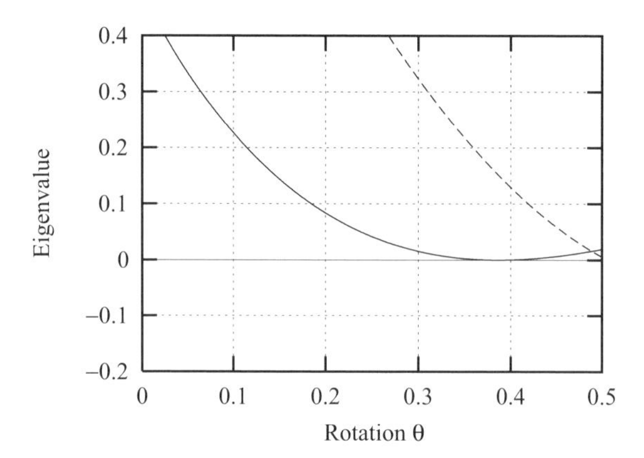

Fig. 4.9 Variations of eigenvalues plotted against rotation θ for $k_{\mathrm{r1}} = k_{\mathrm{r1}}^* = 0.361$ ($k_{\mathrm{r2}} = -k_{\mathrm{r1}} \neq 0$). Solid curve: bifurcation mode; dashed curve: limit point mode.

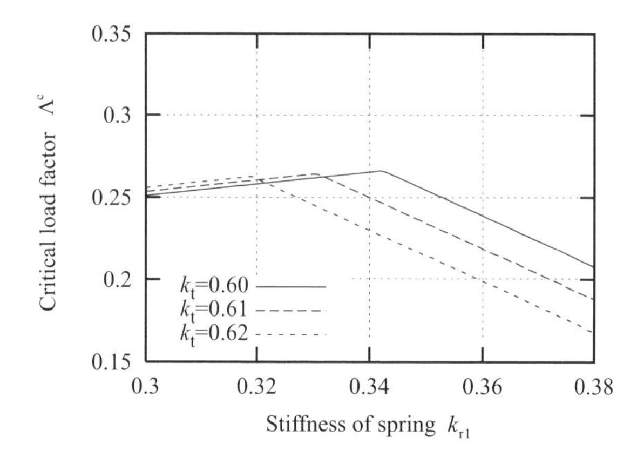

Fig. 4.10 Variation of critical load factor Λ^{c} plotted against spring stiffness k_{r1} ($k_{\mathrm{r2}} = 0$).

Fig. 4.8 shows the variations of the critical load factor Λ^{c} with respect to k_{r1} for three values of $k_{\mathrm{t}} = 0.60, 0.61, 0.62$. For $k_{\mathrm{t}} = 0.60$, Λ^{c} takes the maximum approximately at $k_{\mathrm{r1}} = k_{\mathrm{r1}}^* \simeq 0.361$, which is regarded as the optimal value. It is seen that Λ^{c} is discontinuous with respect to k_{r1} at the optimal solution $k_{\mathrm{r1}} = k_{\mathrm{r1}}^*$, and the sensitivity coefficient of Λ^{c} is unbounded. Fig. 4.9 shows the variation of eigenvalues plotted against rotation θ for $k_{\mathrm{r1}} = k_{\mathrm{r1}}^* \simeq 0.361$. It is observed that a degenerate critical point exists before reaching the limit point (cf., Fig. 4.3). If k_{r1} is decreased from k_{r1}^*, then the bifurcation point disappears, and the limit point turns out to be the first critical point. Hence, Λ^{c} is discontinuous with respect to k_{r1} as observed in Fig. 4.8, which is also observed in the four-bar truss tent with spring in Section 10.3.2.

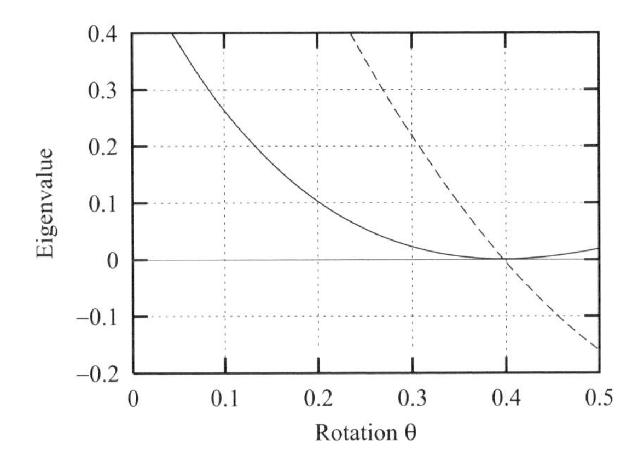

Fig. 4.11 Variations of eigenvalues plotted against rotation θ for $k_{\mathrm{r}1} = k_{\mathrm{r}1}^* = 0.342$ ($k_{\mathrm{r}2} = 0$). Solid curve: bifurcation mode; dashed curve: limit point mode.

4.6.2 Degenerate hilltop

Consider next a symmetric case with $k_{\mathrm{r}2} = 0$. The relation (4.21) is used also for this example. Λ^{c} is plotted against $k_{\mathrm{r}1}$ in Fig. 4.10 for $k_{\mathrm{t}} = 0.60, 0.61, 0.62$. For $k_{\mathrm{t}} = 0.60$, Λ^{c} takes the maximum at $k_{\mathrm{r}1} = k_{\mathrm{r}1}^* \simeq 0.342$, which is regarded as the optimal solution. As can be seen from the plots of the eigenvalues with respect to θ for $k_{\mathrm{r}1} = k_{\mathrm{r}1}^*$ in Fig. 4.11, the critical point is a degenerate hilltop branching point. This corresponds to a special case of vanishing of many derivatives as follows:

$$\begin{aligned} &V_{,111} = V_{,112} = V_{,122} = V_{,1\Lambda} = V_{,2\xi} = V_{,1122} = 0 \\ &V_{,222},\ V_{,1111},\ V_{,2\Lambda}: \quad \text{nonzero} \end{aligned} \tag{4.22}$$

A degenerate hilltop branching point will be studied in Chapter 10.

4.7 Historical Development

There are many studies on design methodologies under buckling constraint [21]. In the early stage of optimum design under buckling constraints, optimal shapes of columns were investigated by analytical approaches. Prager and Taylor [250] derived optimality conditions for columns under linear buckling constraints. Thereafter, a number of works have been published on sensitivity analysis and optimization of column-type structures under linear buckling constraints [230]. Difficulties owing to discontinuity of sensitivity coefficients related to multiple eigenvalues have been extensively discussed [114, 152, 196, 231]. Optimization methods of columns under linear buckling constraints can be found in [113, 271].

Optimization of finite-dimensional structures against buckling started in the 1970's. Linear buckling formulation neglecting prebuckling deformation was first employed. Khot *et al.* [159] presented an optimality criteria approach for trusses and frames. In the 1980's, more practical problems were studied incorporating constraints on displacements and stresses, as well as linear buckling load factors [184]. Difficulties related to multiple eigenvalues were noticed also for finite-dimensional structures [114, 179, 231, 271]. The optimum design with multiple linear buckling load factors can be found without difficulty by successively solving semidefinite programming [152, 173].

The difference of the buckling load factors by linear and nonlinear formulations is magnified for flexible structures, such as arches and shallow trusses [157]. Simple trusses exhibiting limit point instability were studied in the early stage of optimization of geometrically nonlinear finite-dimensional structures [249]. The maximum total potential energy was also used as a performance measure [158]. An NLP approach for maximizing limit point loads was presented [151].

In the 1990's, numerical approaches were presented for optimum designs of moderately large geometrically nonlinear structures. Optimality criteria approaches were mainly used for maximizing the limit point load factor [179, 264]. A method based on parametric programming approach was developed [225].

For building frames, special optimization methods were developed, because they have unique characteristics such as brace buckling and interaction of local and global buckling modes [171]. Numerical methods utilizing the characteristics of building frames were developed [21, 110]. An optimality criteria approach for buckling and displacement constraints under lateral loads was presented [122].

4.8 Summary

In this chapter,

- general formulation of optimization problem has been introduced,
- the optimization problem under stability constraints has been formulated as a foundation of this part, and
- optimal designs under nonlinear buckling constraints have been found for the symmetric shallow truss dome.

The major findings of this chapter are as follows.

- The optimization problem under stability constraint is demonstrated to be useful through the application to a truss dome.
- It is demonstrated that optimization of a symmetric system under nonlinear buckling constraints can lead to a structure with a degenerate hilltop branching point.

5

Optimal Structures Under Snapthrough Constraint

5.1 Introduction

It is vital in the design of latticed domes and shell roofs to assign appropriate members so as to ensure adequate stiffness against instability. For column-type structures, such as transmission towers and highrise buildings, a linear eigenvalue formulation is usually put to use in design because the effect of deformation before buckling is negligible. For shallow dome structures, however, it is necessary to incorporate the effect of prebuckling deformation in evaluating buckling loads, and the optimal dome structures often exhibit snapthrough behavior. The difference between the linear and the nonlinear behaviors is magnified by snapthrough.

In this chapter, we optimize the stiffness distribution of elastic finite-dimensional structures under nonlinear buckling constraints. Three optimization problems are considered for structures undergoing snapthrough. Although these problems appear physically feasible at a first glance, some of them turn out to be spurious. Thus a proper knowledge on snapthrough behavior is vital in successful formulation of an optimization problem, as we will see in consistent formulation of the shape optimization problem for compliant mechanism in Chapter 6.

This chapter is organized as follows. In Section 5.2, three optimization problems for structures undergoing snapthrough are formulated. Structures undergoing snapthrough are optimized in Sections 5.3 and 5.4.

5.2 Optimization Problems for Structures Undergoing Snapthrough

Three different optimization problems of a structure undergoing snapthrough are formulated to demonstrate the necessity of pertinent formulations based on a proper knowledge on geometrically nonlinear behavior. The total structural volume V of this structure is a function of the cross-sectional area A, which is common for all members, and the structure exhibits the displacement u against the external load Λ. We consider three optimization problems:

$$\text{Problem 1:} \begin{cases} \text{minimize } V(A) \\ \text{subject to } u(A, \Lambda^*) \leq \overline{u} \end{cases} \tag{5.1a}$$

$$\text{Problem 2:} \begin{cases} \text{minimize } u(A, \Lambda^*) \\ \text{subject to } V(A) \leq \overline{V} \end{cases} \tag{5.1b}$$

$$\text{Problem 3:} \begin{cases} \text{maximize } \Lambda^{\mathrm{M}}(A) \\ \text{subject to } u(A, \Lambda^{\mathrm{M}}) \leq \overline{u}, \quad V(A) \leq \overline{V} \end{cases} \tag{5.1c}$$

where $\overline{u}$ and $\overline{V}$ are the upper bounds of u and V, respectively, Λ^* is a specified load factor, and Λ^{M} is the maximum of Λ under an upper-bound constraint on u. Note that $u(A, \Lambda^*)$ and $\Lambda^{\mathrm{M}}(A)$ are implicit functions of A.

As we have seen in Chapter 4, the simplest structural optimization problem is to minimize an objective function of the total structural volume V under constraints on structural responses, such as displacements for specified loads. This is formulated as Problem 1 in (5.1a). If we reverse the objective function and constraints, we have Problem 2 in (5.1b): minimization of the displacement at a specified load under the upper-bound constraint $V \leq \overline{V}$ on the total structural volume. The problem can be alternatively formulated as Problem 3 in (5.1c): maximization of the allowable load under constraints on the structural volume and displacement.

If geometrical nonlinearity is not incorporated, optimal solutions for Problems 1–3 turn out to be identical, provided that the upper bounds are appropriately assigned. For geometrically nonlinear structures, especially those undergoing snapthrough, some of the seemingly feasible Problems 1–3 turn out to be spurious; optimization algorithm may diverge, or cannot produce a practically acceptable solution. In the following sections, the optimal solutions of demonstrative examples are investigated in detail.

5.3 Two-Bar Truss

Consider the two-bar truss as shown in Fig. 5.1. The y-directional load Λp is applied at the center node. Young's modulus is $E = 200$ kN/mm^2; $L = 1000$ mm, $H = L/\sqrt{3}$, $p = 1000$ kN. The two truss members have the same cross-sectional area A. The units of force kN and of length mm are suppressed in the following.

The equilibrium path for a specified value of A is shown in Fig. 5.2. The limit point indicated by 'S' is reached as Λ is increased from 0 to the limit point load

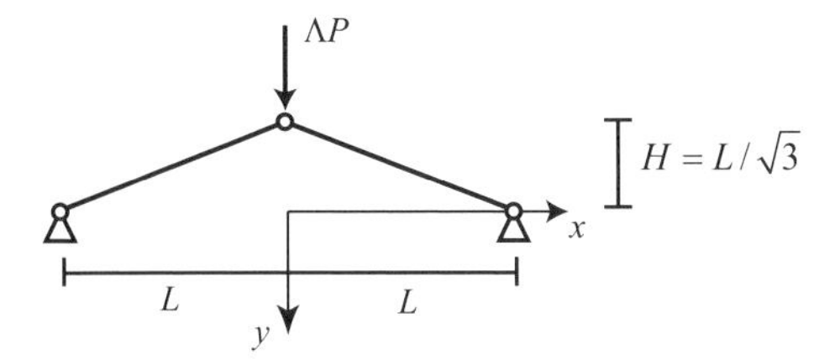

Fig. 5.1 Two-bar truss.

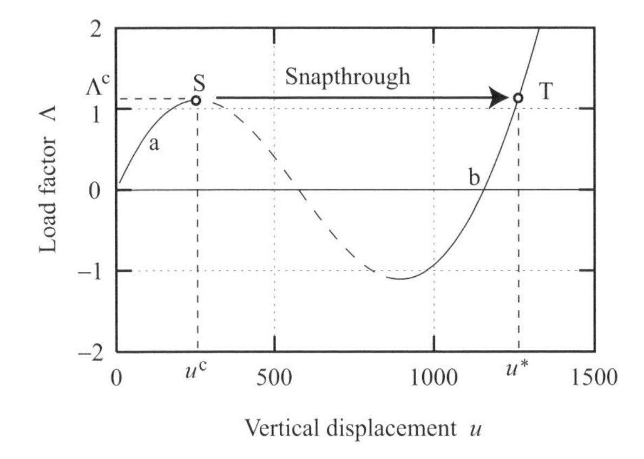

Fig. 5.2 Equilibrium path of the two-bar truss. Solid curve: stable equilibrium path; dashed curve: unstable one; u: y-directional displacement of the center node.

Λ^{c}. The stable equilibrium states before and after the limit point are indicated by 'a' and 'b', respectively. The displacements u^{c} at 'S' and u^* at 'T' play an important role in optimization. If Λ is further increased over Λ^{c}, the equilibrium state jumps dynamically from 'S' to 'T' by snapthrough.

To demonstrate special features to be taken into account in the optimization of structures undergoing snapthrough, consider the three optimization Problems 1–3 in (5.1a)–(5.1c). The cross-sectional area A is chosen to be the only design variable. The total structural volume is $V = (4/\sqrt{3})AL$.

For this two-bar truss, the displacement at the limit point remains $u^{\mathrm{c}} \simeq 240.0$ for any value of A. Therefore, a problem of a displacement constraint $u(\Lambda^{\mathrm{c}}) \leq u^{\mathrm{c}}$ at the limit point load Λ^{c} is ill-defined as it is impossible to control u^{c} by varying the value of design variable A. Such independence of the displacement at the limit point on the design variable, which is not particular for this simple model but occurs generically for realistic structures (cf., Section 5.4), has to be taken into consideration in the formulation of an optimization problem.

Fig. 5.3 illustrates the equilibrium paths for three different values of V. The value of Λ for a specified u is proportional to V. For a large V, Λ reaches a specified value Λ^* before the limit point. However, for a small V, $\Lambda = \Lambda^*$ is satisfied after reaching the limit point.

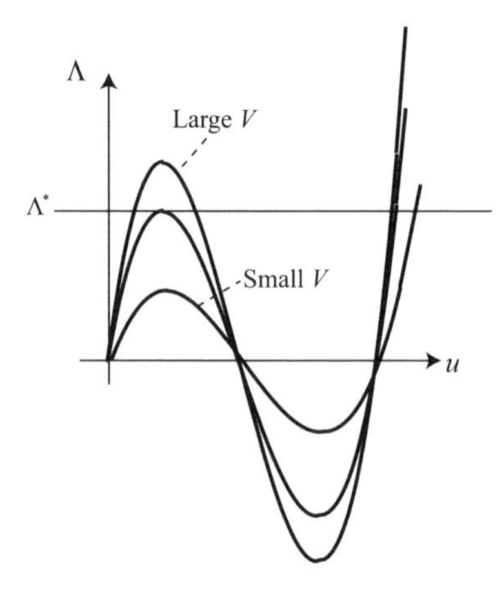

Fig. 5.3 Equilibrium paths for different values of structural volume V.

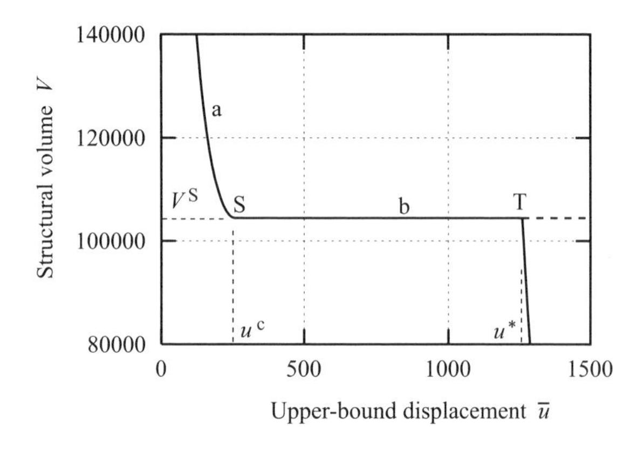

Fig. 5.4 Variation of optimal structural volume plotted against $\overline{u}$ of Problem 1.

For Problem 1, V is to be minimized under the constraint of $u(\Lambda^*) \leq \overline{u}$. Let $\Lambda^* = 1.0$ for simplicity. Since $u(\Lambda^*)$ is a decreasing function of A and $V = (4/\sqrt{3})AL$ is an increasing function of A, the optimal value of V is attained at $u(\Lambda^*) = \overline{u}$, at which the constraint $u(\Lambda^*) \leq \overline{u}$ is satisfied in equality. The variation of the objective function V against $\overline{u}$ is investigated as shown in Fig. 5.4. Note that the curves 'a' and 'b' in this figure correspond to the regions 'a' and 'b', respectively, in Fig. 5.2. There are three regions according to the value of $\overline{u}$.

- For $\overline{u} \leq u^{\mathrm{c}}$, the equilibrium state at $\Lambda = \Lambda^*$ is in the region 'a'.
- For $u^{\mathrm{c}} \leq \overline{u} \leq u^*$, V has a constant optimal value of $V^{\mathrm{S}} = 1.05 \times 10^5$ because the optimal solution is governed by the limit point and $u(\Lambda^*) = u^{\mathrm{c}}$ and $\Lambda^{\mathrm{c}} = \Lambda^*$ are satisfied.
- For $\overline{u} > u^*$, the optimal solution changes according to whether snapthrough from 'S' to 'T' in Fig. 5.2 is allowed or not allowed at $\Lambda < \Lambda^*$. If the snapthrough is allowed, V can decrease in the region $\overline{u} > u^*$ shown by

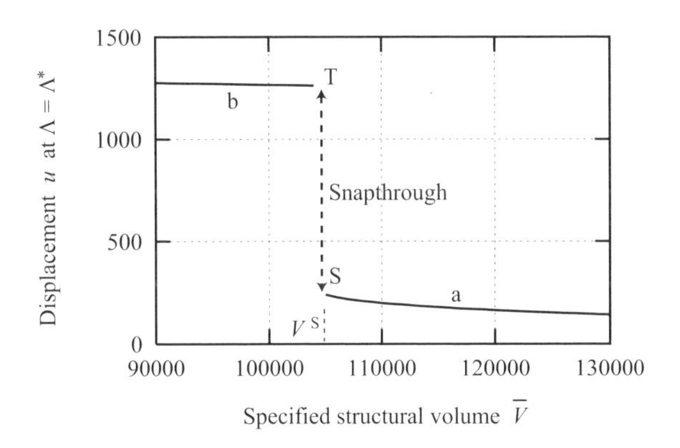

Fig. 5.5 Variation of displacement u of the optimal solution of Problem 2.

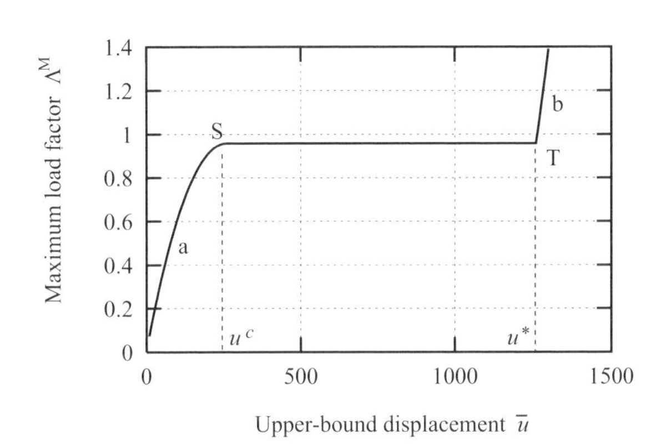

Fig. 5.6 Variation of the maximum load factor Λ^M of the optimal solution of Problem 3 for $V = V^S$.

the solid curve in Fig. 5.4. However, if the snapthrough is not allowed, $u(\Lambda^*) = u^c$ should be satisfied and the optimal V has a constant value of $V^S = 1.05 \times 10^5$ shown by the horizontal dashed line.

For Problem 2, the displacement u at $\Lambda = \Lambda^* = 1.0$ is to be minimized under the constraint $V \leq \overline{V}$. Since $V = (4/\sqrt{3})AL$ is an increasing function of A and $u(\Lambda^*)$ is a decreasing function of A, the optimal value of A is computed from

$$V = \frac{4AL}{\sqrt{3}} = \overline{V} \implies A = \frac{\sqrt{3}\,\overline{V}}{4L} \tag{5.2}$$

- If $\overline{V}$ is sufficiently large, $u(\Lambda^*)$ remains small in the region 'a', and the optimal value of A is almost identical with that of a geometrically linear case. The value of $u(\Lambda^*)$ gradually increases as $\overline{V}$ decreases, as shown in

Fig. 5.5, and the equilibrium state at $\Lambda = \Lambda^*$ reaches the limit point at $V = V^{\rm S} = 1.05 \times 10^5$ denoted by 'S' in Fig. 5.5.

- If $\overline{V}$ is further decreased from $V^{\rm S}$, the equilibrium state jumps to the post-snapthrough state 'T' defined in Fig. 5.2; i.e., $u(\Lambda^*)$ is discontinuous at $\overline{V} = V^{\rm S}$. Therefore, the sensitivity coefficient of $u(\Lambda^*)$ with respect to A also is discontinuous at 'S'. Such discontinuity will result in divergence of an optimization process by a gradient-based optimization algorithm.

For Problem 3, the maximum value $\Lambda^{\rm M}(A)$ of Λ in the range $u \leq \overline{u}$ is to be maximized under the constraint $V \leq \overline{V}$. Fig. 5.6 shows the relation between $\overline{u}$ and $\Lambda^{\rm M}$ for $\overline{V} = 1.0 \times 10^5$.

- For $\overline{u} \leq u^{\rm c}$, $\Lambda^{\rm M} = \Lambda(\overline{u})$ holds and $\Lambda^{\rm M}$ increases as $\overline{u}$ is increased.
- For $u^{\rm c} \leq \overline{u} \leq u^*$, $\Lambda^{\rm M}$ has a constant value $\Lambda^{\rm c}$ at the limit point.
- For $\overline{u} \geq u^*$, Λ has the maximum in post-snapthrough region $u = \overline{u} \geq u^*$ in Fig. 5.2, if snapthrough is allowed, and $\Lambda^{\rm M}$ $(\geq \Lambda^{\rm c})$ increases as $\overline{u}$ is increased.

Note that the curves 'a' and 'b' in Fig. 5.6 are the same as those of the snapthrough behavior in Fig. 5.2.

5.4 Symmetric Shallow Truss Dome

Consider the symmetric shallow truss dome as shown in Fig. 5.7 as a more realistic example (cf., Sections 2.7.2 and 4.4). A concentrated z-directional load Λp ($p = -1$ kN) is applied to the center node 1. Young's modulus of each member is $E = 205.8$. The units kN of force and mm of length are suppressed in the following.

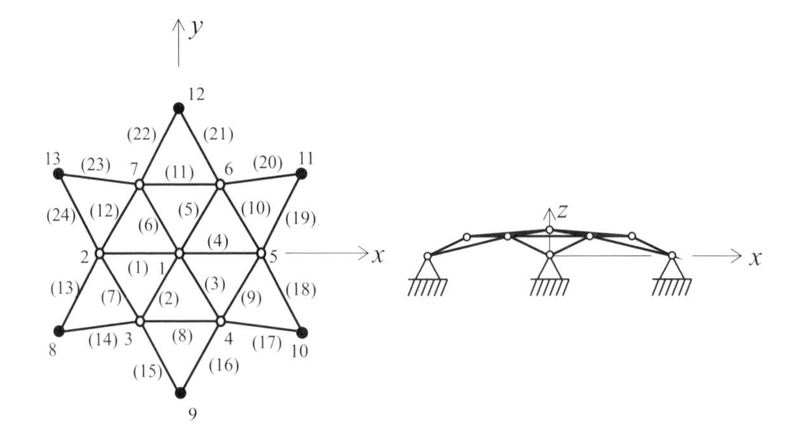

Fig. 5.7 Symmetric shallow truss dome. Nodal coordinates are listed in Table 2.1 in Section 2.7.2. $1, \ldots, 13$ denote node numbers; $(1), \ldots, (24)$ denote member numbers.

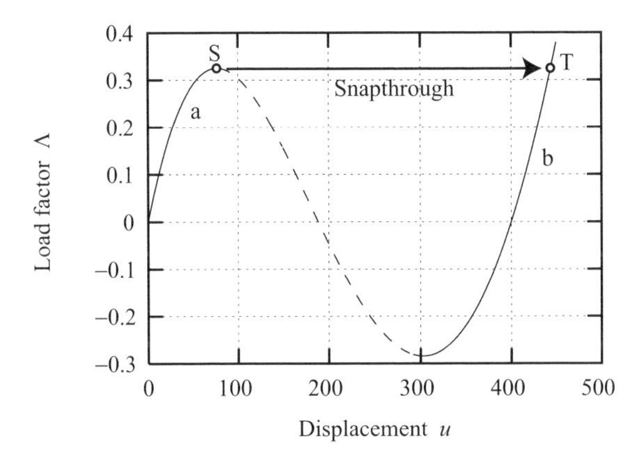

Fig. 5.8 Equilibrium path of the truss dome for $A = 100.0$, $\alpha = 1$. u: displacement in the negative direction of the z-axis of the center node.

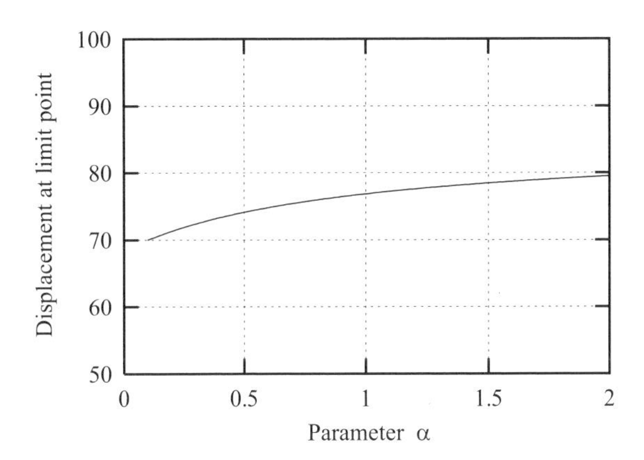

Fig. 5.9 Variation of displacement u^{c} at the limit point of the truss dome with $V = \overline{V} = 7.0 \times 10^6$.

The members are divided into two groups: Group 1 consists of members 1–6 that are connected to the center node, and Group 2 consists of the remaining members. The cross-sectional area A_i of the ith member is assigned by

$$A_i = \begin{cases} A: & \text{for members 1–6 in Group 1} \\ \alpha A: & \text{for members 7–24 in Group 2} \end{cases} \tag{5.3}$$

Note that A and α control the total amount and the distribution of the cross-sectional areas.

The equilibrium path for $\alpha = 1$ is plotted in Fig. 5.8. Here u denote the displacement in the negative direction of the z-axis of the center node. A limit point 'S' is reached as Λ is increased from 0. Fig. 5.9 shows the relation between the design parameter α and the displacement u^{c} at the limit point for $V = \overline{V} = 7.0 \times 10^6$. The variation of u^{c} is small relative to the total height 821.6; i.e., u^{c}

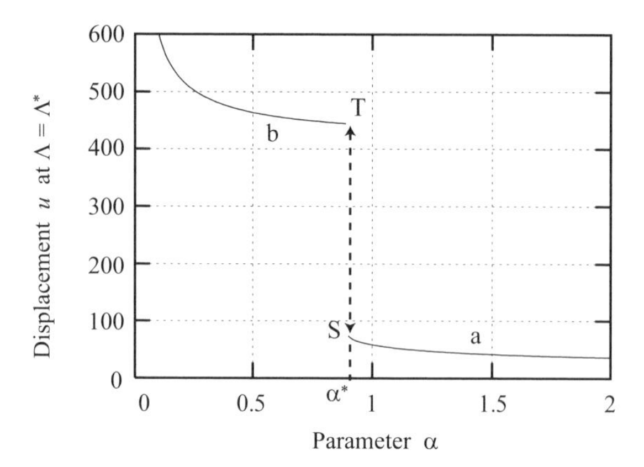

Fig. 5.10 Variation of displacement at $\Lambda = \Lambda^* = 0.3$ of the truss dome with $A = 100.0$ for Problem 1.

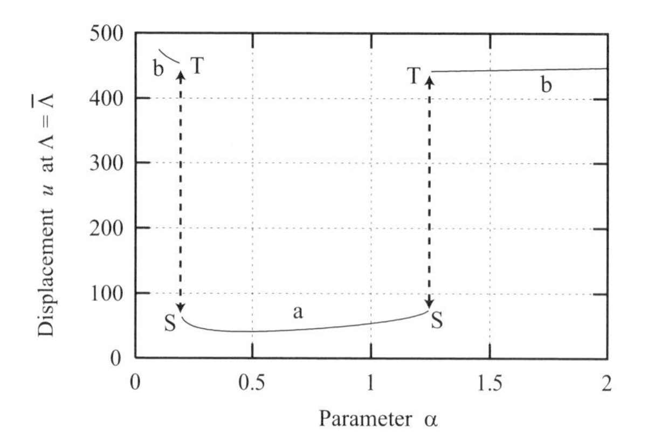

Fig. 5.11 Variation of displacement $u(\Lambda^*)$ at $\Lambda = \Lambda^* = 0.3$ of the truss dome with $V = \overline{V} = 7.0 \times 10^6$ for Problem 2.

is not sensitive to design modification, as was also the case for the two-bar truss in Section 5.3.

Consider the same optimization Problems 1–3 in (5.1a)–(5.1c). The variables V, $u(\Lambda^*)$, Λ^{M}, and so on, are functions of A, but the argument A is suppressed in the sequel for simplicity. Since V is an increasing function of A, the constraint $V \leq \overline{V}$ is satisfied in equality at the optimal solution. In addition to A, the parameter α is chosen as the design variable, and sensitivity coefficients of $u(\Lambda^*)$ and Λ^{M} with respect to α are investigated in the following.

The displacement $u(\Lambda^*)$ for $\Lambda^* = 0.3$ and $A = 100.0$ is discontinuous at $\alpha = \alpha^*$ associated with snapthrough as shown in Fig. 5.10. Accordingly, its sensitivity coefficient with respect to α is discontinuous at $\alpha = \alpha^*$. Therefore, for Problem 1, for which V is minimized under the constraint $u(\Lambda^*) \leq \overline{u}$, the value of V satisfying the constraint $u(\Lambda^*) = \overline{u}$ is discontinuous at $\alpha = \alpha^*$. It can be observed from Fig. 5.10 that α too small suffers from a premature failure owing to uneven

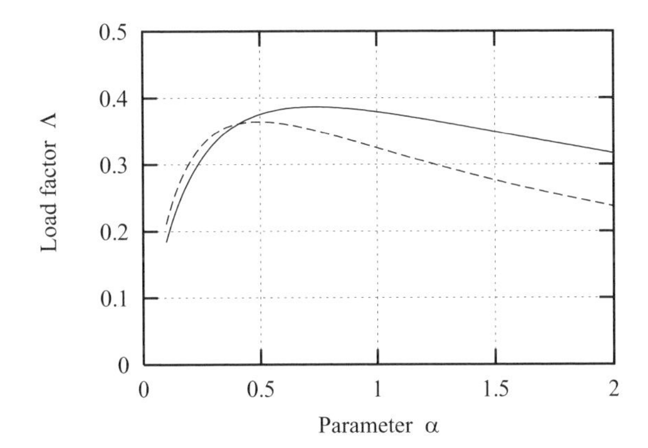

Fig. 5.12 Variation of load factor Λ^{c} at the limit point (solid curve) and load factor $\Lambda(\overline{u})$ at $u = \overline{u} = 45.0$ (dashed curve) for Problem 3.

stiffness distribution that suffers from snapthrough with large deformation in the range 'b', whereas a large value of α restricts the deformation to be small in the range 'a' (cf., the associated curves 'a' and 'b' in Fig. 5.8).

Consider Problem 2, for which $u(\Lambda^*)$ is minimized under the constraint $V \leq \overline{V} = 7.0 \times 10^6$, and let $\Lambda^* = 0.3$. The displacement $u(\Lambda^*)$ for the design satisfying the constraint $V = \overline{V}$ is plotted in Fig. 5.11. Note that A varies with α according to (5.3) and $V = \overline{V}$, whereas A is fixed in Fig. 5.10. As can be seen, $u(\Lambda^*)$ is discontinuous at two values of α, and α too large also suffers from snapthrough. $u(\Lambda^*)$ has a minimum value at $\alpha \simeq 0.47$ that corresponds to the optimum design.

Consider Problem 3, for which Λ^{M} is maximized under the constraints $u(\Lambda^{\mathrm{M}}) \leq \overline{u} = 45.0$ and $V \leq \overline{V} = 7.0 \times 10^6$. The load factor Λ^{c} at the limit point is plotted by the solid curve in Fig. 5.12 and $\Lambda(\overline{u})$ at $u = \overline{u} = 45.0$ by the dashed curve. The maximum load Λ^{M} is given by the larger value of Λ^{c} and $\Lambda(\overline{u})$. Note that these curves intersect at $\alpha \simeq 0.41$, where the sensitivity coefficient of Λ^{M} with respect to α is discontinuous. Note also that the value of u that gives the maximum value of Λ is discontinuous at $\alpha \simeq 0.41$. Therefore, convergence of an optimization algorithm for a gradient-based method will not be guaranteed for the present problem.

5.5 Summary

In this chapter,

- difficulties in optimization for snapthrough behavior have been summarized,
- optimization problems for structures undergoing snapthrough have been presented, and
- structures undergoing snapthrough have been optimized.

The major finding of this chapter is as follows.

- It is vital in the success in optimization of a structure undergoing snapthrough behavior to select a pertinent formulation based on a proper knowledge on this behavior.

6

Shape Optimization of Compliant Mechanisms

6.1 Introduction

An unstable structure called mechanism can exert large deformation for a small input displacement and is used, e.g., for robot arms and retractable structures. This mechanism is also used to change the direction of a motion, e.g., for manipulators, crank shafts and deformation amplifiers for structural control. The crank in Fig. 6.1(a) converts the rotational movement at node 'a' to a translation at node 'b'. Node 'b' of displacement converter in Fig. 6.1(b) moves to right as a result of leftward displacement of node 'a'. For such purposes, it is conventional to employ an unstable bar–joint model called link mechanism [82, 155].

Recently, as an alternative to the unstable mechanism, a compliant mechanism that utilizes flexibility of a mechanical structure has been developed [125, 176, 194], e.g., for switching devices and micro manipulators. Unlike the conventional

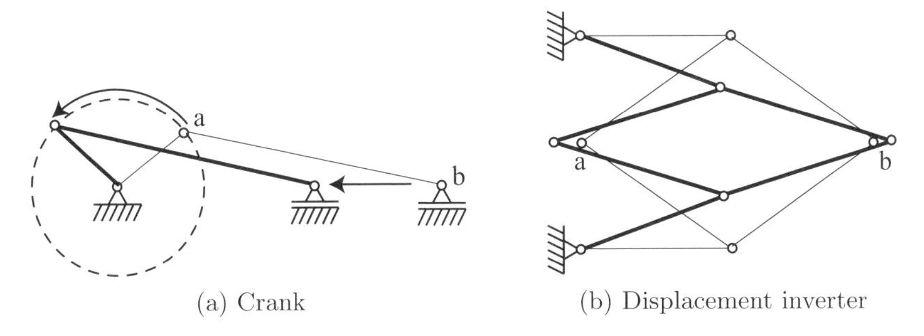

Fig. 6.1 Examples of mechanisms. Thin line: initial shape; thick line: deformed shape.

bar–joint model, desired mechanical properties are realized by appropriately placing flexible units in the structure.

In the general approach for generating a compliant mechanism, a plane domain is divided into a number of finite elements, and unnecessary elements are removed via optimization to obtain an elastically connected beams [212, 274]. However, in most of the studies, external forces should be applied to retain a deformed shape. To overcome this inconvenience, a *bistable structure* with two stable self-equilibrium states [243] is utilized in the compliant mechanism to retain a deformed shape without external forces [194, 226]. Large deformations and catastrophic failures by small additional loads encountered in buckling behavior are not harmful but useful in the new technology.

Optimization considering geometrically nonlinear buckling presented in Chapters 4 and 5 can be effectively combined with a well-developed method of topology optimization of continuum to generate compliant mechanisms. The following studies are conducted to implement snapthrough behavior in optimization:

- The effect of large deformation including snapthrough behavior was incorporated in optimization [39, 240].
- Snapthrough was utilized in topology optimization [38, 270].
- An optimization method was proposed in continuum formulation for generating compliant mechanisms utilizing snapthrough behavior [37].

In this chapter, a general, straightforward and explicit formulation is presented to produce pin-jointed multistable compliant mechanisms that have more than two self-equilibrium states utilizing snapthrough behavior. A truss model is used to avoid numerical instability such as the checkerboard problem in a continuum model [26]. Optimization of the truss model considering geometrical nonlinearity must be constructed with due regard to:

- The presence of many local buckling modes may entail divergence in the analysis and optimization process if the local Euler buckling is considered.
- If bifurcation takes place before reaching the final state, the response may become highly sensitive to imperfections.

This chapter is organized as follows. An illustrative example of the optimization problem of a structure with bistable compliant mechanism is provided in Section 6.2. A general optimization problem is formulated in Section 6.3 for minimizing the total structural volume under constraints on the displacements at the specified nodes, stiffnesses at initial and final states, and load factor at the final state. Cross-sectional areas and nodal locations of trusses are optimized to produce various kinds of compliant mechanisms in Section 6.4.

6.2 Illustrative Examples of Bistable Compliant Mechanisms

Basic properties of a bistable structure and a compliant mechanism are demonstrated by illustrative examples.

6.2.1 Two-bar truss

Snapthrough behavior is utilized to realize a bistable structure, which has two self-equilibrium states: a deformed state and the initial undeformed state. Fig. 6.2 illustrates a relation between the input displacement U_1 and the associated force P_1 that exhibits snapthrough. If the deformation is controlled by the force P_1, it is possible to increase P_1 stably until reaching the limit point 'S', but thereupon the equilibrium state jumps dynamically from 'S' to 'T' by snapthrough.

Alternatively, the deformation of the structure can be controlled by an actuator with displacement control of U_1. In this case, the load P_1 decreases as U_1 is increased beyond the limit point 'S'. Consider the self-equilibrium point 'R' in Fig. 6.2 at $U_1 = U_1^{\mathrm{R}}$ satisfying $P_1 = 0$, and suppose that an obstacle is located to retain the deformation at 'R'. In this case, the final equilibrium state can be stabilized by the stiffness of the actuator, which is locked at the final state. A bistable mechanism with two self-equilibrium states $U_1 = 0$ and $U_1 = U_1^{\mathrm{R}}$ thus have been generated.

For example, consider the two-bar truss as shown in Fig. 6.3, where the solid and dotted lines are the shapes before and after deformation, respectively. If the vertical displacement U_1 of the center node is increased, P_1 increases and reaches a limit point. For a further increase of U_1, $P_1 = 0$ is satisfied when the center node hits a horizontal rigid plate and the two bars become colinear. This final self-equilibrium state can be retained by locking the actuator without any external force. The initial state can be recovered by applying small upward disturbance and releasing the actuator.

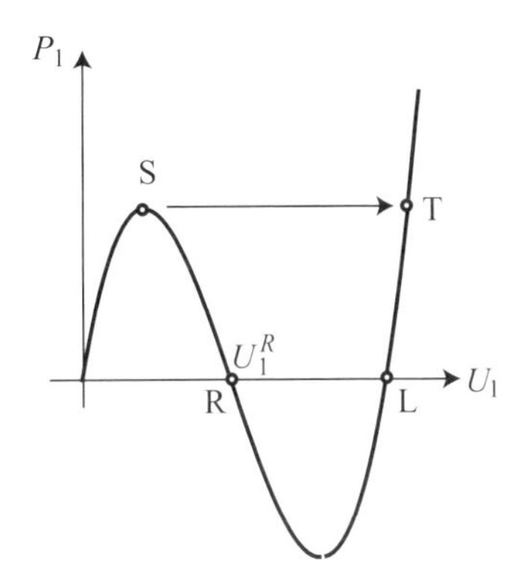

Fig. 6.2 Relation between input displacement U_1 and associated force P_1 of a structure exhibiting snapthrough.

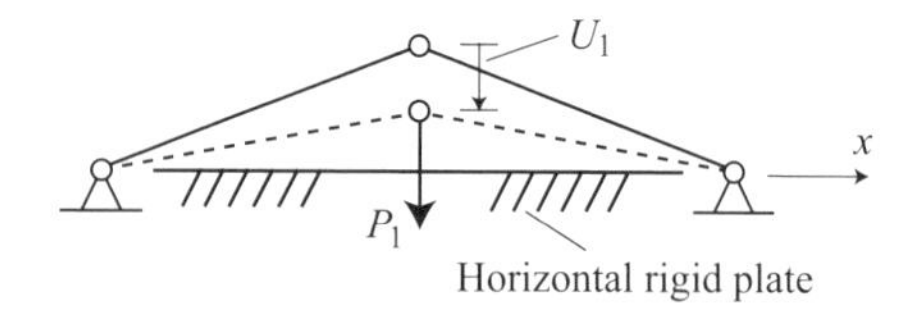

Fig. 6.3 Two-bar truss placed on a horizontal rigid plate. Solid line: initial shape; dashed line: deformed shape.

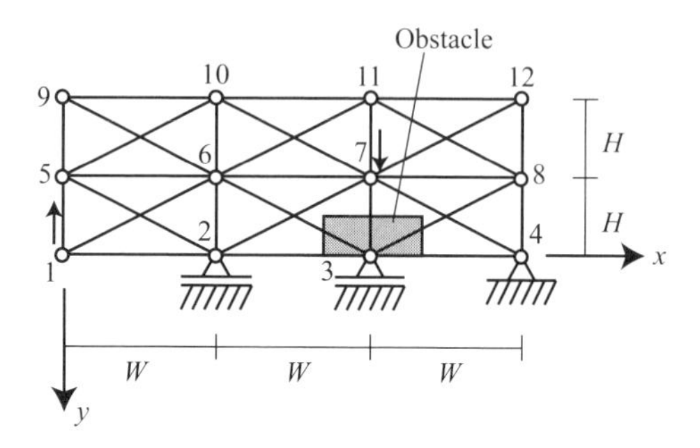

Fig. 6.4 Plane grid truss ($W = 0.2$ m, $H = 0.1$ m). $1, \ldots, 12$ denote node numbers.

Remark 6.2.1 For the two-bar truss, the state 'L' in Fig. 6.2 corresponds to a stable self-equilibrium state that is reverse (reflection symmetric) to the initial state with respect to the x-axis. However, this stable state, which demands no restraint by the actuator, is not employed here to generate a bistable structure, because we assign such a requirement in the general formulation in Section 6.3 that the initial state can be recovered by a small reverse disturbance to the final state. □

6.2.2 Plane grid truss

The framework of the shape optimization to produce a bistable compliant mechanism is illustrated by using the 3×2 plane grid truss as shown in Fig. 6.4 as a ground structure[1]. The pairs of intersecting diagonal members are not connected with each other. Young's modulus is $E = 2.0 \times 10^6$ kN/m^2. The units of force kN and of length m are suppressed in the following. U_{ix} and U_{iy} denote the x- and y-directional displacements at node i, respectively.

By the input y-directional displacement U_{7y} at node 7, the output node 1 should move in the negative direction of the y-axis to the specified position $U_{1y} = -0.05$. The input displacement is terminated by an appropriately placed obstacle as shown in Fig. 6.4.

[1]In the conventional ground structure approach, some members are removed from a highly connected ground structure with fixed nodal locations to arrive at an optimal structure with moderately small number of members. In this section, the nodal locations are also considered as design variables.

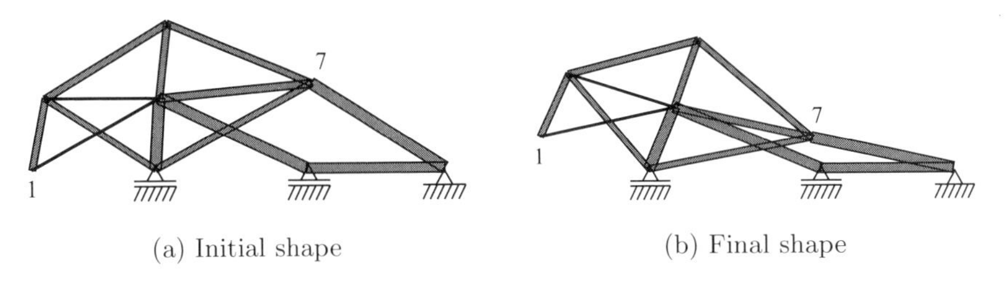

(a) Initial shape (b) Final shape

Fig. 6.5 Optimal grid truss with a bistable mechanism.

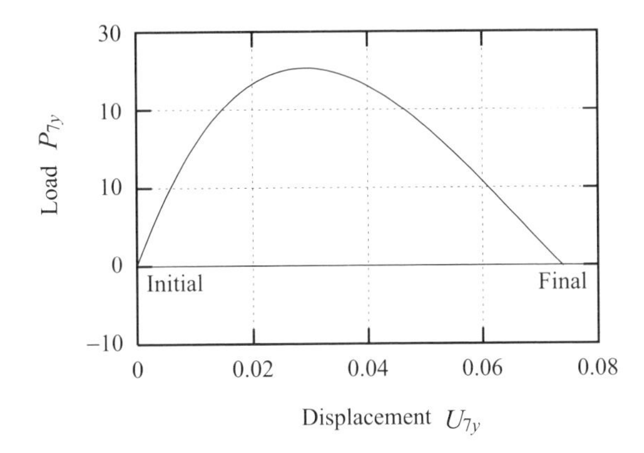

Fig. 6.6 Relation between input displacement U_{7y} and associated force P_{7y} of the optimal truss.

To ensure the stiffness of the structure at the initial and final states, constraints

$$U^0_{7y} \leq 0.01, \quad U^{\mathrm{f0}}_{1x} \leq 0.05, \quad U^{\mathrm{f0}}_{1y} \leq 0.05 \tag{6.1}$$

are imposed. Here U^0_{7y} is the linearly estimated y-directional displacement at node 7 against the y-directional unit load at node 7 at the initial state. U^{f0}_{1x} and U^{f0}_{1y} are the x- and y-directional displacements of node 1, respectively, against the x- and y-directional unit loads at node 1 at the final state, where the tangent stiffness after constraining the input degree-of-freedom is used for estimation.

An optimization problem is formulated to find a bistable compliant mechanism (see Section 6.3 for details). Let (x_i, y_i) denote the coordinates of node i defined in Fig. 6.4. The design variables are the cross-sectional area A_i of each member, x_i $(i = 2, 3, 5, \ldots, 12)$, and y_i $(i = 5, \ldots, 12)$. The lower bound of A_i is set to be a small value of $A^{\mathrm{L}}_i = 1.0 \times 10^{-6}$. The member with $A_i = A^{\mathrm{L}}_i$ after optimization is to be removed from the ground structure. The initial location of node i as shown in Fig. 6.4 is denoted by (x^0_i, y^0_i). The feasible regions for those x_i and y_i chosen as design variables are given as $x^0_i - \Delta x \leq x_i \leq x^0_i + \Delta x$ and $y^0_i - \Delta y \leq y_i \leq y^0_i + \Delta y$, respectively, where $\Delta x = \Delta y = 0.1$.

The initial values of A_i and x_i are randomly generated. The optimal truss is as shown in Fig. 6.5(a), where the width of each member is proportional to its cross-sectional area. Note that the locations of several nodes are modified and several members are removed through the optimization. The deformed final

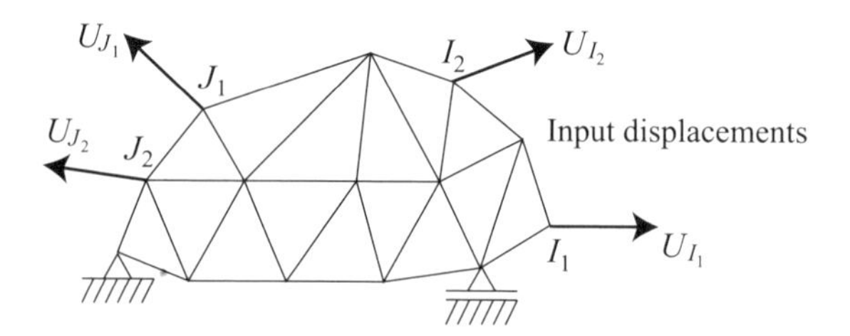

Fig. 6.7 Output displacements U_{J_1} and U_{J_2} respectively by input displacements U_{I_1} and U_{I_2}.

shape is shown in Fig. 6.5(b). The relation between input displacement U_{7y} and the associated force P_{7y} is plotted in Fig. 6.6, which shows that $P_{7y} = 0$ is satisfied at the final state. As can be seen from Fig. 6.5, a large upward output displacement at node 1 is exerted for a downward input displacement at node 7. At the course of this, snapthrough takes place at node 7. This way, a bistable structure can be generated by optimization utilizing snapthrough behavior.

6.3 Shape Optimization Problem for Multistable Compliant Mechanism

A general shape optimization problem for generating a multistable compliant mechanism is presented for a finite-dimensional plane structure.

A multistable compliant mechanism of the structure as shown in Fig. 6.7 is defined as follows:

- A large displacement U_{J_r} at the output node J_r in the specified direction is generated by input displacement U_{I_r} at node I_r of the structure ($r = 1, \dots, n^{\mathrm{p}}$). Here, n^{p} denotes the number of loading conditions that can produce self-equilibrium states.
- An obstacle is placed to terminate the deformation at an unstable self-equilibrium state (point 'R' in Fig. 6.2), and this state is stabilized by locking the actuator without additional external force, i.e., the structure has n^{p} self-equilibrium states in addition to the undeformed initial state. The undeformed initial state can be recovered by reversely applying a small disturbance force and releasing the actuator at each deformed state. For this reversal process, snapthrough behavior is utilized.

Design variables are cross-sectional areas $\mathbf{A} = (A_j) \in \mathbb{R}^{n^{\mathrm{m}}}$ and nodal coordinates $\mathbf{X} = (X_k) \in \mathbb{R}^{n^{\mathrm{x}}}$. The total structural volume is to be minimized to generate a mechanism with small number of members. A final deformed state is defined such that output displacement U_{J_r} for a specified rth forced displacement at input node I_r reaches a specified large value $\overline{U}^{\mathrm{f}}_{J_r}$, i.e.,

$$U_{J_r} = \overline{U}^{\mathrm{f}}_{J_r}, \quad (r = 1, \dots, n^{\mathrm{P}}) \tag{6.2}$$

Note that input displacements and forces are assumed to be positive to simplify the formulation. In the following, the values corresponding to the final state are ascribed with the superscript $(\cdot)^{\mathrm{f}}$.

A series of inequality constraints are introduced:

- Upper and lower bounds $\mathbf{X}^{\mathrm{U}}$ and $\mathbf{X}^{\mathrm{L}}$ are given for $\mathbf{X}$, and very small lower bounds $\mathbf{A}^{\mathrm{L}} = (A^{\mathrm{L}}_j)$ for $\mathbf{A}$ are prescribed for preventing numerical instability. A member with $A_j = A^{\mathrm{L}}_j$ after optimization is to be removed from the initial ground structure with redundant members.
- The rth input force $P^{\mathrm{f}}_{I_r}$ at the final state is constrained to be small so that the final state can approximate a self-equilibrium state:

$$P^{\mathrm{f}}_{I_r} \leq 0, \quad (r = 1, \dots, n^{\mathrm{P}}) \tag{6.3}$$

in which the equality constraint $P^{\mathrm{f}}_{I_r} = 0$ is relaxed to inequality to improve convergence of iterations during optimization. It is shown in the examples that $P^{\mathrm{f}}_{I_r} \simeq 0$ is satisfied for the optimal solution[2].

- The upper bound of the rth input displacement $U^{\mathrm{f}}_{I_r}$ at the final state is incorporated in the constraints as (see Remark 6.3.1 below)

$$U^{\mathrm{f}}_{I_r} \leq \overline{U}^{\mathrm{f}}_{I_r} \tag{6.4}$$

- The stiffness at the initial state is ensured by constraining the displacements

$$U^{0}_{I_r} \leq \overline{U}^{0}_{I_r}, \quad (r = 1, \dots, n^{\mathrm{P}}) \tag{6.5}$$

where $U^{0}_{I_r}$ denotes the linearly estimated displacement U_{I_r} for the unit load $P_{I_r} = 1$ at the initial state. Note that the constraints on stiffness together with that of the final load (6.3) produce a solution that exhibits snapthrough behavior; without the stiffness constraints, minimization of the total structural volume leads to an unstable structure for which $P_{I_r} = 0$ holds throughout the equilibrium path.

- The requirement on the stiffness at the final state is given by the displacement constraints as

$$U^{\mathrm{f0}}_{J_r x} \leq \overline{U}^{\mathrm{f0}}_{J_r x}, \quad U^{\mathrm{f0}}_{J_r y} \leq \overline{U}^{\mathrm{f0}}_{J_r y}, \quad (r = 1, \dots, n^{\mathrm{P}}) \tag{6.6}$$

Here $U^{\mathrm{f0}}_{J_r x}$ denotes the x-directional displacement of node J_r at the final state evaluated by the tangent stiffness for the x-directional unit load at node J_r after constraining the rth input degree-of-freedom. $U^{\mathrm{f0}}_{J_r y}$ is defined similarly.

[2]If $P^{\mathrm{f}}_{I_r} < 0$, the final state terminated by an obstacle is stable without locking the actuator. However, if $P^{\mathrm{f}}_{I_r}$ has a large negative value, a large reversal force is needed to recover the initial state.

- To further improve convergence to a structure with snapthrough, the load at an intermediate state is constrained by

$$P^{\rm m}_{I_r} \geq P^{\rm f}_{I_r} \tag{6.7}$$

Here $P^{\rm m}_{I_r}$ denotes the input load at the intermediate state with $U_{J_r} = \overline{U}_{J_r}/2$.

Remark 6.3.1 In the customary formulation of the optimization problem of compliant mechanism, the absolute value of output versus input displacement ratio $|U^{\rm f}_{J_r}/U^{\rm f}_{I_r}|$ is maximized [212, 274]. However, the maximization of the output/input ratios for multiple loading conditions becomes a multiobjective optimization problem that is difficult to solve. □

Finally, the optimization problem for minimizing the total structural volume $V(\mathbf{A}, \mathbf{X})$ is stated as

$$\begin{aligned}
\text{minimize} \quad & V(\mathbf{A}, \mathbf{X}) && (6.8a)\\
\text{subject to} \quad & P^{\rm f}_{I_r}(\mathbf{A}, \mathbf{X}) \leq 0, \quad (r = 1, \dots, n^{\rm p}) && (6.8b)\\
& U^{\rm f}_{I_r}(\mathbf{A}, \mathbf{X}) \leq \overline{U}^{\rm f}_{I_r}, \quad (r = 1, \dots, n^{\rm p}) && (6.8c)\\
& U^{0}_{I_r}(\mathbf{A}, \mathbf{X}) \leq \overline{U}^{0}_{I_r}, \quad (r = 1, \dots, n^{\rm p}) && (6.8d)\\
& U^{\rm f0}_{J_r x}(\mathbf{A}, \mathbf{X}) \leq \overline{U}^{\rm f0}_{J_r x}, \quad (r = 1, \dots, n^{\rm p}) && (6.8e)\\
& U^{\rm f0}_{J_r y}(\mathbf{A}, \mathbf{X}) \leq \overline{U}^{\rm f0}_{J_r y}, \quad (r = 1, \dots, n^{\rm p}) && (6.8f)\\
& P^{\rm m}_{I_r}(\mathbf{A}, \mathbf{X}) \geq P^{\rm f}_{I_r}, \quad (r = 1, \dots, n^{\rm p}) && (6.8g)\\
& A^{\rm L}_i \leq A_i, \quad (i = 1, \dots, n^{\rm m}) && (6.8h)\\
& X^{\rm L}_i \leq X_i \leq X^{\rm U}_i, \quad (i = 1, \dots, n^{\rm x}) && (6.8i)
\end{aligned}$$

Obviously, the final state satisfying (6.2) cannot be found if the direction of U_{J_r} at the initial solution is opposite to $\overline{U}_{J_r}$. Furthermore, both structural analyses and optimization problems to be solved involve strong nonlinearity. Therefore, initial solutions are randomly generated [160], and several local optimal solutions are found to generate various kinds of compliant mechanisms. The optimization procedure is summarized as follows:

Step 1: Assign random initial values to **A** and **X**.

Step 2: Trace the equilibrium path for each loading condition by adopting the input displacement U_{I_r} as the path parameter.

Step 3: Go to Step 1 if U_{J_r} at the first incremental step is opposite to $\overline{U}_{J_j}$. Otherwise, trace the path until (6.2) is satisfied.

Step 4: Compute sensitivity coefficients of the objective and constraint functions with respect to **A** and **X**.

Step 5: Update **A** and **X** in accordance with an optimization algorithm.

Step 6: Go to Step 2 if the iteration does not converge.

Step 7: Go to Step 1 to find another mechanism.

Note that no human judgment is needed in the proposed method, and several mechanisms can be found by a straightforward optimization from randomly generated initial ground structures.

Remark 6.3.2 When the final value $\overline{U}^{\mathrm{f}}_{J_r}$ of the output displacement at node J_r and the upper bound $\overline{U}^{\mathrm{f}}_{I_r}$ of the input displacement are given, all the parameters in the problem and the cross-sectional areas can be scaled after obtaining an optimal solution without modifying the nodal locations. Suppose that the upper-bound displacements $\overline{U}^{0}_{I_r}$, $\overline{U}^{\mathrm{f0}}_{J_r x}$ and $\overline{U}^{\mathrm{f0}}_{J_r y}$ for stiffness constraints are scaled by the same factor α as

$$U^0_{I_r} \rightarrow \alpha U^0_{I_r}, \quad U^{\mathrm{f0}}_{J_r x} \rightarrow \alpha U^{\mathrm{f0}}_{J_r x}, \quad U^{\mathrm{f0}}_{J_r y} \rightarrow \alpha U^{\mathrm{f0}}_{J_r y} \tag{6.9}$$

Then the optimal cross-sectional areas and other variables at optimal solution are scaled as

$$P^{\mathrm{f}}_{I_r} \rightarrow P^{\mathrm{f}}_{I_r}/\alpha, \quad P^{\mathrm{m}}_{I_r} \rightarrow P^{\mathrm{m}}_{I_r}/\alpha, \quad \mathbf{A} \rightarrow (1/\alpha)\mathbf{A}, \quad V \rightarrow V/\alpha \tag{6.10}$$

Hence, the total structural volume and the global stiffness of an optimal solution obtained can be controlled by the parameter α. □

6.4 Examples of Multistable Compliant Mechanisms

Consider the plane grid truss as shown in Fig. 6.8. The units of force kN and of length m are suppressed in the following.

This truss represents thc uppcr half of an equipment shown in Fig. 6.9(a), which illustrates the following usage of the mechanism:

- If the workpiece is smaller than the initial gap as shown in Fig. 6.9(b), pull node 1 in Fig. 6.8, and release the input force slightly beyond the final state to grip the workpiece without external loads.
- If it is larger than the initial gap as shown in Fig. 6.9(c), pull node 2 in Fig. 6.8 to open the gap, stop at the final position, insert the workpiece, and release the input force to close the gap.

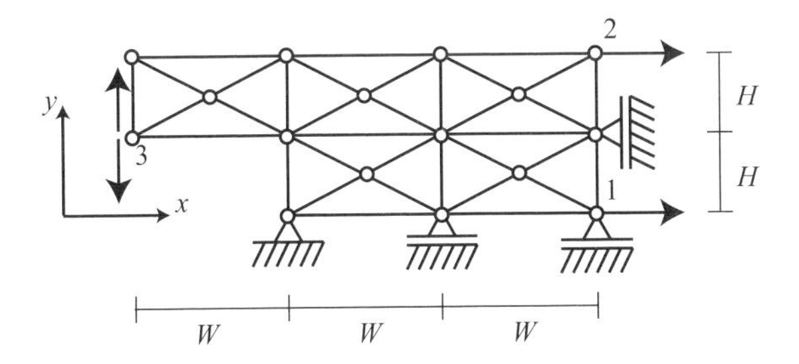

Fig. 6.8 Plane grid truss ($W = 0.2$ m, $H = 0.1$ m).

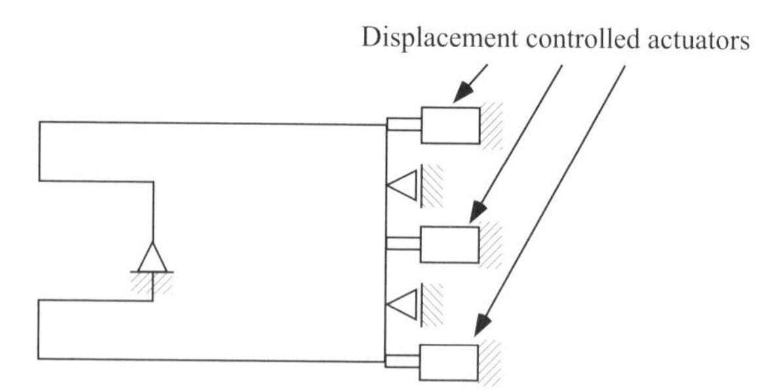

(a) Boundary conditions and locations of the actuators

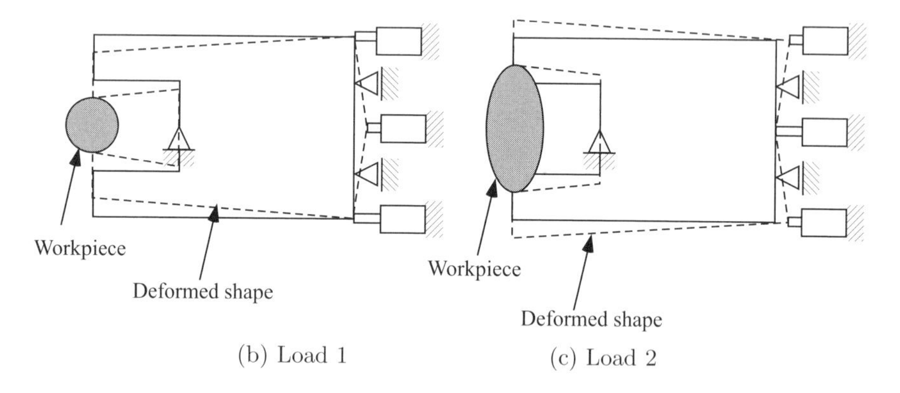

(b) Load 1 (c) Load 2

Fig. 6.9 Equipment with tri-stable mechanism.

We generate a tri-stable mechanism. Consider two loading conditions with input x-directional displacements U_1 and U_2 applied at nodes 1 and 2, respectively, which are called loads 1 and 2 for brevity. The final self-equilibrium states are defined such that the y-directional displacements of node 3 for loads 1 and 2 reach the specified values -0.02 and 0.02, respectively. Other constraint parameters are $\overline{U}^{\mathrm{f}}_{I_r} = \overline{U}^{0}_{I_r} = \overline{U}^{\mathrm{f0}}_{J_r x} = \overline{U}^{\mathrm{f0}}_{J_r y} = 0.02$.

The cross-sectional areas $\mathbf{A} = (A_i)$ of all members are chosen to be independent variables with the lower bound $A_i^{\mathrm{L}} = 1.0 \times 10^{-6}$, and the members with the lower bound cross-sectional areas after optimization are removed. The coordinates of the output node 3, the pin support, and the roller supports in the constrained directions are fixed during optimization. Let x_i^0 denote the initial x-coordinate of an unconstrained node in the grid shown in Fig. 6.8. The upper and lower bounds of x_i are given by $x_i^0 \pm 0.02$. The feasible regions of the y-coordinates are defined similarly. Optimization is carried out by IDESIGN Ver. 3.5 [14] that utilizes SQP in Section 4.3.2. Sensitivity coefficients are computed by the forward finite difference method.

Initial values of A_i are given randomly by

$$A_i = 0.001 + 0.002(R_i - 0.5) \tag{6.11}$$

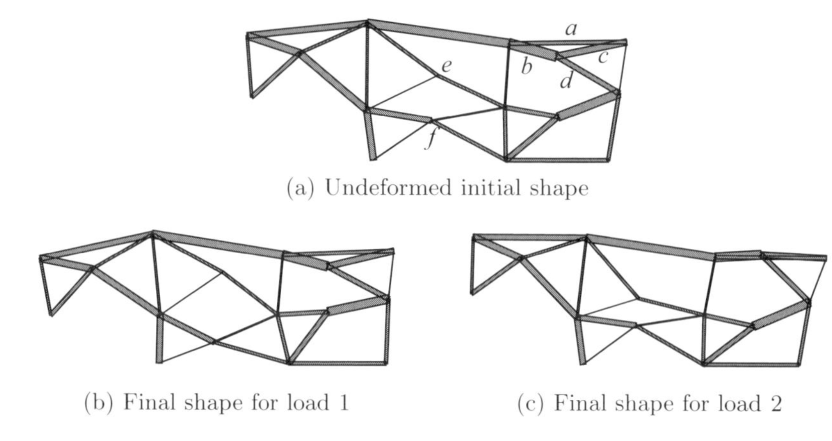

(a) Undeformed initial shape

(b) Final shape for load 1

(c) Final shape for load 2

Fig. 6.10 Type 1 optimum design ($V^{\mathrm{opt}} = 1.3479 \times 10^{-2}$).

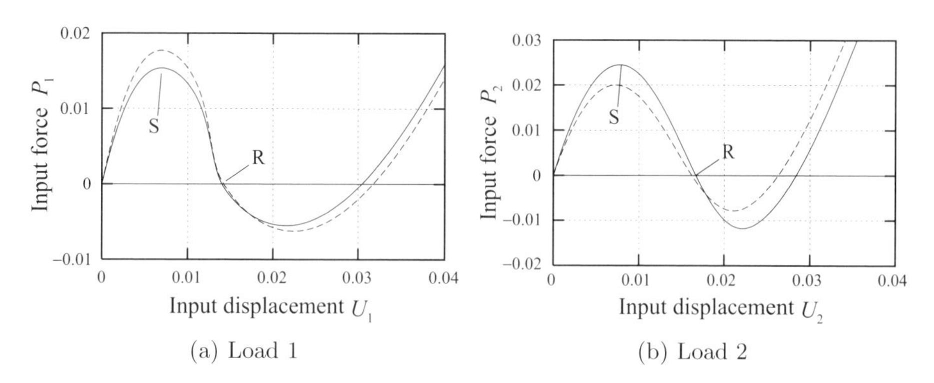

(a) Load 1

(b) Load 2

Fig. 6.11 Relation between input displacements and the associated forces of the Type 1 optimum design for the loads 1 and 2. Solid curve: perfect structure; dashed curve: imperfect structure.

with the use of uniform random values $R_i \in [0, 1)$. For the unconstrained nodal locations, the initial value of x_i is given as

$$x_i = x_i^0 + 0.05(R_i - 0.5) \tag{6.12}$$

The y-coordinates are defined similarly.

Two local optimal solutions have been found starting from different random initial solutions. Type 1 solution is as shown in Fig. 6.10(a), where the width of each member is proportional to its cross-sectional area. The optimal total structural volume is $V^{\mathrm{opt}} = 1.3479 \times 10^{-2}$. The final deformed shapes for the loads 1 and 2 are as shown in Figs. 6.10(b) and (c), respectively. It can be observed from Figs. 6.10(a) and (c) that snapthrough takes place for the load 2 at the triangular unit formed by members a, b and c. Local buckling is also observed around the nodes e and f for the load 1 (cf., Fig. 6.10(b)).

The relation between the input displacement U_1 and the associated force P_1 of Type 1 solution is shown by the solid curve in Fig. 6.11(a). P_1 increases as U_1 is increased until reaching a limit point ‘S’. P_1 decreases by further increasing U_1 to

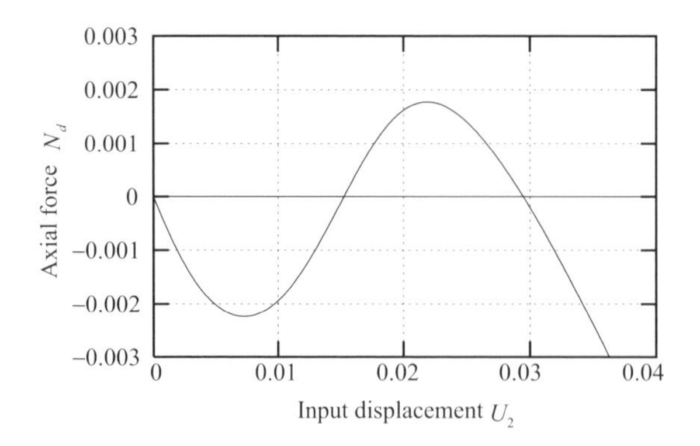

Fig. 6.12 Relation between input displacement U_2 and axial force N_d of the optimum design Type 1.

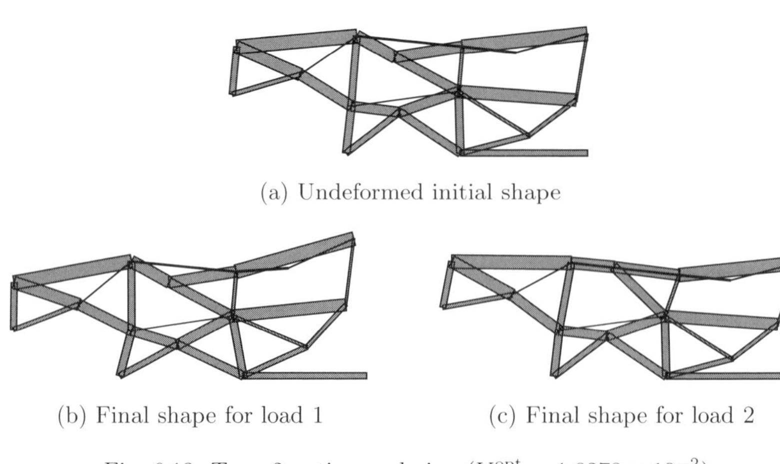

Fig. 6.13 Type 2 optimum design ($V^{\rm opt} = 1.8278 \times 10^{-2}$).

reach the final state 'R' with $P_1 = 0$. The relation between the input displacement U_2 and the associated force P_2 is plotted by the solid curve in Fig. 6.11(b), where snapthrough behavior is observed. The relation between U_2 and the axial force N_d of member d is shown in Fig. 6.12, which also indicates the presence of a limit point 'S' approximately at $U_2 \simeq 0.007$.

At the course of this, it has been confirmed that no bifurcation point exists along the equilibrium path, and the equilibrium path is not sensitive to initial imperfections.

To further investigate imperfection sensitivity, the nodal locations of imperfect structures are given randomly using the uniform random values $R_i \in [0, 1)$ as

$$x_i = x_i^{\rm opt} + 0.002(R_i - 0.5) \tag{6.13}$$

where $x_i^{\rm opt}$ are the x-coordinates of the optimal solution in Fig. 6.10(a). The y-coordinates are defied similarly.

Equilibrium paths are traced for 10 cases of imperfect structures. For the ith imperfect structure, e_i denotes the mean absolute value of deviation of the load from that of the perfect structure throughout the incremental step of path-tracing before reaching the final state. The worst case is associated with the maximum value of e_i among the 10 cases for loads 1 and 2, which are 1.4779×10^{-3} and 3.4687×10^{-3}, respectively; and the equilibrium paths corresponding to these worst cases for the loads 1 and 2 are plotted by the dashed curves in Figs. 6.11(a) and (b) to demonstrate that the equilibrium paths of the optimal solutions are not sensitive to initial imperfections.

Type 2 optimal solution obtained from a different initial solution is shown in Fig. 6.13, where the value of the objective function is $V^{\mathrm{opt}} = 1.8278 \times 10^{-2}$, which is larger than that of Type 1. Recall that our objective is not to find the global optimal solution, but to generate various possible mechanisms from randomly generated initial solutions.

6.5 Summary

In this chapter,

- multistable compliant mechanism has been introduced,
- a new formulation has been presented for generating multistable compliant mechanisms consisting of bar elements, and
- optimum shape design of a plane grid truss has been conducted to generate tri-stable mechanisms.

The major findings of this chapter are as follows.

- A multistable structure that has more than two self-equilibrium states can be found by the ground structure approach considering geometrical nonlinearity. The final state is defined such that output displacements reach specified values, and the input displacements as well as the stiffness at the initial and final states are included in the constraints. This way, multistable mechanisms with small number of members can be successfully found by minimizing the total structural volume.
- Various types of mechanisms can be found by optimization from randomly generated initial solutions; i.e., the proposed method does not require multistage procedures that demand human judgment. The difficulties in continuum formulations can be avoided successfully by using a truss model.
- The multistable mechanism with large deformation can be realized by local snapthrough behavior. Therefore, members irrelevant to snapthrough can be removed without disturbing the snapthrough.

7

Optimal Braced Frames with Coincident Buckling Loads

7.1 Introduction

Elastic buckling plays a pivotal role in designing slender steel frames, such as the symmetric braced frame as shown in Fig. 7.1(a). The frame without braces as shown in Fig. 7.1(b) exhibits a so-called sway buckling, which is classified as symmetric bifurcation and global buckling. The sway buckling can be avoided by attaching slender braces [172]. Then the braced frame as shown in Fig. 7.1(c) exhibits a non-sway buckling, which is classified as slightly asymmetric bifurcation and member buckling without joint displacements as the prebuckling deformation is negligibly small[1]. Bažant and Cedolin [19] investigated buckling of frames, such as the L-shaped frame as shown in Fig. 7.2, and showed that an asymmetric frame undergoing non-sway asymmetric bifurcation buckling is imperfection-sensitive, as was demonstrated by Roorda [257].

It is possible to evaluate the buckling load factor of a symmetric braced frame accurately by the linear eigenvalue analysis because its prebuckling deformation is very small. However, the deformation of an imperfect system becomes very large before reaching a critical point; therefore, geometrical nonlinearity should be incorporated in order to evaluate imperfection sensitivity. Even for the perfect system, the non-sway bifurcation point often disappears owing to small symmetric prebuckling deformation, and the sway bifurcation is stable as demonstrated in the structural examples in Section 7.4. Then the maximum allowable load factor should be evaluated based on the local and global deformations, or plastic failure [189], instead of the bifurcation load.

---

[1] The non-sway buckling is not local buckling in a strict sense because bending moment and rotation at the end of a buckled member lead to deformation of other members.

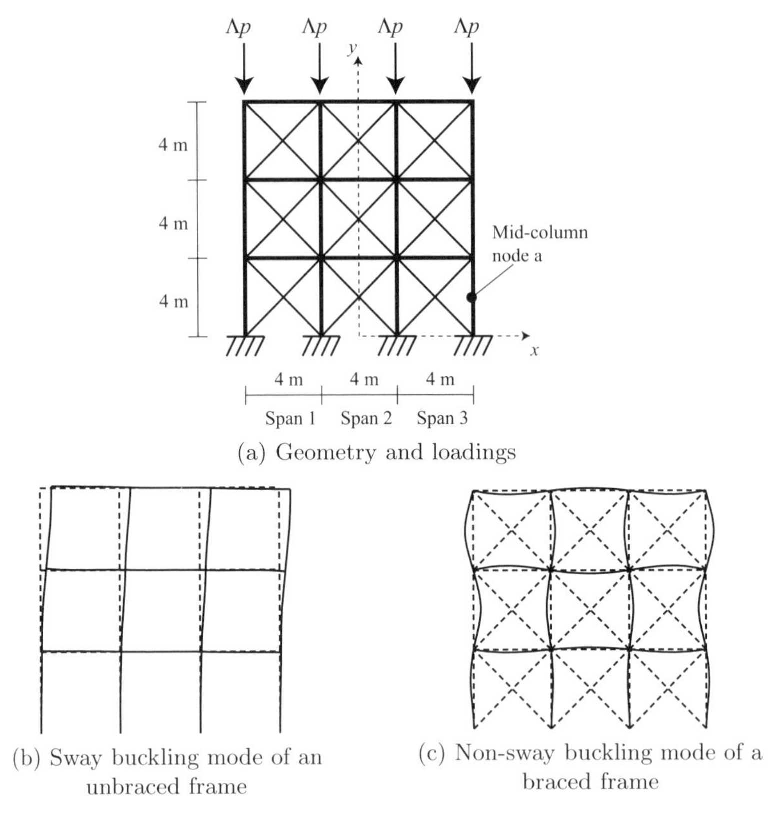

(a) Geometry and loadings

(b) Sway buckling mode of an unbraced frame

(c) Non-sway buckling mode of a braced frame

Fig. 7.1 Three-story three-span braced frame.

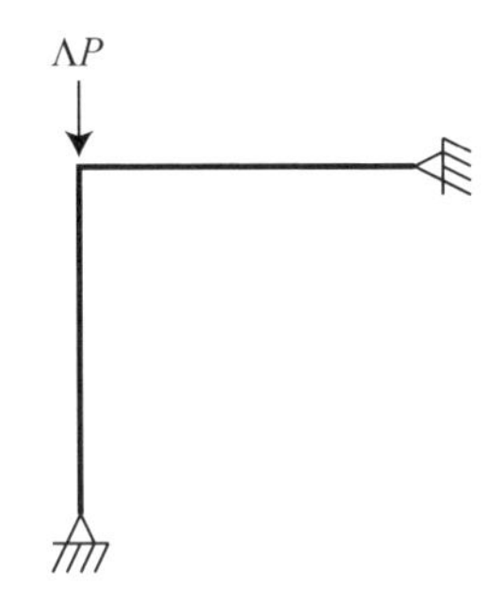

Fig. 7.2 L-shaped frame.

A straightforward optimization of the frame under stability constraint will produce an optimal frame with coincident bifurcation. In a customary structural analysis, a coincident critical point serves as a rare exceptional case that is generated only through an elaborate selection of structural parameters, as seen in Section 2.7.3. Nonetheless amazing mechanism in optimization of structures is its capability to produce coincident critical points in a systematic manner. This encourages the readers to study the theory of coincident critical points presented in Section 3.3.2. Again a proper knowledge on elastic stability plays a pivotal role in the evaluation of the performance of optimal structures.

An optimal braced frame often has an asymmetric (non-sway) bifurcation point slightly above a symmetric (sway) bifurcation point on the fundamental path to produce a nearly coincident bifurcation point [19]. Thus the theory of a semi-symmetric bifurcation point presented in Section 3.6.2 serves as a foundation of this chapter. At this point, imperfect bifurcation behaviors become very complex in association with the existence of multiple bifurcation modes and the mode interaction among them. For this reason, the influence of imperfections varies greatly according to their modes.

In this chapter, imperfection sensitivity of the nonlinear buckling load factors of non-optimal and optimal symmetric frames is investigated[2]. Higher differential coefficients are computed for quantitative evaluation of imperfection sensitivity and mode interaction. It is shown that the linear buckling load factors corresponding to sway and non-sway modes of an optimum design are coincident. However, the asymmetric bifurcation point with non-sway type deformation disappears for geometrically nonlinear analysis. For optimal frames, the non-sway bifurcation is shown to be nearly symmetric, and the interaction between sway and non-sway modes does not always enhance imperfection sensitivity. This supports the major claim of this book that the imperfection sensitivity at the coincident bifurcation is not necessarily enhanced.

This chapter is organized as follows. An optimization problem is formulated under constraints on linear buckling load factors in Section 7.2. Imperfection sensitivity laws and associated imperfect behaviors are studied in Section 7.3. Numerical examples of non-optimal and optimal frames are presented in Section 7.4.

7.2 Optimization Problem of a Braced Frame

Optimization of the braced frame as shown in Fig. 7.1(a) subjected to proportional vertical loads is conducted.

7.2.1 Problem formulation

Optimization of a braced frame in terms of linear buckling load factor is conducted as follows:

- The total volume $V = \sum_{i=1}^{\nu} A_i L_i$ is chosen as the objective function, where A_i is the cross-sectional area and L_i is the length of the ith member ($i = 1, \ldots, \nu$).
- Linear buckling analysis (cf., Section 1.5.1) is conducted to ensure the stability of the frame; to be precise, the lower bound $\overline{\Lambda}$ for the rth lowest

[2] Only vertical loads are applied to frames in this chapter, although multiple loading conditions including horizontal loads are considered in practical design [122].

positive linear buckling loads Λ_{Lr} $(r = 1, \ldots, h)$ is specified, where h is the largest possible multiplicity of the buckling load.

- Side constraints are specified for $\mathbf{A} = (A_i)$.
- The second moment of inertia I_i of the ith member is considered to be a function of the cross-sectional area A_i. Note that the braces are modeled by truss members without member buckling.

To sum up, the optimization problem is formulated as

$$\text{minimize} \quad \sum_{i=1}^{\nu} A_i L_i \tag{7.1a}$$

$$\text{subject to} \quad \Lambda_{Lr}(\mathbf{A}) \geq \overline{\Lambda}, \quad (r = 1, \ldots, h) \tag{7.1b}$$

$$A_i^{\mathrm{L}} \leq A_i \leq A_i^{\mathrm{U}}, \quad (i = 1, \ldots, \nu) \tag{7.1c}$$

It is well known that the optimum design under constraint on linear buckling loads often has coincident buckling loads [231]. Then the sensitivity coefficients of $\Lambda_{Lr}(\mathbf{A})$ with respect to $\mathbf{A}$ become discontinuous at the optimum design with coincident buckling loads (cf., Section 2.6.2). A difficulty arising from discontinuity in sensitivity coefficient can be avoided by using the line search in optimization algorithm (cf. Section 4.3.2).

7.2.2 Definition of maximum load factor

The non-sway bifurcation point of the braced frame obtained by linear buckling analysis disappears if geometrically nonlinear analysis is conducted as shown in the following examples. Since deformation at the load level near the linear buckling load becomes very large for its imperfect systems, the maximum allowable load factor is defined in view of global and local deformations.

The global displacement Δ_j is defined by the interstory displacement

$$\Delta_j = U_j - U_{j-1}, \quad (j = 1, \ldots, n^{\mathrm{s}}) \tag{7.2}$$

where U_j denotes the horizontal displacement in the jth floor level, and n^{s} is the number of stories. We employ the standard assumption of a rigid floor, in which the horizontal displacement of the nodes in a floor is assigned with the same value. The local deformation δ_i is defined as

$$\delta_i = \frac{1}{2}(u_i^{\mathrm{u}} + u_i^{\mathrm{b}}) - u_i^{\mathrm{m}}, \quad (j = 1, \ldots, m^{\mathrm{c}}) \tag{7.3}$$

where horizontal displacements of the upper, middle and bottom nodes of the ith column are denoted by u_i^{u}, u_i^{m} and u_i^{b}, respectively, and m^{c} is the number of columns.

The upper bounds $\overline{\Delta}_j$ and $\overline{\delta}_i$ for $|\Delta_j|$ and $|\delta_i|$ are specified, respectively. The maximum load factor Λ^{M} under the constraints on global and local displacements is defined as the value of Λ at which one of the following conditions is first satisfied in equality as Λ is increased from 0:

$$\begin{cases} |\Delta_j| \leq \overline{\Delta}_j, & (j = 1, \dots, n^{\mathrm{s}}) \\ |\delta_i| \leq \overline{\delta}_i, & (j = 1, \dots, m^{\mathrm{c}}) \end{cases} \tag{7.4}$$

7.3 Imperfection Sensitivity of Semi-Symmetric Bifurcation Point

We investigate theoretically imperfection sensitivity and associated imperfect behaviors of a semi-symmetric double bifurcation point, which emerges as a coincidence of a simple symmetric bifurcation point and a simple asymmetric one (cf., Section 3.3.2).

Suppose the critical eigenmodes $\boldsymbol{\Phi}_1^{\mathrm{c}}$ and $\boldsymbol{\Phi}_2^{\mathrm{c}}$ are related to symmetric and asymmetric bifurcation points, respectively. Since imperfect behaviors of a semi-symmetric system in general are very complicated, we focus only on the imperfection mode in the direction of $\boldsymbol{\Phi}_2^{\mathrm{c}}$, and assume [129, 284]

$$\begin{aligned} &V_{,1\xi} = 0, \quad V_{,2\xi} < 0 \\ &V_{,111} = V_{,122} = 0, \quad V_{,222} < 0, \quad V_{,112} \neq 0 \end{aligned} \tag{7.5}$$

By the further assumption that the fundamental path is stable until reaching this point, we have

$$V_{,11\Lambda} < 0, \quad V_{,22\Lambda} < 0 \tag{7.6}$$

The imperfection sensitivity for the major imperfection under consideration is given by the 1/2-power law [129] (cf., Appendix A.7 for derivation)

$$\Lambda^{\mathrm{c}}(\xi) = \Lambda^{\mathrm{c}0} + C_8(\kappa)|V_{,2\xi}\xi|^{\frac{1}{2}} \tag{7.7}$$

where

$$\kappa = \frac{V_{,112}V_{,22\Lambda}}{V_{,222}V_{,11\Lambda}} \tag{7.8}$$

is a structural parameter, which is independent of ξ, and

$$C_8(\kappa) = \begin{cases} \dfrac{|2V_{,222}|^{\frac{1}{2}}}{V_{,22\Lambda}} : & \text{limit point load for } \xi > 0 \\ -\mathrm{sign}\,\xi\, \dfrac{V_{,112}}{V_{,11\Lambda}} \left| \dfrac{2}{V_{,222}(2\kappa - 1)} \right|^{\frac{1}{2}} : & \text{bifurcation load for} \\ & \quad \kappa > 1/2 \text{ and } \xi > 0; \\ & \quad \text{or } \kappa < 1/2 \text{ and } \xi < 0 \\ \text{non-existent} : & \text{for } \kappa > 1/2 \text{ and } \xi < 0 \end{cases} \tag{7.9}$$

In the application of this law, care must be taken for the associated imperfect behaviors that vary with the value of the structural parameter κ. The imperfect behaviors are classified to the following three cases:

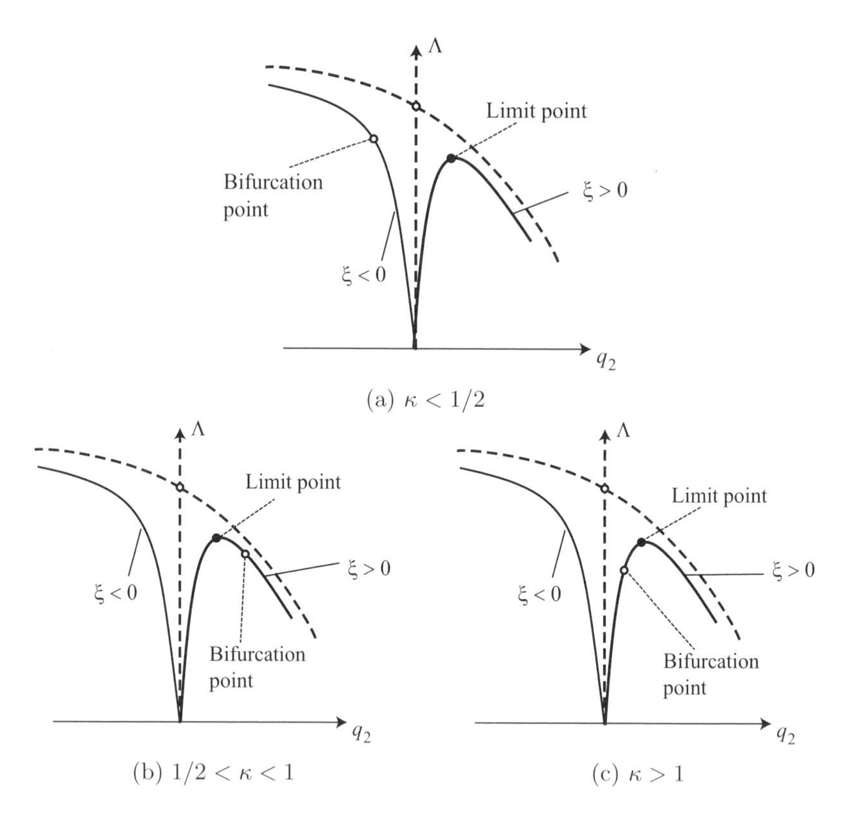

Fig. 7.3 Classification of imperfect behaviors of a semi-symmetric bifurcation point for an imperfection in the direction of $\mathbf{\Phi}_2^{\mathrm{c}}$ corresponding to an asymmetric bifurcation. Dashed curve: perfect equilibrium path; solid curve: imperfect equilibrium path; ○: bifurcation point; •: limit point.

- For $\kappa < 1/2$, a limit point exists for $\xi > 0$, and a bifurcation point exists for $\xi < 0$ (cf., Fig. 7.3(a)).
- For $1/2 < \kappa < 1$, a limit point exists for $\xi > 0$ to govern the critical load, a bifurcation point exists beyond this limit point, and no critical point exists for $\xi < 0$ (cf., Fig. 7.3(b)).
- For $\kappa > 1$, a bifurcation point exists before reaching the limit point for $\xi > 0$ owing to mode interaction, and no critical point exists for $\xi < 0$ (cf., Fig. 7.3(c)). The existence of this bifurcation point on the fundamental path leads to a severe reduction of the critical load.

The structural parameter κ in (7.8) involves the third-order cross-term $V_{,112}$ between the two critical modes $\mathbf{\Phi}_1^{\mathrm{c}}$ and $\mathbf{\Phi}_2^{\mathrm{c}}$. Thus the *third-order interaction* between these modes is influential on κ and, in turn, on resulting imperfect behaviors. By (7.5), (7.6) and (7.8),

$$-V_{,112} \text{ increases} \iff \kappa \text{ increases} \tag{7.10}$$

i.e., $|\kappa|$ is an increasing function of the modal interaction term $|V_{,112}|$. An imperfection in the direction of $\mathbf{\Phi}_2^{\mathrm{c}}$ under consideration serves as a minor imperfection for the symmetric bifurcation associated with $\mathbf{\Phi}_1^{\mathrm{c}}$. Therefore, the critical bifurcation load corresponding to the symmetric bifurcation for $\mathbf{\Phi}_1^{\mathrm{c}}$ is reduced by the deformation in the direction of $\mathbf{\Phi}_2^{\mathrm{c}}$. For this reason, for $\kappa > 1$, a bifurcation point may exist before reaching the limit point on the equilibrium path of an imperfect system. Although the 1/2-power law (7.7) for the bifurcation point is of the same form as the simple asymmetric bifurcation in (3.48), the semi-symmetric point is more imperfection-sensitive in that the absolute value $|C_8(\kappa)|$ of the coefficient is enhanced by mode interaction [129].

7.4 Non-Optimal and Optimal Frames

Consider the three-story three-span frame as shown in Fig. 7.1(a). A proportional vertical load Λp is applied at each node on the roof ($p = 1000$ kN). Young's modulus is $E = 200.0$ kN/mm^2. The beams and columns have sandwich cross-sections; namely, I_i is proportional to A_i as $I_i = h^2 A_i/4$, where h is the distance between the two flanges. Green's strain is used. The units of force kN and of length mm are suppressed in the following.

The beams and columns are modeled by a beam element with bending and axial deformations, and a truss element is used for braces. Each column is divided into two elements. The assumption of rigid floor is employed; i.e., the nodes in the same story have the same horizontal displacement. The upper bounds for global and local displacements are given as $\overline{\Delta}_j = 400$ and $\overline{\delta}_j = 200$, which are 1/10 and 1/20 of the column height, respectively.

Only the imperfections of nodal locations in the direction of a linear combination of the buckling modes are considered. Imperfection sensitivity is investigated for non-optimal unbraced frames and optimal braced frames with two different cross-sectional areas with $h = 200$ and 500.

7.4.1 Non-optimal unbraced frames

The non-optimal unbraced frames with $h = 200$ and 500 are considered. The cross-sectional areas of beams and columns are $A = 10000$.

First, for $h = 200$, the two lowest linear buckling load factors are $\Lambda_{\mathrm{L}1} = 6.6527$ and $\Lambda_{\mathrm{L}2} = 8.9118$ and both correspond to sway buckling. The nonlinear critical load factor $\Lambda^{\mathrm{c}0} = 6.6524$ obtained by path-tracing analysis agrees within good accuracy with $\Lambda_{\mathrm{L}1}$. The effect of prebuckling deformation on the critical load factor, accordingly, is negligibly small.

Mode interaction can be investigated with reference to higher differential coefficients of the total potential energy as discussed in Section 3.6.2. The differential coefficients of Π^{V} with respect to the active coordinate q_1 at the bifurcation point are $V_{,111} = 0$ and $V_{,1111} = 0.42058 > 0$; therefore, the critical point is a stable-symmetric bifurcation point (cf., (3.28) and (3.29)). The total potential energy is also symmetric with respect to q_2.

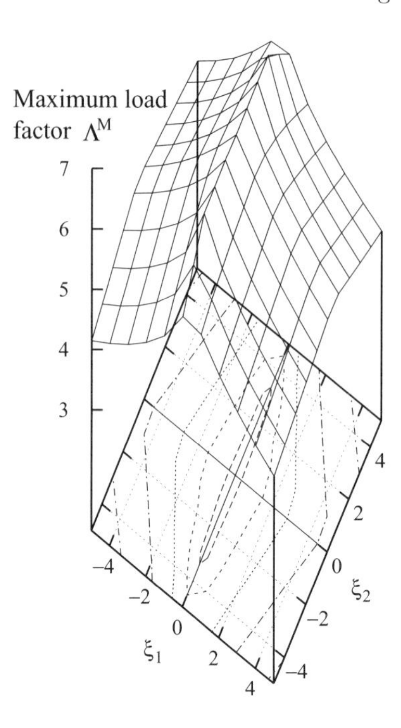

Fig. 7.4 Variation of maximum load factor Λ^{M} of imperfect systems of the non-optimal unbraced frame ($h = 200$) plotted against ξ_1 and ξ_2.

Let ξ_1 and ξ_2 denote the imperfection parameters in the directions of $\mathbf{\Phi}_1^{c0}$ and $\mathbf{\Phi}_2^{c0}$, respectively. Since the bifurcation point is stable, an imperfect system remains stable beyond the buckling load factor Λ^{c0} of the perfect system; accordingly, its maximum load factor Λ^{M} is to be defined by the displacement constraints (7.4). The maximum load factor Λ^{M} is plotted against the imperfection parameters ξ_1 and ξ_2 in Fig. 7.4; Λ^{M} is not sensitive to ξ_2 but sensitive to ξ_1. The most sensitive direction for a constant square norm $\sqrt{(\xi_1)^2 + (\xi_2)^2}$ almost coincides with $\mathbf{\Phi}_1^{c0}$, but has a small component of $\mathbf{\Phi}_2^{c0}$. The minimum of Λ^{M} in the range of $-5.0 \leq \xi_1 \leq 5.0$, $-5.0 \leq \xi_2 \leq 5.0$ is $\Lambda^{M}_{\min} = 4.1426$, and the reduction ratio is $\Lambda^{M}_{\min}/\Lambda^{c0} = 0.62269$ (cf., $\Lambda^{c0} = 6.6524$).

Next, for $h = 500$, the two lowest linear buckling load factors are $\Lambda_{L1} = 39.567$ and $\Lambda_{L2} = 55.232$ and both correspond to sway buckling again. The nonlinear critical load factor is $\Lambda^{c0} = 39.565$, which is close to Λ_{L1}.

The differential coefficients of Π^{V} are $V_{,111} = V_{,222} = 0$ and $V_{,1111} = 0.43298 > 0$. Similarly to the case of $h = 200$, the bifurcation point is stable and symmetric. The second bifurcation point is also symmetric. It is observed from the variation of maximum load factor Λ^{M} plotted against the imperfection parameters ξ_1 and ξ_2 in Fig. 7.5 that the most sensitive direction is given by the mixture of $\mathbf{\Phi}_1^{c0}$ and $\mathbf{\Phi}_2^{c0}$. The reduction ratio of the maximum load is $\Lambda^{M}_{\min}/\Lambda^{c0} = 0.44929$.

Finally, the performances for $h = 200$ and 500 are compared. The lowest linear buckling load factors are $\Lambda_{L1} = 6.6527$ and 39.567 for $h = 200$ and 500, respectively. Thus $h = 500$ with larger I_i has much greater buckling load. Yet $h = 500$

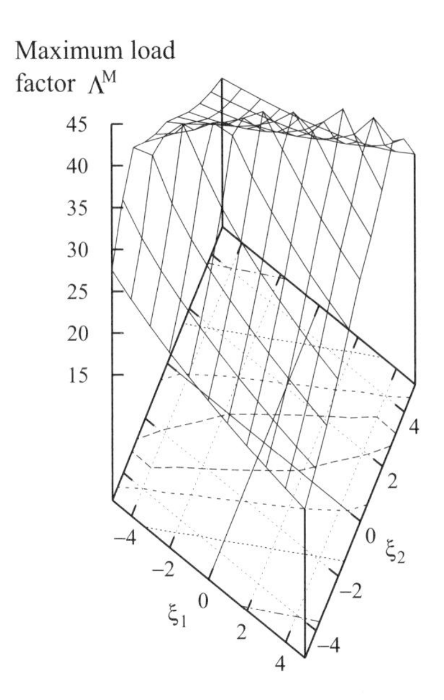

Fig. 7.5 Variation of maximum load factor Λ^{M} of imperfect systems of the non-optimal unbraced frame ($h = 500$) plotted against ξ_1 and ξ_2.

with $\Lambda^{M}_{\min}/\Lambda^{c0} = 0.44929$ has a much larger rate of reduction of the maximum load than $h = 200$ with $\Lambda^{M}_{\min}/\Lambda^{c0} = 0.62269$. Thus the apparently stronger design with $h = 500$ suffers from higher imperfection sensitivity.

7.4.2 Optimal braced frames

Optimal braced frames are obtained and their imperfection sensitivity is investigated. The same cross-sectional area is assigned for each of the following four groups of members: beams, columns, external braces in spans 1 and 3 defined in Fig. 7.1(a), and internal braces in span 2. Therefore, the number of design variables is 4. The lower bound of cross-sectional area is $A_i^{\mathrm{L}} = 100$ for columns and beams, and is $A_i^{\mathrm{L}} = 0$ for braces. A_i^{U} is set sufficiently large so that upper-bound constraints are inactive at all the members of the optimal frames. The specified buckling load factor is $\overline{\Lambda} = 16.0$. The method of modified feasible directions with line search presented in Section 4.3.2 is used for optimization and the design sensitivity coefficients are approximately computed by the finite difference approach.

We consider below two cases:

- $h = 200$ for a semi-symmetric bifurcation point, and
- $h = 500$ for a nearly semi-symmetric bifurcation point due to disappearance of non-sway asymmetric bifurcation by prebuckling deformation.

Table 7.1 Optimal cross-sectional areas and the total structural volume.

	$h = 200$	$h = 500$
Total volume	5.6797×10^8	1.0205×10^8
Columns	10304.0	1482.9
Beams	1469.8	406.32
External braces	301.21	167.87
Internal braces	0.0	143.74

Table 7.2 Differential coefficients of the optimal frames.

	$h = 200$	$h = 500$
$V_{,111}$	0.0	0.0
$V_{,222}$	-2.0088×10^{-4}	-3.7803×10^{-2}
$V_{,112}$	-2.1970×10^{-4}	-4.8820×10^{-2}
$V_{,122}$	0.0	0.0
$V_{,1111}$	0.69500	9.9120×10^{-2}
$V_{,2222}$	0.50535	7.9012×10^{-2}
$V_{,1122}$	0.54968	8.2986×10^{-2}
$V_{,11\Lambda}$	-1.0806×10^3	—
$V_{,22\Lambda}$	-1.2132×10^3	—

Semi-symmetric bifurcation point

First, for $h = 200$, the optimal cross-sectional areas are listed in the middle column of Table 7.1, which shows that the cross-sectional areas of the columns are very large compared with those of beams and braces. The cross-sectional areas of the internal braces vanish; accordingly, these braces have been removed. The nonlinear buckling load factor corresponding to a symmetric bifurcation point with sway buckling mode is $\Lambda^{c0} = 15.969$, which is very close to $\overline{\Lambda} = 16.0$.

Mode interaction at coincident critical points should be investigated in reference to higher differential coefficients of the total potential energy with due regard to the facts:

- Nonzero third-order term $V_{,112}$ characterizes the mode interaction at a semi-symmetric bifurcation.
- A negative fourth-order cross-term $V_{,1122}$ can lead to instability of a completely-symmetric double bifurcation point even though the two simple bifurcation points are stable if they are slightly separated [287] (cf., Section 3.6.3).

The values of differential coefficients of Π^{V} are listed in the middle column of Table 7.2. $V_{,222} \neq 0$ indicates that $\mathbf{\Phi}_2^{c0}$ corresponds to an asymmetric bifurcation.

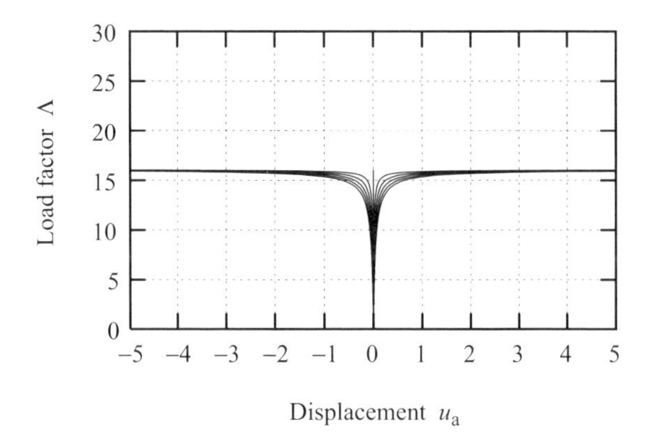

Fig. 7.6 Equilibrium paths of the optimal frame ($h = 200$).

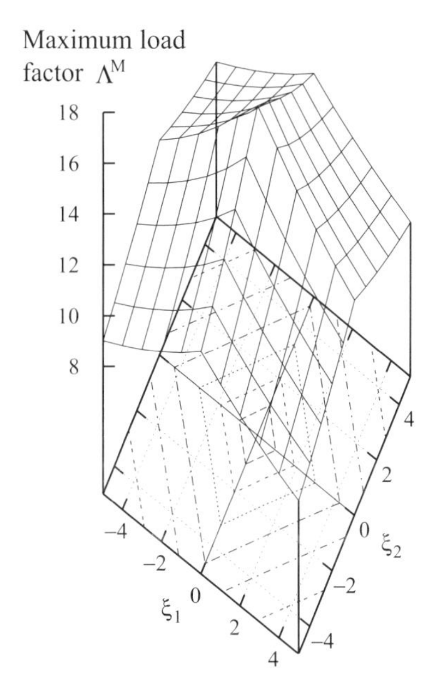

Fig. 7.7 Variation of maximum load factor Λ^M of imperfect systems of the optimal frame ($h = 200$) plotted against ξ_1 and ξ_2.

This is a semi-symmetric double bifurcation point with $V_{,112} \neq 0$, and interaction exists between the modes 1 and 2, as discussed in Section 3.6.2.

However, the absolute value of $V_{,112}$ is very small compared with those of the fourth-order terms. Therefore, the asymmetry of mode 2 is very small, and the double critical point is nearly completely symmetric. Since $V_{,1122} > 0$, there will be no instability related to the fourth-order terms.

Imperfect behaviors for an imperfection pattern in the direction of $\mathbf{\Phi}_2^{c0}$ are investigated. The structural parameter in (7.8) is $\kappa = 1.2149 > 1$, which means that

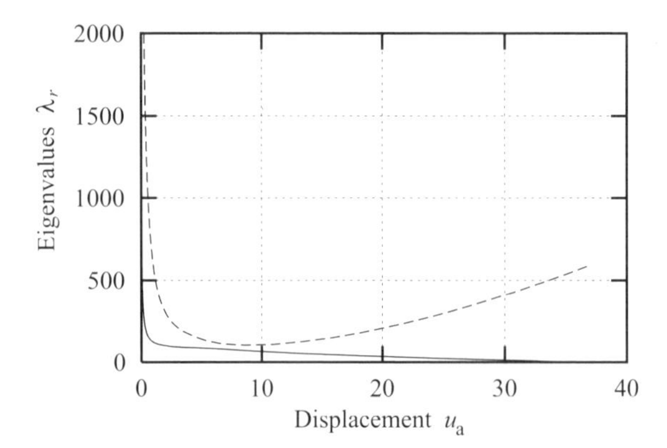

Fig. 7.8 The two lowest eigenvalues λ_1 and λ_2 of the optimal frame ($h = 500$) plotted against u_{a}. Solid curve: λ_1 corresponding to sway mode; dashed curve: λ_2 corresponding to non-sway mode.

the first critical point of an imperfect system is a bifurcation point or non-existent depending on the sign of ξ_2. Fig. 7.6 shows (cf., Fig. 7.3(c)) equilibrium paths of imperfect systems with $\xi_2 = \pm 1, \pm 2, \dots, \pm 5$ that are almost symmetric with respect to ξ_2, and are similar to those of a stable-symmetric bifurcation point. Here u_{a} denotes the horizontal displacement of node 'a' defined in Fig. 7.1(a) that represents the local deformation of the column since the beam-column joints do not move in the horizontal direction in the prebuckling deformation.

The maximum load factor Λ^{M} under displacement constraints (7.4) is plotted in Fig. 7.7, which is almost symmetric with respect to two imperfection parameters. The reduction ratio is $\Lambda^{\mathrm{M}}_{\mathrm{min}}/\Lambda^{\mathrm{c0}} = 0.56393$, which is slightly smaller than $\Lambda^{\mathrm{M}}_{\mathrm{min}}/\Lambda^{\mathrm{c0}} = 0.59625$ for the non-optimal frame. Therefore, imperfection sensitivity is not severely increased via optimization.

Nearly semi-symmetric bifurcation point

Next for $h = 500$, the optimal cross-sectional areas are listed in the last column of Table 7.1. It is observed that the total structural volume decreases due to an increase of h. The nonlinear critical load factor corresponding to a symmetric bifurcation point is $\Lambda^{\mathrm{c0}} = 16.227$, which is slightly larger than $\overline{\Lambda} = 16.0$.

The two lowest eigenvalues λ_1 and λ_2 of the tangent stiffness matrix are plotted against u_{a} in Fig. 7.8. As can be seen, λ_2 corresponding to a non-sway mode shown by the dashed curve decreases to a small value but does not reach zero; accordingly, the asymmetric bifurcation point is non-existent even for the perfect system. Although Λ^{c0} is close to the two linear buckling load factors, the frame does not have a coincident critical point. The prebuckling deformation is very large contrary to the case of $h = 200$. In the study of mode interaction, it is pertinent to regard the point at which λ_2 has the minimum value as an approximate critical point. The tangent stiffness matrix is constructed from the linearly estimated prebuckling deformation, and the eigenmodes are computed to approximate $\mathbf{\Phi}_1^{\mathrm{c0}}$ and $\mathbf{\Phi}_2^{\mathrm{c0}}$. The values of differential coefficients of Π^{V} are as

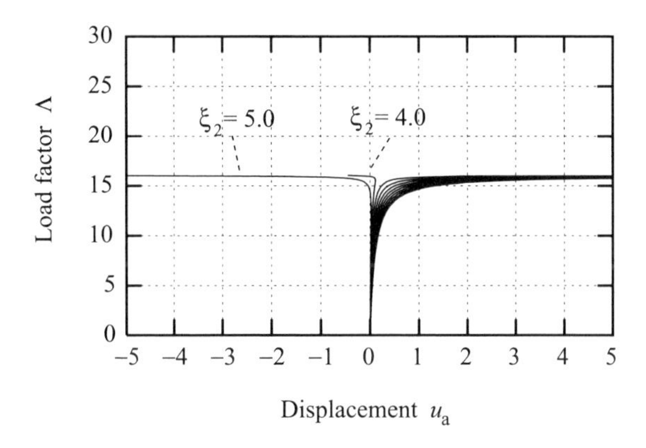

Fig. 7.9 Equilibrium paths of the optimal frame ($h = 500$).

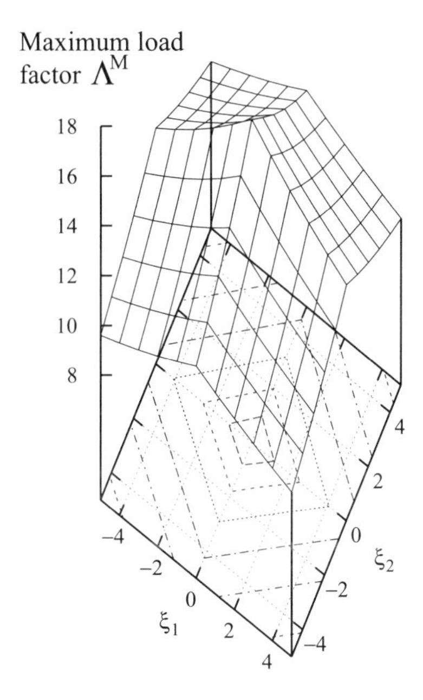

Fig. 7.10 Maximum load factor Λ^{M} of imperfect systems of the optimal frame of ($h = 500$) plotted against ξ_1 and ξ_2.

listed in the last column of Table 7.2; $V_{,111} = 0$ and $V_{,222} \neq 0$ indicate that the first and second buckling modes correspond to symmetric and asymmetric bifurcation points, respectively. The absolute values of the third- and fourth-order terms are of the same order. The nonzero value of $V_{,112}$ indicating possible mode interaction is larger than that of $h = 200$. Since $V_{,1122} > 0$, there is no instability related to the fourth-order terms.

Yet it is to be noted that the maximum mid-column displacement of the non-sway mode $\mathbf{\Phi}_2^{\mathrm{c}0}$ is 0.47688, and, therefore, is very small as a practically acceptable initial imperfection in comparison with the column height 4000.0. For example, if we choose a larger practically acceptable imperfection of $\xi_2 = 10$, the mid-column imperfection becomes 4.7688; the third- and fourth-order terms are computed as $V_{,222}\xi^3/6 = 6.3005$ and $V_{,2222}\xi^4/24 = 20.342$, and the fourth-order term thus is predominant in the range of imperfection.

The equilibrium paths of imperfect systems are investigated in the directions of $\mathbf{\Phi}_1^{c0}$ and $\mathbf{\Phi}_2^{c0}$ that are approximated by the linear buckling modes. Fig. 7.9 shows equilibrium paths of the imperfect systems with $\xi_2 = \pm 1, \pm 2, \ldots, \pm 5$. The reduction ratio is $\Lambda^{\rm M}_{\rm min}/\Lambda^{\rm c0} = 0.60011$ (cf., Fig. 7.10), which is much larger than 0.44929 for the non-optimal unbraced frame. Therefore, imperfection sensitivity does not increase and even decreases via optimization.

7.5 Summary

In this chapter,

- an optimization problem of a braced frame has been formulated, and
- imperfection sensitivity of braced frames has been investigated to demonstrate that imperfection sensitivity does not strongly increase via optimization.

The major findings of this chapter are as follows.

- Although the buckling load factors of symmetric frames can be estimated accurately by linear eigenvalue analysis, the asymmetric bifurcation point corresponding to a non-sway mode that can be obtained from the linear eigenvalue formulation disappears if geometrically nonlinear analysis is carried out. Therefore, the maximum load factor of the braced frame exhibiting non-sway buckling should be determined in view of local deformation of the columns.
- The non-sway asymmetric bifurcation point is nearly symmetric if the bending stiffnesses of the columns and beams are small. The imperfection sensitivity of the three-story three-span frames is also nearly symmetric.
- A frame with a semi-symmetric double bifurcation point is generated by optimization. Effect of mode interaction can be estimated based on the differential coefficients of the total potential energy. For the case where the non-sway bifurcation point disappears due to prebuckling deformation, the load factor that leads to a rapid increase of non-sway deformation can be estimated by linear buckling analysis. It has been confirmed that interaction between sway and non-sway modes does not always enhance imperfection sensitivity.

8

Hilltop Branching Point I: Simple Bifurcation

8.1 Introduction

Optimization of a structure for a specified nonlinear buckling load factor often engenders a coincident critical point. Designing such a structure is potentially dangerous because of increased imperfection sensitivity compared with that of a structure with moderately spaced critical points. The danger in naive optimization cautioned by Thompson [287] might have overshadowed the importance of structural optimization, as it is certainly ironic to produce a dangerous structure in the attempt to optimize its performance. Yet a profound meaning of the word "naive" has to be understood in the light of theory of elastic stability, a fraction of which was introduced in Part I. It is to be rephrased as "without knowing the type of coincident criticality you encountered." He did not unconditionally reject structures with coincident critical points, while it is often believed that it is meaningless to optimize a structure for a specified nonlinear buckling load factor. The imperfection sensitivity is severe for the semi-symmetric bifurcation point with the 1/2-power law, but is not that severe for the nondegenerate hilltop branching point with the piecewise linear law (cf., Section 3.6). The danger of naive optimization may be rephrased as "In the optimization of structures, one must check if the critical point in question is imperfection-sensitive or not."

A simple example of a hilltop branching point, the four-bar truss, was studied in Chapter 3.7. Yet it is to be emphasized that a hilltop branching is not limited to such a simple example but can be generated for realistic structures via optimization. For dome trusses, as will be demonstrated in Section 10.4, optimization produces hilltop branching, and the maximum loads can be effectively increased by optimizing the perfect system against buckling [216]. By virtue of the optimization, their imperfection sensitivity is improved from the 2/3-power law for simple symmetric bifurcation to the piecewise linear law for hilltop branching that is less imperfection-sensitive.

Hilltop branching points have many variants, including

- that with simple bifurcation, for which a limit point coincides with a simple bifurcation point,
- that with multiple bifurcations, for which a limit point coincides with multiple bifurcation points, and
- degenerate hilltop branching point.

We start with the hilltop point with simple bifurcation in this chapter. Hilltop point with multiple bifurcation is treated in Chapter 9, and the degenerate hilltop point in Chapter 10.

This chapter is organized as follows. In Section 8.2, imperfection sensitivity laws for nondegenerate hilltop points with simple bifurcation are introduced. A hilltop branching point with asymmetric bifurcation is produced for the bar–spring model in Section 8.3.

8.2 Imperfection Sensitivity Laws

In Section 3.6.1, the imperfection sensitivity laws (3.56) and (3.57) for hilltop points for a general asymmetric imperfection ξ were introduced. Recall that the hilltop point with simple bifurcation is classified to that with symmetric bifurcation and that with asymmetric bifurcation. Recall also that the sensitivity law for nondegenerate hilltop points with symmetric bifurcation reads

$$\Lambda^{\rm c}(\xi) = \Lambda^{\rm c0} + C_4 |V_{,1\xi}\xi| + C_5 V_{,2\xi}\xi \tag{8.1}$$

and that with asymmetric bifurcation reads

$$\Lambda^{\rm c}(\xi) = \begin{cases} \Lambda^{\rm c0} + C_6 |V_{,1\xi}\xi| + C_5 V_{,2\xi}\xi : & \text{for } \xi > 0 \\ \Lambda^{\rm c0} + C_7 |V_{,1\xi}\xi| + C_5 V_{,2\xi}\xi : & \text{for } \xi < 0 \end{cases} \tag{8.2}$$

where $\Lambda^{\rm c0} = \Lambda^{\rm c}(0)$. Here we have the classification:

$$\begin{cases} \boldsymbol{\Phi}_1^{\rm c} : \text{ bifurcation mode} \\ \boldsymbol{\Phi}_2^{\rm c} : \text{ limit point mode} \end{cases} \tag{8.3}$$

and

$$\begin{cases} C_5 V_{,2\xi}\xi : & \text{minor imperfection} \\ C_4 |V_{,1\xi}\xi|,\ C_6 |V_{,1\xi}\xi|,\ C_7 |V_{,1\xi}\xi| : & \text{major imperfection} \end{cases} \tag{8.4}$$

Although these imperfections are of the same order, the words *major* and *minor* are used to be consistent with other critical points. In this section, imperfection sensitivity laws are derived for these hilltop points to arrive at the explicit forms of the coefficients $C_4, \ldots, C_7$.

8.2.1 General formulation

The V-formulation is employed to derive imperfection sensitivity laws of non-degenerate hilltop branching points. The derivatives of the potential Π^{V} at this hilltop point satisfy:

- The conditions of a double critical point read

$$V_{,1} = V_{,2} = 0 \tag{8.5a}$$
$$V_{,11} = V_{,12} = V_{,21} = V_{,22} = 0 \tag{8.5b}$$

- By the classification (8.3) of $\mathbf{\Phi}_1^{\mathrm{c}}$ and $\mathbf{\Phi}_2^{\mathrm{c}}$, we have

$$V_{,1\Lambda} = 0, \quad V_{,2\Lambda} \neq 0 \tag{8.6}$$

- Minor and major imperfections are characterized by

$$\begin{cases} \text{minor (symmetric) imperfection:} & V_{,1\xi} = 0, \quad V_{,2\xi} \neq 0 \\ \text{major (antisymmetric) imperfection:} & V_{,1\xi} \neq 0, \quad V_{,2\xi} = 0 \end{cases} \tag{8.7}$$

and assume that the perfect system has the trivial solution $q_1 = 0$, namely, the first-order terms of Π^{V} with respect to q_1 vanish as follows:

$$V_{,122} = V_{,12\Lambda} = V_{,1\Lambda\Lambda} = \cdots = 0 \tag{8.8}$$

The total potential energy at the hilltop point of the perfect system $(\mathbf{q}^{\mathrm{a}}, \widetilde{\Lambda}, \xi) = (\mathbf{0}, 0, 0)$ is expanded as

$$\begin{aligned} \Pi^{\mathrm{V}}(\mathbf{q}^{\mathrm{a}}, \widetilde{\Lambda}, \xi) = {} & \Pi^{\mathrm{V}}(\mathbf{0}, 0, 0) + V_{,2\Lambda} q_2 \widetilde{\Lambda} + V_{,1\xi} q_1 \xi + V_{,2\xi} q_2 \xi \\ & + \frac{1}{6} V_{,111} (q_1)^3 + \frac{1}{2} V_{,112} (q_1)^2 q_2 + \frac{1}{6} V_{,222} (q_2)^3 \\ & + \frac{1}{2} V_{,11\Lambda} (q_1)^2 \widetilde{\Lambda} + \frac{1}{2} V_{,22\Lambda} (q_2)^2 \widetilde{\Lambda} \\ & + \frac{1}{2} V_{,11\xi} (q_1)^2 \xi + V_{,12\xi} q_1 q_2 \xi + \frac{1}{2} V_{,22\xi} (q_2)^2 \xi \\ & + \frac{1}{24} V_{,1111} (q_1)^4 + \frac{1}{4} V_{,1122} (q_1)^2 (q_2)^2 \\ & + \frac{1}{6} V_{,1112} (q_1)^3 q_2 + \frac{1}{2} V_{,112\Lambda} (q_1)^2 q_2 \widetilde{\Lambda} + \text{h.o.t.} \end{aligned} \tag{8.9}$$

Higher-order terms, h.o.t., are often suppressed in the sequel. Recall that all derivatives are evaluated at the hilltop point of the perfect system $(\mathbf{q}^{\mathrm{a}}, \widetilde{\Lambda}, \xi) = (\mathbf{0}, 0, 0)$.

The bifurcation equations are obtained as

$$\frac{\partial \Pi^{\mathrm{V}}}{\partial q_1} = V_{,1\xi} \xi + \frac{1}{2} V_{,111} (q_1)^2 + V_{,112} q_1 q_2 + V_{,11\Lambda} q_1 \widetilde{\Lambda} + \frac{1}{6} V_{,1111} (q_1)^3 = 0 \tag{8.10}$$

$$\frac{\partial \Pi^{\mathrm{V}}}{\partial q_2} = V_{,2\Lambda} \widetilde{\Lambda} + V_{,2\xi} \xi + \frac{1}{2} V_{,112} (q_1)^2 + \frac{1}{2} V_{,222} (q_2)^2 + V_{,22\Lambda} q_2 \widetilde{\Lambda} = 0 \tag{8.11}$$

in which the leading-order terms of q_1, q_2, Λ and ξ are contained. For example, the leading order term of q_1 in (8.10) is

$$\begin{cases} \frac{1}{2}V_{,111}(q_1)^2 : & \text{for } V_{,111} \neq 0 \\ \frac{1}{6}V_{,1111}(q_1)^3 : & \text{for } V_{,111} = 0,\ V_{,1111} \neq 0 \end{cases} \tag{8.12}$$

The criticality condition is given as

$$\det \mathbf{S}^{\mathrm{V}}(q_1, q_2, \widetilde{\Lambda}, \xi) = 0 \tag{8.13}$$

with the expression of the stability (tangent stiffness) matrix of the bifurcation equation

$$\mathbf{S}^{\mathrm{V}}(q_1, q_2, \widetilde{\Lambda}, \xi) = \begin{pmatrix} S_{11}^{\mathrm{V}} & S_{12}^{\mathrm{V}} \\ S_{21}^{\mathrm{V}} & S_{22}^{\mathrm{V}} \end{pmatrix} \tag{8.14}$$

with

$$S_{11}^{\mathrm{V}} = V_{,111}q_1 + V_{,112}q_2 + V_{,11\Lambda}\widetilde{\Lambda} + V_{,11\xi}\xi + \frac{1}{2}V_{,1122}(q_2)^2 \tag{8.15a}$$

$$S_{12}^{\mathrm{V}} = S_{21}^{\mathrm{V}} = V_{,112}q_1 + V_{,12\xi}\xi + V_{,1122}q_1q_2 + V_{,112\Lambda}q_1\widetilde{\Lambda} \tag{8.15b}$$

$$S_{22}^{\mathrm{V}} = V_{,222}q_2 + V_{,22\Lambda}\widetilde{\Lambda} + V_{,22\xi}\xi + \frac{1}{2}V_{,1122}(q_1)^2 \tag{8.15c}$$

8.2.2 Trivial fundamental path

Consider the perfect system with $\xi = 0$. Then the bifurcation equations (8.10) and (8.11) have the trivial fundamental path

$$q_1 = 0, \quad \widetilde{\Lambda} = -\frac{V_{,222}}{2V_{,2\Lambda}}(q_2)^2 \tag{8.16}$$

Note that $\widetilde{\Lambda}$ is of the order of $(q_2)^2$ and the fifth term $V_{,22\Lambda}q_2\widetilde{\Lambda}$ in (8.11) of the order of $(q_2)^3$ has been suppressed in (8.16).

The stability matrix $\mathbf{S}^{\mathrm{V}}$ in (8.14) on this trivial path with $q_1 = 0$ reduces to

$$\mathbf{S}^{\mathrm{V}}(0, q_2, \widetilde{\Lambda}, 0) = \begin{pmatrix} \lambda_\alpha & 0 \\ 0 & \lambda_\beta \end{pmatrix} \tag{8.17}$$

with two eigenvalues

$$\lambda_\alpha = V_{,112}q_2 + V_{,11\Lambda}\widetilde{\Lambda} + \frac{1}{2}V_{,1122}(q_2)^2 = V_{,112}q_2 + C_\alpha(q_2)^2 \tag{8.18a}$$

$$\lambda_\beta = V_{,222}q_2 + V_{,22\Lambda}\widetilde{\Lambda} = V_{,222}q_2 + \text{h.o.t.} \tag{8.18b}$$

Here

$$C_\alpha = \frac{1}{2V_{,2\Lambda}}(V_{,2\Lambda}V_{,1122} - V_{,222}V_{,11\Lambda}) \tag{8.19}$$

and $V_{,22\Lambda}\widetilde{\Lambda}$, of the order of $(q_2)^2$, is a higher-order term in λ_β.

Remark 8.2.1 The eigenvalue λ_α in (8.18a) becomes degenerate for $V_{,112} = 0$ in the sense that λ_α is tangential to the q_2-axis at $q_2 = 0$. Then the coefficient C_α plays an important role (cf., Appendices A.8 and A.9). □

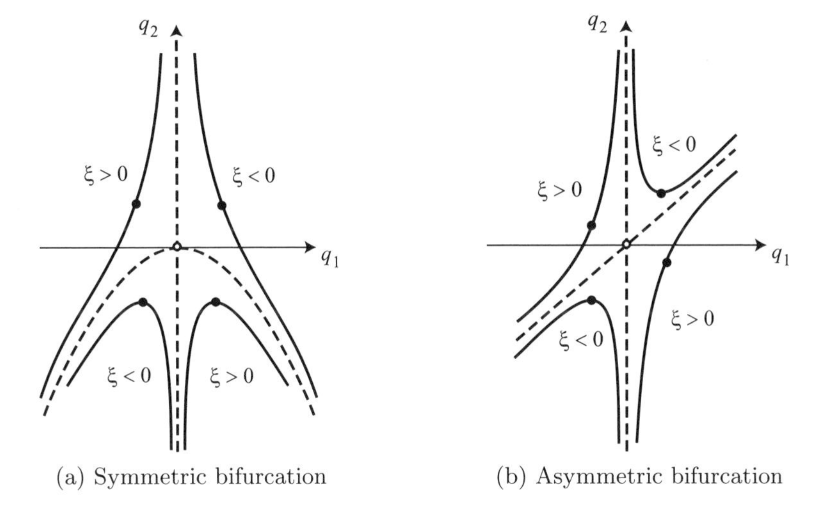

(a) Symmetric bifurcation (b) Asymmetric bifurcation

Fig. 8.1 Plan views of asymptotic behaviors in the vicinity of nondegenerate hilltop branching points. ∘: hilltop point; •: limit point; dashed curve: perfect equilibrium path; solid curve: imperfect equilibrium path.

8.2.3 Perfect and imperfect behaviors

Consider the most customary case in practice where the system has a rising fundamental path and becomes unstable at the hilltop point. Then from (8.16), (8.18a) and (8.18b), we have

$$V_{,112} < 0, \quad V_{,222} < 0, \quad V_{,2\Lambda} < 0 \tag{8.20}$$

For this case, plan views of perfect and imperfect behaviors are shown in Fig. 8.1(a) for the hilltop point with symmetric bifurcation and in Fig. 8.1(b) for that with asymmetric bifurcation. In addition, the associated bird's-eye views for symmetric bifurcation are shown in Fig. 8.2. It is noteworthy that the values of $\widetilde{\Lambda}$ are decreasing along all the perfect equilibrium paths from the hilltop point; therefore, this point literally is located at a hilltop. Each imperfect path has a limit point near the hilltop point.

It makes a sharp contrast with the associated behaviors of simple bifurcation points as shown in Fig. 8.3. It is noteworthy that perfect equilibrium paths are not all descending from the bifurcation point; accordingly, not all imperfect paths have a limit point.

8.2.4 Imperfection sensitivity for minor imperfection

Consider the minor imperfection with $V_{,1\xi} = 0$ and $V_{,2\xi} \neq 0$ in (8.7). Then (8.10) gives (cf., Fig. 8.2)

$$\begin{cases} q_1 = 0 : & \text{fundamental path} \\ \frac{1}{2}V_{,111}q_1 + \frac{1}{6}V_{,1111}(q_1)^2 + V_{,112}q_2 + V_{,11\Lambda}\widetilde{\Lambda} = 0 : & \text{bifurcation path} \end{cases} \tag{8.21}$$

On the fundamental path with $q_1 = 0$, there exists a limit point. Since the criticality condition (8.13) with (8.14) is satisfied for $q_2 = -(V_{,22\Lambda}/V_{,222})\widetilde{\Lambda}$, the

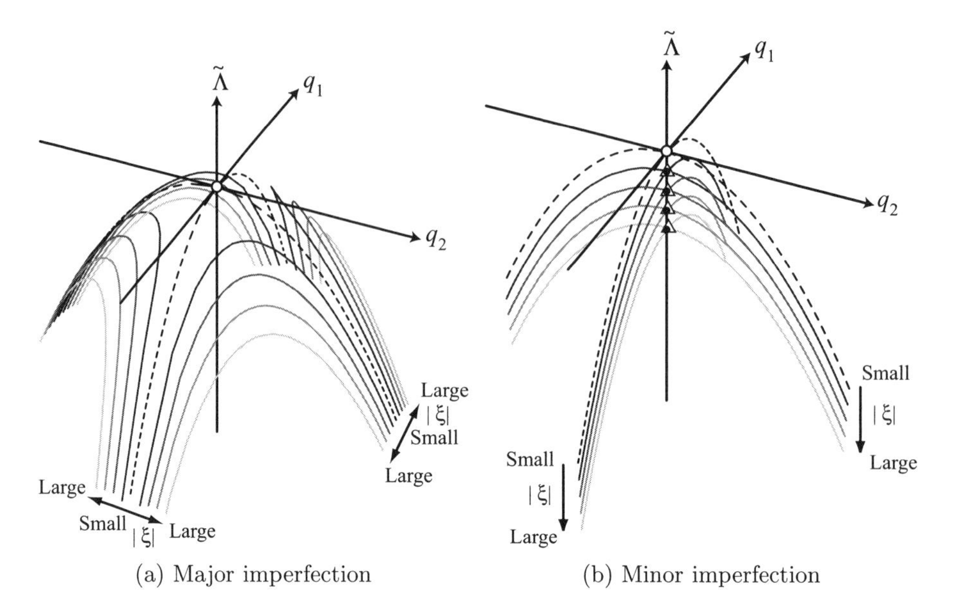

Fig. 8.2 Bird's-eye views of asymptotic behaviors in the vicinity of nondegenerate hilltop branching points with symmetric bifurcation. ∘: hilltop point; •: limit point of an imperfect system; △: simple bifurcation point; dashed curve: perfect equilibrium path; solid curve: imperfect equilibrium path.

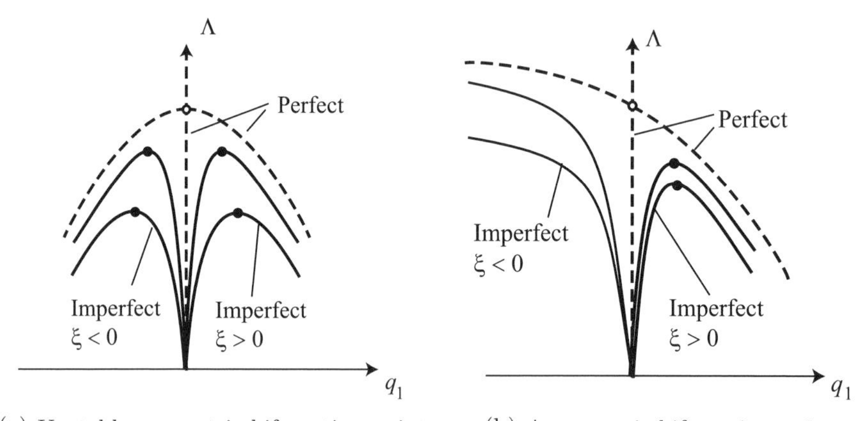

Fig. 8.3 Asymptotic behaviors in the vicinity of simple bifurcation points. ∘: simple bifurcation point; •: limit point of an imperfect system; dashed curve: perfect equilibrium path; solid curve: imperfect equilibrium path.

limit point load is obtained from (8.11) as

$$\widetilde{\Lambda}^{\mathrm{c}}(\xi) = -\frac{V_{,2\xi}}{V_{,2\Lambda}}\xi + \text{h.o.t.} \tag{8.22}$$

On the bifurcation path, there exists a bifurcation point. Then from the simultaneous solution of (8.11), (8.13), and (8.21), the bifurcation load $\widetilde{\Lambda}^{\rm c}$ turns out to be asymptotically identical with the limit point load $\widetilde{\Lambda}^{\rm c}$ in (8.22) up to the order of ξ.

To sum up, the sensitivity coefficient in (8.1) is given by

$$C_5 = -\frac{1}{V_{,2\Lambda}} > 0 \tag{8.23}$$

8.2.5 Imperfection sensitivity for major imperfection

Consider the major imperfection with $V_{,1\xi} \neq 0$ and $V_{,2\xi} = 0$ in (8.7) of a nondegenerate hilltop point.

Hilltop point with symmetric bifurcation

Consider the case of symmetric bifurcation with $V_{,111} = 0$, $V_{,1111} \neq 0$ (cf., (3.32)). The locations of limit points of an imperfect system are determined by simultaneously solving the bifurcation equations (8.10) and (8.11) and criticality condition (8.13). We see the existence of solutions of the order (cf., Remark 8.2.2)

$$\Lambda = O(\xi), \quad q_1 = O(\xi^{\frac{1}{2}}), \quad q_2 = O(\xi^{\frac{1}{2}}) \tag{8.24}$$

where $O(\,\cdot\,)$ denotes the order of the term in the parentheses. Then (8.10)–(8.13) reduce to

$$V_{,1\xi}\xi + V_{,112}q_1q_2 = 0 \tag{8.25a}$$

$$V_{,2\Lambda}\widetilde{\Lambda} + \frac{1}{2}V_{,112}(q_1)^2 + \frac{1}{2}V_{,222}(q_2)^2 = 0 \tag{8.25b}$$

$$V_{,112}V_{,222}(q_2)^2 - (V_{,112})^2(q_1)^2 = 0 \tag{8.25c}$$

Note that (8.25a)–(8.25c) are consistent in order in that they all consist of the terms of the order ξ. The solutions of this set of equations (8.25a)–(8.25c) are

$$(q_1^{\rm c}(\xi))^2 = \frac{V_{,222}}{V_{,112}}(q_2^{\rm c}(\xi))^2 = -\frac{1}{V_{,112}}\left(\frac{V_{,222}}{V_{,112}}\right)^{\frac{1}{2}}|V_{,1\xi}\xi| \tag{8.26}$$

$$\widetilde{\Lambda}^{\rm c}(\xi) = C_4|V_{,1\xi}\xi| \tag{8.27}$$

where

$$C_4 = \frac{1}{V_{,2\Lambda}}\left(\frac{V_{,222}}{V_{,112}}\right)^{\frac{1}{2}} < 0 \tag{8.28}$$

The explicit form of the coefficient C_4 in the sensitivity law (8.1) has thus been obtained.

Remark 8.2.2 In the power series expansion method, it is essential to correctly assume the orders of the terms. If these orders are consistent, a symbolic computation software can produce the explicit forms of imperfection sensitivity law in a systematic manner. In fact, Maple 9 [193] has been utilized to derive these forms. □

Remark 8.2.3 The sensitivity law (8.27) with (8.28) does not contain $V_{,ij\xi}$, $V_{,ij\Lambda}$ or the fourth-order terms; accordingly, there is no contamination by passive coordinates, and all the terms in this law in V-formulation can be written in the D-formulation by replacing $V_{,\dots}$ simply by the corresponding $D_{,\dots}$ (cf., (3.19a)–(3.19c)). □

Hilltop point with asymmetric bifurcation

Consider the case of asymmetric bifurcation with $V_{,111} \neq 0$, $V_{,112} < 0$ (cf., (3.32)). The locations of limit points of an imperfect system are determined by simultaneously solving the bifurcation equations (8.10) and (8.11) and the criticality condition (8.13). We see the existence of solutions of the order (cf., Remark 8.2.2)

$$\Lambda = O(\xi), \quad q_1 = O(\xi^{\frac{1}{2}}), \quad q_2 = O(\xi^{\frac{1}{2}}) \tag{8.29}$$

Then (8.10)–(8.13) reduce to

$$V_{,1\xi}\xi + \frac{1}{2}V_{,111}(q_1)^2 + V_{,112}q_1q_2 = 0 \tag{8.30a}$$

$$V_{,2\Lambda}\widetilde{\Lambda} + \frac{1}{2}V_{,112}(q_1)^2 + \frac{1}{2}V_{,222}(q_2)^2 = 0 \tag{8.30b}$$

$$(V_{,111}q_1 + V_{,112}q_2)V_{,222}q_2 - (V_{,112})^2(q_1)^2 = 0 \tag{8.30c}$$

Note that (8.30a)–(8.30c) are consistent in order in that they all consist of the terms of the order of ξ.

The following set of solutions is obtained from (8.30a)–(8.30c):

$$q_1^{\rm c}(\xi) = \pm(4\alpha)^{\frac{1}{4}}|V_{,1\xi}\xi|^{\frac{1}{2}} \tag{8.31}$$

$$q_2^{\rm c}(\xi) = \mp\frac{1 + \mathrm{sign}(V_{,1\xi}\xi)V_{,111}\alpha^{\frac{1}{2}}}{V_{,112}\alpha^{\frac{1}{4}}}\,\mathrm{sign}(V_{,1\xi}\xi)|V_{,1\xi}\xi|^{\frac{1}{2}} \tag{8.32}$$

$$\widetilde{\Lambda}^{\rm c}(\xi) = -\frac{V_{,222}(1 + \mathrm{sign}(V_{,1\xi}\xi)V_{,111}\alpha^{\frac{1}{2}})}{2(V_{,112})^2V_{,2\Lambda}\alpha^{\frac{1}{2}}}|V_{,1\xi}\xi| \tag{8.33}$$

where the double signs take the same order and

$$\alpha = \frac{V_{,222}}{V_{,222}(V_{,111})^2 + 4(V_{,112})^3} > 0 \tag{8.34}$$

From (8.33), the explicit forms of the coefficients C_6 and C_7 in the sensitivity law (8.2) are given as

$$C_6 = -\frac{V_{,222}(1 + \mathrm{sign}(V_{,1\xi})V_{,111}\alpha^{\frac{1}{2}})}{2(V_{,112})^2V_{,2\Lambda}\alpha^{\frac{1}{2}}} < 0: \ \text{for } \xi > 0 \tag{8.35a}$$

$$C_7 = -\frac{V_{,222}(1 - \mathrm{sign}(V_{,1\xi})V_{,111}\alpha^{\frac{1}{2}})}{2(V_{,112})^2V_{,2\Lambda}\alpha^{\frac{1}{2}}} < 0: \ \text{for } \xi < 0 \tag{8.35b}$$

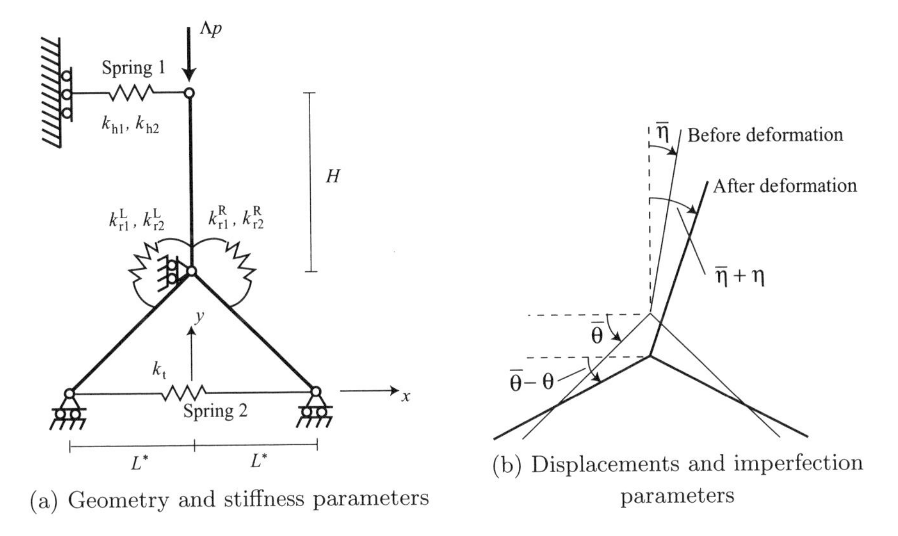

(a) Geometry and stiffness parameters

(b) Displacements and imperfection parameters

Fig. 8.4 Bar–spring model.

8.3 Bar–Spring Model: Hilltop with Asymmetric Bifurcation

Consider the bar–spring model as shown in Fig. 8.4(a), which consists of rigid bars and elastic springs. The perfect behavior of this model has been studied in Section 4.6. The length of the diagonal bars is given as $L = L^*/\cos\overline{\theta} = 1$; we set $H = 1.5$ and $p = 1$. In the following, the units of length and force are omitted for brevity. Spring 1 has a nonlinear relation $N = k_{\mathrm{h}1}d + \frac{1}{2}k_{\mathrm{h}2}d^2$ between the extension d and the axial force N. The presence of the nonlinear quadratic term $\frac{1}{2}k_{\mathrm{h}2}d^2$ makes the system asymmetric. Spring 2 is linear with the extensional stiffness k_{t}. The two rotational springs have a nonlinear relation $M = k_{\mathrm{r}1}r + \frac{1}{2}k_{\mathrm{r}2}r^2$ between the rotation r and the moment M. The properties for the right and left springs are indicated by $(\,\cdot\,)^{\mathrm{R}}$ and $(\,\cdot\,)^{\mathrm{L}}$, respectively. The model is symmetric with respect to the y-axis, if $k_{\mathrm{h}2} = 0$, $k_{\mathrm{r}1}^{\mathrm{R}} = k_{\mathrm{r}1}^{\mathrm{L}}$, and $k_{\mathrm{r}2}^{\mathrm{R}} = k_{\mathrm{r}2}^{\mathrm{L}}$ are satisfied.

Define the displacement vector by $\mathbf{U} = (\eta, \theta)^\top$ as shown in Fig. 8.4(b), where the thin and thick solid lines are the states before and after deformation, respectively. The initial imperfections are defined by $\mathbf{v} = (\overline{\eta}, \overline{\theta})^\top$, where $\overline{\theta}$ and $\overline{\eta}$ correspond to minor (symmetric) and major (asymmetric) imperfections, respectively. For the perfect system, $\mathbf{v}^0 = (0, (2/9)\pi)^\top$.

Suppose the stiffness of a rotational spring is inversely proportional to the angle spanned by the spring, and $k_{\mathrm{r}1}^{\mathrm{R}}$, $k_{\mathrm{r}2}^{\mathrm{R}}$, $k_{\mathrm{r}1}^{\mathrm{L}}$, and $k_{\mathrm{r}2}^{\mathrm{L}}$ can be defined in terms of the constants $\widehat{k}_{\mathrm{r}1}$ and $\widehat{k}_{\mathrm{r}2}$ as

$$k_{\mathrm{r}i}^{\mathrm{R}} = \frac{1}{\overline{\theta} - \overline{\eta} + \pi/2}\widehat{k}_{\mathrm{r}i}, \quad k_{\mathrm{r}i}^{\mathrm{L}} = \frac{1}{\overline{\theta} + \overline{\eta} + \pi/2}\widehat{k}_{\mathrm{r}i}, \quad (i = 1, 2) \tag{8.36}$$

The total potential energy Π is then given as

$$\begin{aligned}\Pi(\mathbf{U},\Lambda,\mathbf{v}) \;=\;& \frac{1}{2}k_{h1}H^2[\sin(\overline{\eta}+\eta)-\sin\overline{\eta}]^2+\frac{1}{6}k_{h2}H^3[\sin(\overline{\eta}+\eta)-\sin\overline{\eta}]^3\\ &+\frac{1}{2}k_{\mathrm{t}}[2L(\cos(\overline{\theta}-\theta)-\cos\overline{\theta})]^2\\ &+\frac{1}{2}k^{\mathrm{L}}_{\mathrm{r1}}(\theta-\eta)^2+\frac{1}{6}k^{\mathrm{L}}_{\mathrm{r2}}(\theta-\eta)^3+\frac{1}{2}k^{\mathrm{R}}_{\mathrm{r1}}(\theta+\eta)^2+\frac{1}{6}k^{\mathrm{R}}_{\mathrm{r2}}(\theta+\eta)^3\\ &-\Lambda pL[\sin\overline{\theta}-\sin(\overline{\theta}-\theta)]-\Lambda pH[\cos\overline{\eta}-\cos(\overline{\eta}+\eta)] \end{aligned} \tag{8.37}$$

Here the extension of spring 1 is written as $d = H[\sin(\overline{\eta}+\eta)-\sin\overline{\eta}]$. The derivatives of Π with respect to U_1 $(=\eta)$ and of U_2 $(=\theta)$ are

$$\begin{aligned}S_{,1} =\;& [k_{h1}Ha+\frac{1}{2}k_{h2}(Ha)^2]H\cos(\overline{\eta}+\eta)-k^{\mathrm{L}}_{\mathrm{r1}}(\theta-\eta)-\frac{1}{2}k^{\mathrm{L}}_{\mathrm{r2}}(\theta-\eta)^2\\ &+k^{\mathrm{R}}_{\mathrm{r1}}(\theta+\eta)+\frac{1}{2}k^{\mathrm{R}}_{\mathrm{r2}}(\theta+\eta)^2-\Lambda pH\sin(\overline{\eta}+\eta)\end{aligned} \tag{8.38}$$

$$\begin{aligned}S_{,2} =\;& 4k_{\mathrm{t}}L^2[\cos(\overline{\theta}-\theta)-\cos\overline{\theta}]\sin(\overline{\theta}-\theta)+k^{\mathrm{L}}_{\mathrm{r1}}(\theta-\eta)+\frac{1}{2}k^{\mathrm{L}}_{\mathrm{r2}}(\theta-\eta)^2\\ &+k^{\mathrm{R}}_{\mathrm{r1}}(\theta+\eta)+\frac{1}{2}k^{\mathrm{R}}_{\mathrm{r2}}(\theta+\eta)^2-\Lambda pL\cos(\overline{\theta}-\theta)\end{aligned} \tag{8.39}$$

where $a=\sin(\overline{\eta}+\eta)-\sin\overline{\eta}$. Higher derivatives are derived in Appendix A.6.

The eigenmode $\mathbf{\Phi}_r$ and eigenvalue λ_r are defined by the eigenvalue problem:

$$\begin{bmatrix} S_{,11} & S_{,12}\\ S_{,21} & S_{,22}\end{bmatrix}\mathbf{\Phi}_r=\lambda_r\mathbf{\Phi}_r,\quad (r=1,2) \tag{8.40}$$

where $S_{,12}=S_{,21}$.

All the quantities are evaluated at the perfect system with $(\overline{\eta},\overline{\theta})=(0,(2/9)\pi)$, where the tangent stiffness matrix $\mathbf{S}$ is diagonal with $S_{,12}=S_{,21}=0$ in (8.40); $\mathbf{\Phi}_1=(1,0)^\top$ and $\mathbf{\Phi}_2=(0,1)^\top$ correspond to the bifurcation point and limit point, respectively.

For the perfect system, the equilibrium equation $S_{,1}=0$ is satisfied by the trivial solution $\eta=0$ even for the asymmetric case with $k_{\mathrm{h2}}\neq 0$, i.e., the equilibrium state before buckling is symmetric with respect to the y-axis. Therefore, θ is the only displacement component along the fundamental path, and Λ can be computed for a given θ from the equilibrium equation $S_{,2}=0$.

Consider an asymmetric case where $k_{\mathrm{t}}=0.1$, $k_{\mathrm{h2}}=2k_{\mathrm{h1}}/L$ and $\widehat{k}_{\mathrm{r2}}=-2\widehat{k}_{\mathrm{r1}}$. Independent design variables are k_{h1} and k_{r1}. To numerically find a solution that has a hilltop branching point satisfying $S_{,11}=S_{,22}=0$, an optimization library DOT Ver. 5.0 [308] was used for obtaining k_{h1} and k_{r1} that minimize the error e defined by

$$e=(S_{,11})^2+(S_{,22})^2 \tag{8.41}$$

A hilltop branching is attained by $k_{\mathrm{h1}}=0.018483$ and $\widehat{k}_{\mathrm{r1}}=0.10778$.

Fig. 8.5 shows the relations between θ and the eigenvalues. From $\mathbf{\Phi}_1=(1,0)^\top$ and $\mathbf{\Phi}_2=(0,1)^\top$, q_1 and q_2 are the generalized coordinates in the directions of U_1 $(=\eta)$ and U_2 $(=\theta)$, respectively. Consider the major imperfection $\xi=\overline{\eta}$. The

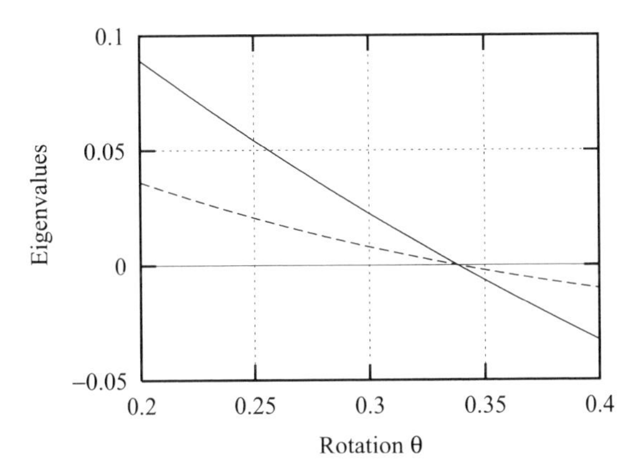

Fig. 8.5 Variation of eigenvalues plotted against rotation θ near the hilltop branching point. Solid curve: bifurcation mode; dashed curve: limit point mode.

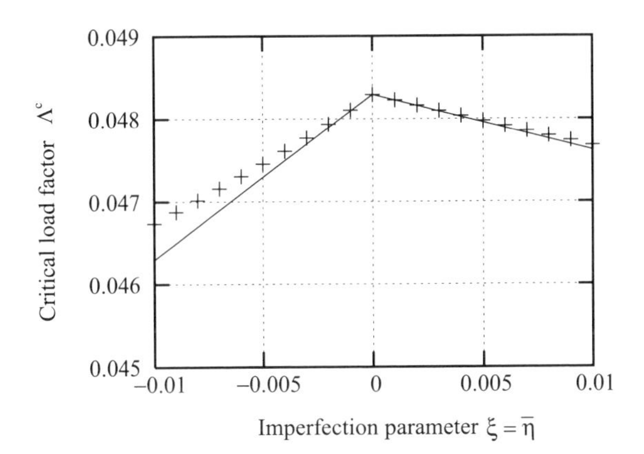

Fig. 8.6 Imperfection sensitivity of the asymmetric model. +: result of path-tracing analysis; solid line: sensitivity law (8.43).

values of the derivatives at the hilltop point are

$$\begin{aligned} &V_{,111} = 0.12476, \quad V_{,222} = -0.56460, \quad V_{,112} = 0.19001, \quad V_{,122} = 0.0 \\ &V_{,1\Lambda} = 0.0, \quad V_{,2\Lambda} = -0.93581, \quad V_{,1\xi} = -0.06307, \quad V_{,2\xi} = 0 \end{aligned} \tag{8.42}$$

The imperfection sensitivity law (8.2) with (8.35a) and (8.35b) is evaluated to

$$\Lambda^{\rm c}(\xi) = \begin{cases} \Lambda^{\rm c0} - 0.067144\xi : & \text{for } \xi \geq 0 \\ \Lambda^{\rm c0} + 0.19923\xi : & \text{for } \xi < 0 \end{cases} \tag{8.43}$$

The piecewise linear relation (8.43) between $\xi = \overline{\eta}$ and $\Lambda^{\rm c}$ is plotted in Fig. 8.6 by the solid lines to show good correlation with the results of numerical path-tracing shown by '+'.

8.4 Summary

In this chapter,

- explicit forms of imperfection sensitivity laws for nondegenerate hilltop points with simple bifurcation have been introduced, and
- imperfection sensitivity of an optimal structure with a hilltop branching point with asymmetric bifurcation has been investigated.

The major finding of this chapter is as follows.

- Imperfection sensitivity laws for nondegenerate hilltop points enjoy the piecewise linear law that is less severe than the 2/3-power or 1/2-power law for a simple bifurcation point. An optimized structure with hilltop branching is not imperfection-sensitive, and optimization does not always lead to a dangerous structure. It encourages the readers to carry out optimization of structures with confidence.

9

Hilltop Branching Point II: Multiple Bifurcations

9.1 Introduction

Hilltop branching points with simple bifurcation were studied in Chapter 8 to demonstrate the inherent mechanism of optimization to produce coincident critical points. In this chapter, the imperfection sensitivity of hilltop points with multiple bifurcations is studied, while details are to be consulted with [145, 224].

It is noteworthy that, for a pin-jointed truss, member buckling can occur almost independently from global buckling [224, 242]. Therefore, it is possible to create a hilltop branching point at which arbitrary many symmetric bifurcation points exist at a limit point; i.e., many members buckle simultaneously with global snapthrough. In this chapter, hilltop points with as many as four bifurcation points are actually created for a pin-jointed truss with simultaneously buckling members. Some of the bifurcation modes are symmetric and contain components of the limit point mode, therefore, have infinitesimally small third-order interactions. The orthogonality conditions among bifurcation modes and the limit point mode are relaxed by ignoring such interactions to extract pure member buckling modes.

The system with regular-polygonal symmetry, the most common symmetry in structures, has group-theoretic double bifurcation points (cf., Section 3.3.2). For example, it will be observed in Section 10.5 for a spherical truss dome with regular-hexagonal symmetry that six critical points are very closely located along the fundamental path near the peak, which can be approximated by a hilltop point with many bifurcation points. In this chapter, a triple hilltop branching point, which occurs as a coincidence of a limit point and a double bifurcation point, is produced for regular-polygonal truss tents.

This chapter is organized as follows. The imperfection sensitivity laws of hilltop points are investigated theoretically in Section 9.2. The hilltop point with four nearly symmetric branches of the pin-jointed truss is studied in Section 9.3. Triple hilltop branching points of regular-polygonal trusses are investigated in Section 9.4.

9.2 Imperfection Sensitivity

Imperfection sensitivity laws for hilltop points with multiple bifurcations with multiplicity $m \geq 3$ are presented [145, 224].

9.2.1 Hilltop point with many symmetric bifurcations

Consider a hilltop point with $m-1$ completely-symmetric bifurcation points and a limit point mode. Assumptions employed are

$$\begin{aligned} &V_{,ijk} = 0, \quad (i,j,k = 1,\dots,m-1) \\ &V_{,imm} = 0, \quad (i = 1,\dots,m-1) \end{aligned} \tag{9.1}$$

where $\boldsymbol{\Phi}_1^{\mathrm{c}},\dots,\boldsymbol{\Phi}_{m-1}^{\mathrm{c}}$ correspond to bifurcation modes and $\boldsymbol{\Phi}_m^{\mathrm{c}}$ to limit point mode. We consider the most customary case of

$$\begin{aligned} &V_{,m\Lambda} < 0, \quad V_{,mmm} < 0 \\ &V_{,iim} < 0, \quad (i = 1,\dots,m-1) \end{aligned} \tag{9.2}$$

Then the imperfection sensitivity law is written as [224]

$$\Lambda^{\mathrm{c}}(\xi) = \Lambda^{\mathrm{c}0} + C_{13}\sqrt{-\beta_m}\,|\xi| + C_{14}V_{,m\xi}\xi \tag{9.3}$$

with

$$C_{13} = \frac{\sqrt{-V_{,mmm}}}{V_{,m\Lambda}}, \quad C_{14} = -\frac{1}{V_{,m\Lambda}} \tag{9.4}$$

and β_m is defined by

$$\beta_m = \sum_{i=1}^{m-1}\sum_{j=1}^{m-1} V_{,ijm}\widehat{q}_i\widehat{q}_j \tag{9.5}$$

Here $\widehat{q}_j$ is obtained by solving the $m-1$ simultaneous linear equations

$$\sum_{j=1}^{m-1} V_{,ijm}\widehat{q}_j + V_{,i\xi} = 0, \quad (i = 1,\dots,m-1) \tag{9.6}$$

In particular, for a hilltop point with two symmetric bifurcation points; i.e., $m = 3$, and $V_{,123} = 0$,

$$\widehat{q}_j = -\frac{V_{,j\xi}}{V_{,jj3}}, \quad (j = 1,2) \tag{9.7}$$

is obtained from (9.6). Therefore, β_m in (9.5) is expressed explicitly by $V_{,ijm}$ and $V_{,j\xi}$, and, in turn, (9.3) reduces to

$$\Lambda^{\rm c}(\xi) = \Lambda^{\rm c0} + \frac{1}{V_{,3\Lambda}} \sqrt{V_{,333} \left[\frac{(V_{,1\xi})^2}{V_{,113}} + \frac{(V_{,2\xi})^2}{V_{,223}} \right]} \, |\xi| - \frac{V_{,3\xi}}{V_{,3\Lambda}} \xi \tag{9.8}$$

which agrees with the results in [145, 284]. for the most customary case of (9.2), namely,

$$V_{,3\Lambda} < 0, \ V_{,333} < 0, \ V_{,113} < 0, \ V_{,223} < 0 \tag{9.9}$$

9.2.2 Hilltop point for a system with dihedral-group symmetry

Structures with dihedral-group symmetry appear generically for shells of revolution and reticulated regular polygonal truss domes. These structures have particular double bifurcation points due to symmetry, in addition to limit points and simple symmetric bifurcation points.

In particular, Ikeda *et al.* [145] considered a triple hilltop branching point, which occurs as a coincidence of a limit point and a double bifurcation point. The power series expansion form in the V-formulation was determined through exploitation of the dihedral-group symmetry. The perfect and imperfect bifurcation behaviors in the neighborhood of this hilltop branching point were thoroughly investigated.

To sum up, through complex derivations, the sensitivity law is given by a piecewise linear law

$$\Lambda^{\rm c}(\xi) = \begin{cases} \Lambda^{\rm c0} + C_{15}\sqrt{(V_{,1\xi})^2 + (V_{,2\xi})^2}\,|\xi| + C_{16}V_{,3\xi}\xi : & \text{for } \ \widehat{n} \geq 4 \\ \Lambda^{\rm c0} + C_{17}(\psi)\sqrt{(V_{,1\xi})^2 + (V_{,2\xi})^2}\,|\xi| + C_{16}V_{,3\xi}\xi : & \text{for } \ \widehat{n} = 3 \end{cases} \tag{9.10}$$

with $\psi = \tan^{-1}(V_{,2\xi}/V_{,1\xi})$, as is the case for other hilltop points, where $\widehat{n}$ is the number of bifurcation paths, and $\tan^{-1}(\,\cdot\,)$ takes the principal value in the range $[-\pi/2, \pi/2]$. The coefficients C_{15} and C_{16} enjoy explicit expressions

$$C_{15} = \frac{1}{V_{,3\Lambda}}\sqrt{\frac{V_{,333}}{V_{,113}}}, \quad C_{17} = -\frac{1}{V_{,3\Lambda}} \tag{9.11}$$

Here $V_{,113} = V_{,223}$ holds by symmetry.

9.3 Arch-Type Truss: Hilltop with Multiple Symmetric Bifurcations

Consider the arch-type truss as shown in Fig. 9.1 to investigate interaction between member buckling modes and the limit point mode. The units of force kN and of length mm are suppressed in the following. The y-directional load Λp is applied to the center node. We set Young's modulus $E = 1$ and $p = 0.001$ for simplicity. The members are divided into four groups as shown in Fig. 9.1. Let A_i and I_i, respectively, denote the cross-sectional area and second moment of

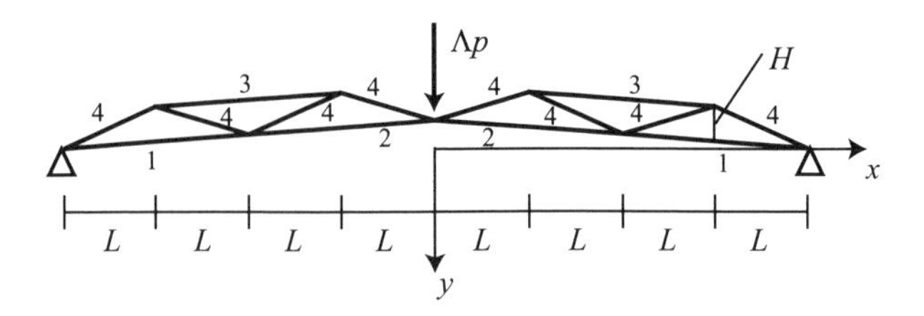

Fig. 9.1 Arch-type truss ($L = 1000$, $H = 200$). Rise of the center node is 100; $1, \ldots, 4$ denote member groups.

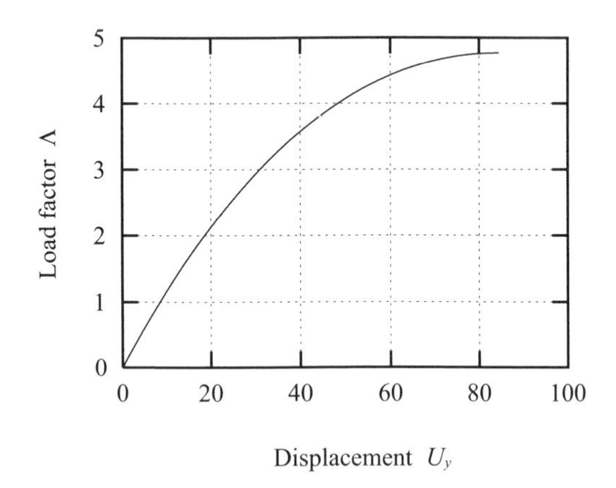

Fig. 9.2 Equilibrium path of the arch-type truss. U_y: the y-directional displacement of the center node.

inertia of the members in the ith group. A sandwich cross-section is assumed, where $I_i = h_i^2 A_i/4$ and h_i is the distance between the flanges that is independent of A_i. Green's strain is used for simplicity. Each member is divided into four elements. Differentiation and mathematical manipulation are carried out by a symbolic computation package Maple 9 [193].

Fig. 9.2 shows an equilibrium path for the perfect system with $(A_1, A_2, A_3, A_4) = (100, 100, 1000, 300)$ and $(h_1, h_2, h_3, h_4) = (36.30, 36.10, 60.0, 200.0)$. A limit point is attained at $\Lambda^{\mathrm{c}} = 4.7681$.

Fig. 9.3 shows the variation of the five lowest eigenvalues λ_r $(r = 1, \ldots, 5)$ of $\mathbf{S}$ plotted against the y-directional displacement U_y of the center node. The five eigenvalues vanish simultaneously at the hilltop point. Note that, along the fundamental path, λ_1 and λ_2 are exactly coincident, λ_3 and λ_4 are nearly coincident, and λ_5 corresponds to the limit point mode.

The eigenmodes $\mathbf{\Phi}_1^{\mathrm{c}0}, \ldots, \mathbf{\Phi}_5^{\mathrm{c}0}$ at the hilltop point of the perfect system are shown in Fig. 9.4(a); $V_{,iii} = 0$ is not always satisfied for member buckling modes. For example, $\mathbf{\Phi}_3^{\mathrm{c}0}$ and $\mathbf{\Phi}_4^{\mathrm{c}0}$ in Fig. 9.4(a) are symmetric with respect to the y-axis, and nodal displacements should exist to satisfy orthogonality with the limit point mode $\mathbf{\Phi}_5^{\mathrm{c}0}$, which is also symmetric. Hence $\mathbf{\Phi}_3^{\mathrm{c}0}$ and $\mathbf{\Phi}_4^{\mathrm{c}0}$ are the mixture of the limit point mode and the member buckling mode. Note that $\mathbf{\Phi}_1^{\mathrm{c}0}$ and $\mathbf{\Phi}_2^{\mathrm{c}0}$ are

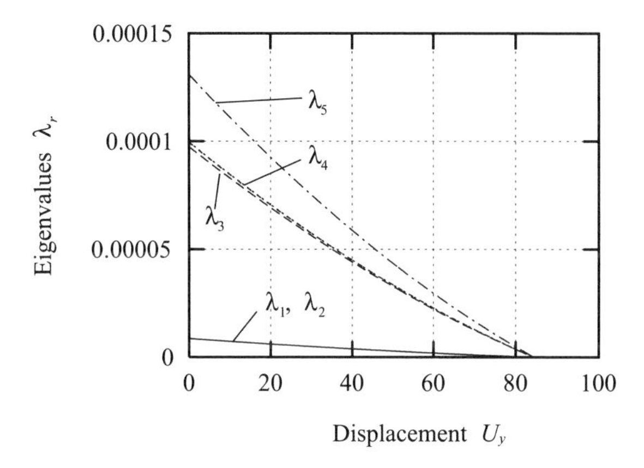

Fig. 9.3 Variation of eigenvalues $\lambda_1, \ldots, \lambda_5$ plotted against the y-directional displacement U_y of the center node.

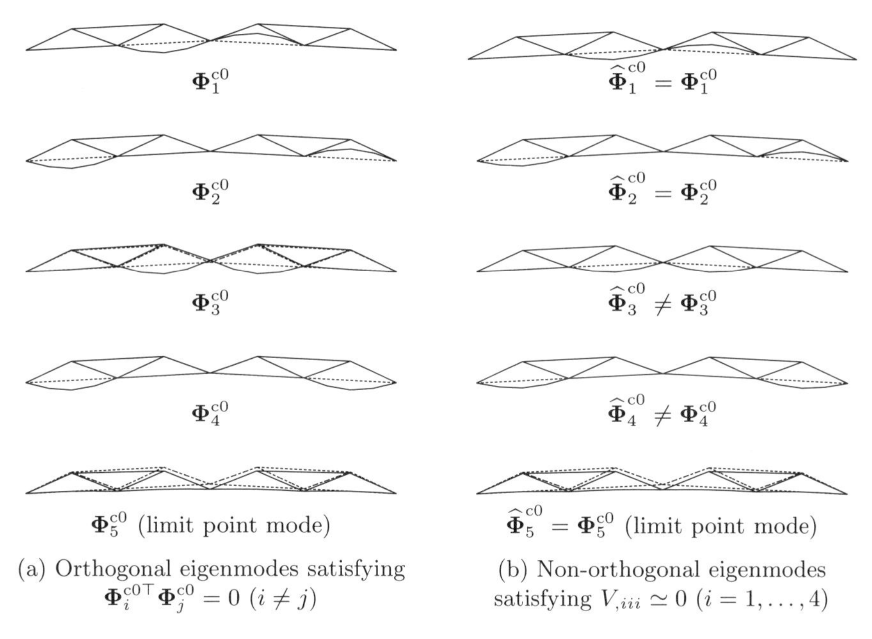

Fig. 9.4 Eigenmodes corresponding to the five zero eigenvalues of $\mathbf{S}$.

antisymmetric with respect to the y-axis, and the y-directional displacement of the center node vanishes; therefore, $V_{,111} = V_{,222} = 0$ is satisfied.

In order to separate limit point mode and member buckling modes, the limit point mode $\mathbf{\Phi}_5^{c0}$ is first defined as the incremental displacement vector at the limit point. Then the components of $\mathbf{\Phi}_5^{c0}$ are subtracted from $\mathbf{\Phi}_3^{c0}$ and $\mathbf{\Phi}_4^{c0}$ to arrive at pure member buckling modes $\widehat{\mathbf{\Phi}}_3^{c0}$ and $\widehat{\mathbf{\Phi}}_4^{c0}$ shown in Fig. 9.4(b). Note that symmetric modes $\widehat{\mathbf{\Phi}}_3^{c0}$, $\widehat{\mathbf{\Phi}}_4^{c0}$ and $\widehat{\mathbf{\Phi}}_5^{c0}$ do not satisfy the orthogonality condition, but $V_{,333} = V_{,444} \simeq 0$ hold.

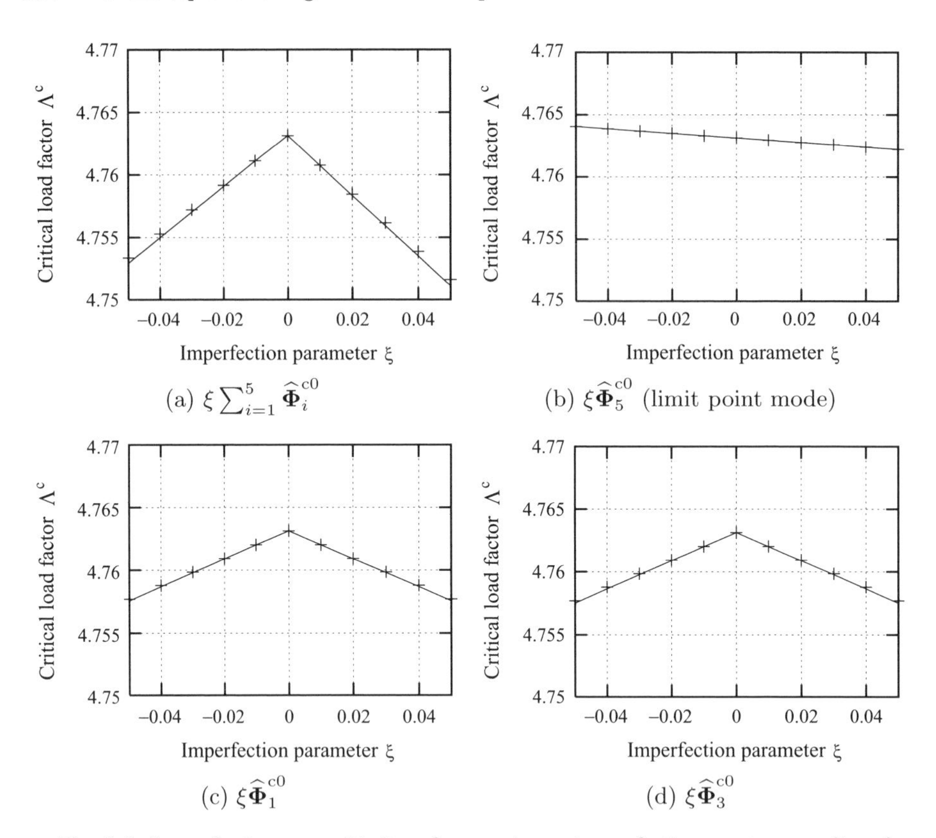

(a) $\xi \sum_{i=1}^{5} \widehat{\boldsymbol{\Phi}}_i^{\mathrm{c}0}$ (b) $\xi \widehat{\boldsymbol{\Phi}}_5^{\mathrm{c}0}$ (limit point mode)

(c) $\xi \widehat{\boldsymbol{\Phi}}_1^{\mathrm{c}0}$ (d) $\xi \widehat{\boldsymbol{\Phi}}_3^{\mathrm{c}0}$

Fig. 9.5 Imperfection sensitivity for various imperfections. +: result of path-tracing analysis; solid line: sensitivity laws (9.14)–(9.17).

With the use of the modes $\widehat{\boldsymbol{\Phi}}_1^{\mathrm{c}0}, \dots, \widehat{\boldsymbol{\Phi}}_5^{\mathrm{c}0}$, the nonzero third-order differential coefficients of V are obtained as

$$\begin{aligned} &V_{,115} = V_{,225} = -2.0842 \times 10^{-7}, \quad V_{,335} = V_{,445} = -2.1415 \times 10^{-7} \\ &V_{,345} = -2.2262 \times 10^{-9}, \quad V_{,555} = -1.8214 \times 10^{-8} \end{aligned} \tag{9.12}$$

Other coefficients are

$$\begin{aligned} &V_{,1\xi} = V_{,2\xi} = -9.7979 \times 10^{-5}, \quad V_{,3\xi} = V_{,4\xi} = -9.7963 \times 10^{-5} \\ &V_{,5\xi} = -4.7352 \times 10^{-5} \\ &V_{,1\Lambda} = \cdots = V_{,4\Lambda} = 0, \quad V_{,5\Lambda} = -7.0748 \times 10^{-11} \end{aligned} \tag{9.13}$$

First consider an asymmetric imperfection in the direction of the sum $\xi \sum_{i=1}^{5} \widehat{\boldsymbol{\Phi}}_i^{\mathrm{c}0}$ of five modes shown in Fig. 9.4(b). The use of (9.12) and (9.13) in (9.3) leads to

$$\Lambda^{\mathrm{c}}(\xi) = \Lambda^{\mathrm{c}0} - 0.22176|\xi| - 1.8393 \times 10^{-2}\xi \tag{9.14}$$

which is plotted by the solid line in Fig. 9.5(a), and correlates well with '+' that denotes Λ^{c} of an imperfect system attained at a limit point obtained by path-tracing analysis.

For the minor imperfection $\xi \widehat{\boldsymbol{\Phi}}_5^{\mathrm{c}0}$ in the direction of limit point mode, the first term in the right-hand-side of (9.14) vanishes, and the relation between ξ and Λ^{c} is given by a linear law

$$\Lambda^{\mathrm{c}}(\xi) = \Lambda^{\mathrm{c}0} - 1.8393 \times 10^{-2}\xi \tag{9.15}$$

which is plotted in Fig. 9.5(b).

For the major imperfection $\xi \widehat{\boldsymbol{\Phi}}_1^{\mathrm{c}0}$, the second term in the right-hand-side of (9.14) vanishes, and the relation between ξ and Λ^{c} is given by a symmetric piecewise linear law

$$\Lambda^{\mathrm{c}}(\xi) = \Lambda^{\mathrm{c}0} - 0.11134|\xi| \tag{9.16}$$

which is plotted in Fig. 9.5(c). For the third mode $\xi \widehat{\boldsymbol{\Phi}}_3^{\mathrm{c}0}$, the relation is

$$\Lambda^{\mathrm{c}}(\xi) = \Lambda^{\mathrm{c}0} - 0.11249|\xi| \tag{9.17}$$

which is also symmetric as plotted in Fig. 9.5(d). Although the mode $\xi \widehat{\boldsymbol{\Phi}}_3^{\mathrm{c}0}$ corresponds to a slightly asymmetric bifurcation, the linear term is negligible.

The critical loads have thus been accurately estimated by linear and piecewise linear relations for all the cases to assess the validity of the proposed formula (9.3).

9.4 Regular-Polygonal Truss Tents: Hilltop with Group-Theoretic Double Point

Consider the regular-polygonal n-bar truss tents shown in Fig. 9.6 subjected to the z-directional load Λ. Green's strain is used; set $EA = 1.0$ kN for simplicity. The length of each bar projected to the xy-plane is set constant as $L = 1000.0$ mm, and H is varied to find a triple hilltop branching point at $H = 1224.72$ mm, at which a group-theoretic double bifurcation point coincides with the limit point, irrespective of the number n of members ($n = 3, \dots, 6$). The units of force kN and of length mm are suppressed in the following. Differentiation and mathematical manipulations are carried out by a symbolic computation package Maple V [116].

The critical loads at the hilltop point for the perfect system are $\Lambda^{\mathrm{c}0} = 0.26832$, 0.35776, 0.44720, and 0.53664 for $n = 3, \dots, 6$, respectively. The number of bifurcation paths is equal to $\widehat{n} = n$. The imperfection vector is the (x, y, z) coordinates of the center node with $\mathbf{v}^0 = (0, 0, -H)^\top$ and the imperfection pattern vector is given as $\mathbf{d} = (1, 1, 1)^\top$. Critical loads Λ^{c} for imperfect trusses shown by '+' are plotted against imperfection parameter ξ in Fig. 9.7.

For $n = \widehat{n} = 4, 5, 6$, the law (9.8) becomes

$$\Lambda^{\mathrm{c}}(\xi) = \begin{cases} \Lambda^{\mathrm{c}0} - 3.5055 \times 10^{-4}|\xi| + 4.9572 \times 10^{-4}\xi : & \text{for } n = \widehat{n} = 4 \\ \Lambda^{\mathrm{c}0} - 6.1965 \times 10^{-4}|\xi| + 4.3819 \times 10^{-4}\xi : & \text{for } n = \widehat{n} = 5 \\ \Lambda^{\mathrm{c}0} - 5.2579 \times 10^{-4}|\xi| + 7.4358 \times 10^{-4}\xi : & \text{for } n = \widehat{n} = 6 \end{cases} \tag{9.18}$$

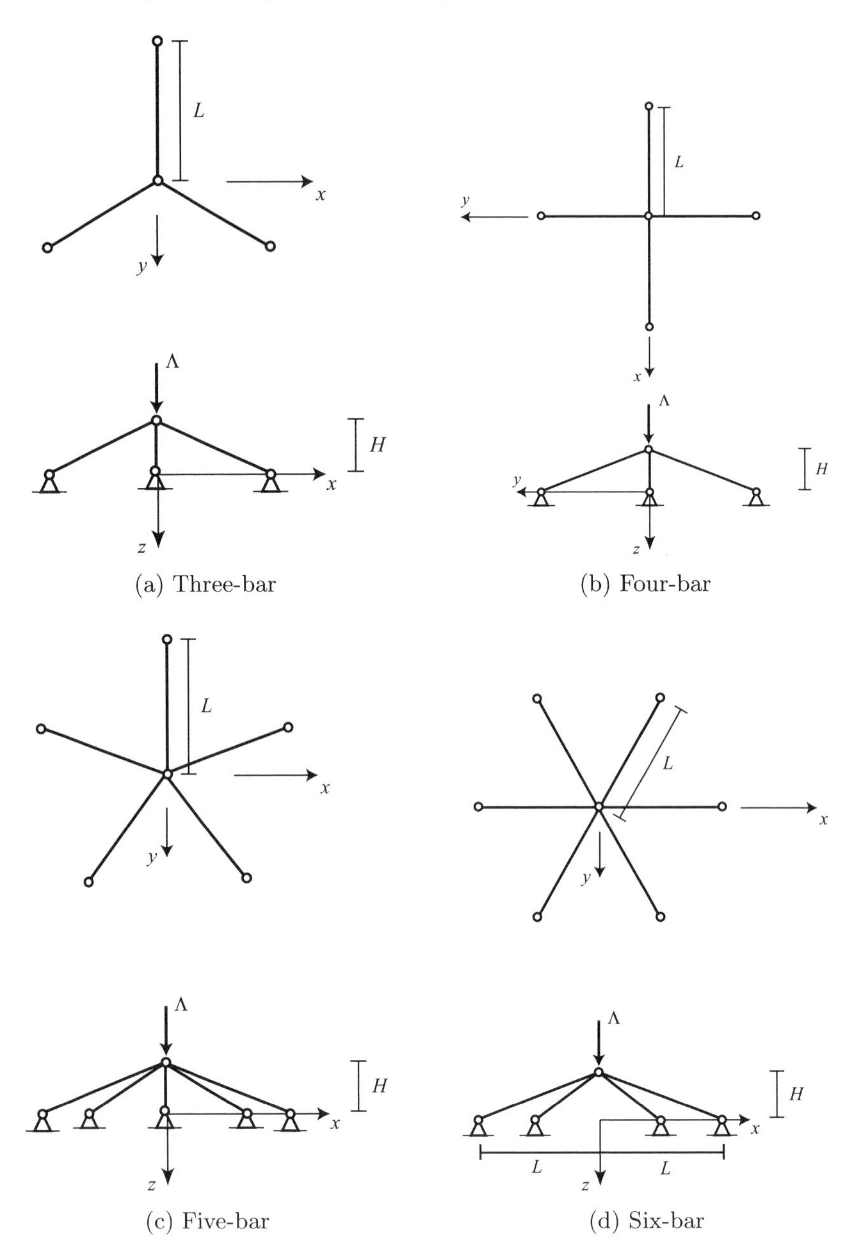

Fig. 9.6 Regular-polygonal n-bar truss tents.

Here coefficients for the law (9.8) are as listed in Table 9.1 for $n = 4, 5, 6$, and these laws are shown by the solid lines in Figs. 9.7(b)–(d). As can be seen, the critical loads of imperfect systems can be predicted accurately by the asymptotic formula (9.10) for $\widehat{n} \geq 4$. For $n = 3$, although explicit derivation of imperfection sensitivity law is not possible, a piecewise linear relation between ξ and Λ^{c} can be observed in Fig. 9.7(a).

Table 9.1 Coefficients for sensitivity law of regular-polygonal n-bar truss tents.

	$n = 4$	$n = 5$	$n = 6$
$V_{,113}$	-6.1572×10^{-7}	-8.9464×10^{-7}	-1.0736×10^{-6}
$V_{,223}$	-6.1572×10^{-7}	-8.9464×10^{-7}	-1.0736×10^{-6}
$V_{,333}$	-1.0736×10^{-6}	-2.6840×10^{-6}	-3.2207×10^{-6}
$V_{,3\Lambda}$	-1.0	-1.0	-1.0
$V_{,1\xi}$	-2.0238×10^{-4}	-2.5297×10^{-4}	-3.0357×10^{-4}
$V_{,2\xi}$	-2.0238×10^{-4}	-2.5297×10^{-4}	-3.0357×10^{-4}
$V_{,3\xi}$	-3.5055×10^{-4}	-4.3819×10^{-4}	-5.2582×10^{-4}

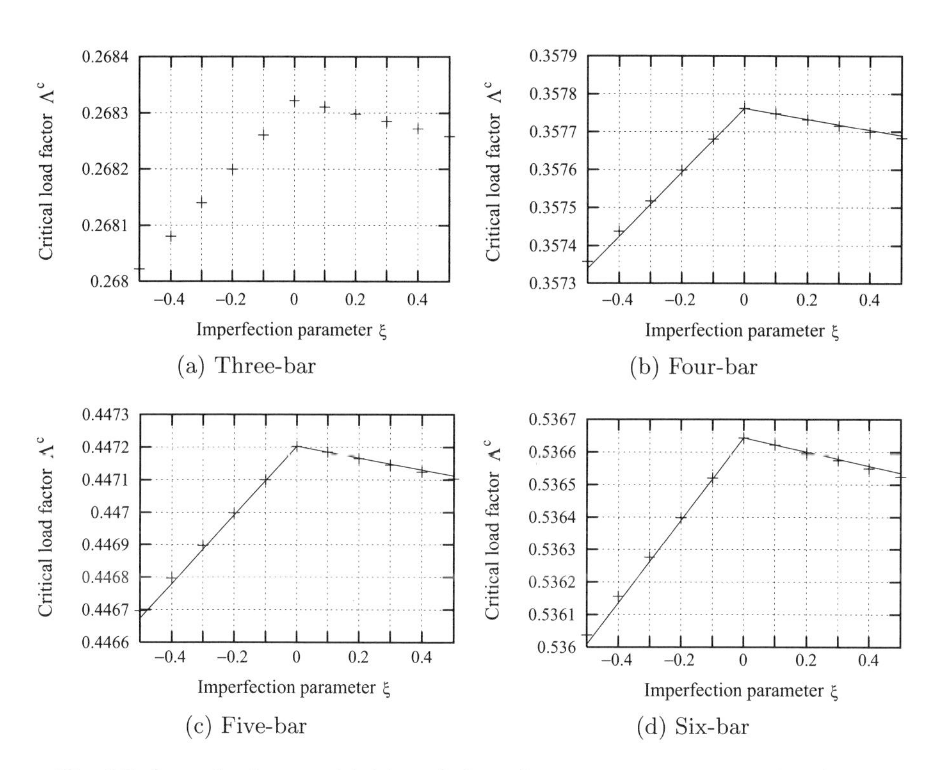

(a) Three-bar

(b) Four-bar

(c) Five-bar

(d) Six-bar

Fig. 9.7 Imperfection sensitivities of the n-bar truss tents. $+$: result of path-tracing analysis; solid line: sensitivity law (9.8).

9.5 Summary

In this chapter,

- imperfection sensitivity laws for hilltop branching point with many symmetric bifurcation points have been derived, and

- hilltop branching points with many symmetric bifurcation points have been produced for the arch-type truss and the regular-polygonal truss tents, and have been shown to be imperfection-insensitive.

The major findings of this chapter are as follows.

- The arch-type truss studied herein has practical importance, because it is related to interaction between global snapthrough and local member buckling. The maximum loads of imperfect systems are piecewise linear functions of an imperfection parameter. Therefore, the existence of member buckling at the limit point is not dangerous in view of imperfection sensitivity. The "simultaneous mode design" for this case is not that pessimistic as was cautioned the "erosion of optimization by compound bifurcation."
- The reduction of critical load due to a symmetric imperfection that is classified as a minor imperfection is of the same order as that due to an antisymmetric major imperfection. Minor imperfection, therefore, should be properly considered in evaluating the critical loads of imperfect systems.

10

Hilltop Branching Point III: Degenerate

10.1 Introduction

A degenerate critical point has somewhat been set aside up to this chapter as a rare exceptional case that can arise from accidental vanishing of some higher differential coefficients of the potential function. Moreover, much care is not paid to this critical point in stability design, because the point does not have any negative effect on the stability of the structure; there is no bifurcation path, and the equilibrium path of an imperfect system makes only a slight detour around the point [217].

Yet an amazing feature of a symmetric structure is its inherent mechanism to annihilate systematically higher differential coefficients. In fact, truss domes with regular-polygonal symmetry that are studied in this chapter do encounter degenerate hilltop points generically as a result of optimization. Moreover, these points often are imperfection sensitive as worked out in Appendix A.9, unlike the nondegenerate hilltop points that enjoy the piecewise-linear law and are not imperfection-sensitive (cf., Chapters 8 and 9). Cautions, accordingly, must be exercised on a degenerate hilltop branching point, which has somewhat been overlooked, despite its importance.

This chapter is organized as follows. Asymptotic behaviors at a degenerate hilltop point are investigated theoretically in Section 10.2. Degenerate hilltop points are generated via optimization for a four-bar truss tent in Section 10.3, a symmetric shallow truss in Section 10.4, and a spherical truss in Section 10.5.

10.2 Degenerate Behaviors

A degenerate hilltop branching point that occurs as a coincidence of a limit point and a simple unstable-symmetric or an asymmetric bifurcation point is studied theoretically. This point is characterized by $V_{,112} = 0$ (cf., Remark 8.2.1 in Section 8.2). There are many variants of sensitivity laws for this point according to the vanishing of $V_{,111}$, $V_{,1122}$, and so on. While the explicit forms of these laws are derived in Appendices A.8 and A.9, we focus on degenerate behaviors of the perfect system in this section.

Consider a degeneracy of a hilltop point that is characterized by

$$V_{,112} = 0, \quad V_{,1122} \neq 0 \tag{10.1}$$

Recall the results of Section 8.2.2 that on the trivial path (8.16), $\mathbf{S}^{\mathrm{V}}(q_1, q_2, \widetilde{\Lambda}, \xi)$ in (8.17) enjoys a diagonal form

$$\mathbf{S}^{\mathrm{V}}(0, q_2, \widetilde{\Lambda}, 0) = \begin{pmatrix} \lambda_\alpha & 0 \\ 0 & \lambda_\beta \end{pmatrix} \tag{10.2}$$

with two eigenvalues ($V_{,112} = 0$)

$$\begin{aligned} \lambda_\alpha &= S_{11}^{\mathrm{V}} = V_{,11\Lambda}\widetilde{\Lambda} + \frac{1}{2}V_{,1122}(q_2)^2 = C_\alpha (q_2)^2 \\ \lambda_\beta &= S_{22}^{\mathrm{V}} = V_{,222}q_2 + V_{,22\Lambda}\widetilde{\Lambda} = V_{,222}q_2 + \text{h.o.t.} \end{aligned} \tag{10.3}$$

and

$$C_\alpha = \frac{1}{2V_{,2\Lambda}}(V_{,2\Lambda}V_{,1122} - V_{,222}V_{,11\Lambda}) \tag{10.4}$$

We focus on the most customary case in practice where the system has a rising fundamental path and becomes unstable at the hilltop point. Hence, we have (cf., (8.20))

$$V_{,2\Lambda} < 0, \quad V_{,222} < 0 \tag{10.5}$$

and assume

$$C_\alpha > 0 \implies V_{,1122} > \frac{V_{,222}V_{,11\Lambda}}{V_{,2\Lambda}} \tag{10.6}$$

We encounter complex behavior in the vicinity of this degenerate hilltop point due to vanishing of many differential coefficients. The two eigenvalues λ_α and λ_β in (10.3) behave as shown in Fig. 10.1. The curve of λ_α is tangential to the q_2-axis, and λ_α is positive except for the hilltop point. In the presence of a small design modification or imperfection, λ_α ceases to vanish at the degenerate hilltop point to trigger catastrophic change that entails several difficulties in design sensitivity analysis, including a discontinuity in the critical point and its unbounded sensitivity coefficients with respect to a design modification, as discussed in Section 4.4.2.

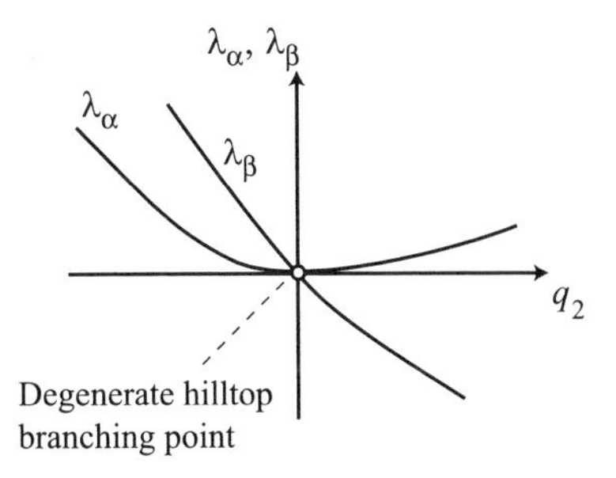

Fig. 10.1 Variation of the two lowest eigenvalues λ_α and λ_β at a degenerate hilltop branching point.

10.3 Four-Bar Truss Tent

Optimum designs are found for the four-bar truss tent with a spring as shown in Fig. 10.2, and its imperfection sensitivity is investigated. The truss has four symmetrically located bars with the same cross-sectional area A. A vertical spring with extensional stiffness K is attached at the center node. Let $L = 1000$ mm, $E = 200$ kN/mm^2, $p = 1000$ kN. The units of force kN and of length mm are suppressed in the following.

The total structural volume V is to be minimized under the constraint on the critical load factor Λ^c. The height H and the cross-sectional area A are chosen as design variables. The critical point is a limit point if H is sufficiently small, and is a double bifurcation point if H is moderately large. We consider two cases:

- A truss without spring, $K = 0$, to demonstrate the emergence of a hilltop branching by optimization.
- A truss with a spring, $K \neq 0$, to demonstrate discontinuities in optimum design with respect to the problem parameters and in sensitivity coefficients with respect to the design variables.

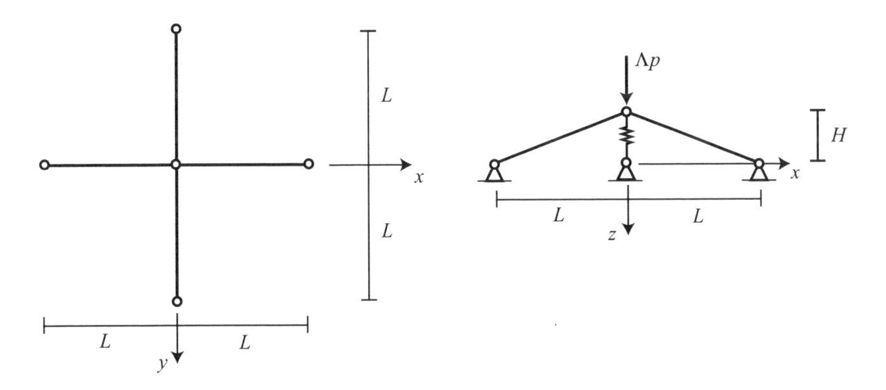

Fig. 10.2 Four-bar truss tent with a spring.

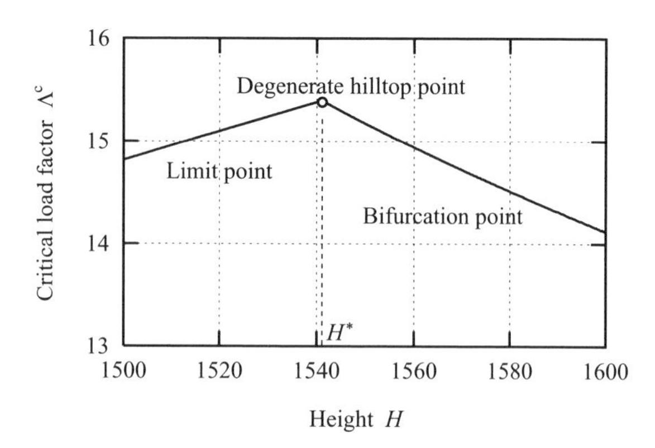

Fig. 10.3 Variation of critical load factor Λ^{c} of the four-bar truss without spring plotted against height H ($K = 0$, $A = 100$). ∘: degenerate hilltop point.

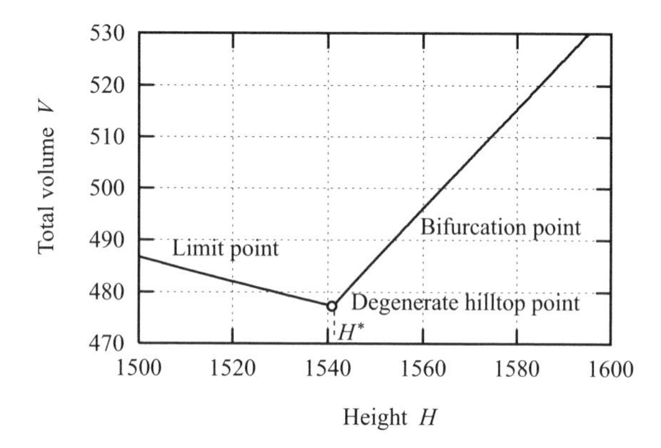

Fig. 10.4 Variation of optimal structural volume V of the four-bar truss without spring plotted against height H ($K = 0, \Lambda^{\mathrm{c}} = 1$). ∘: degenerate hilltop point.

10.3.1 Without spring

Consider the truss without spring. First we fix $A = 100$ to arrive at the plot of critical load factor Λ^{c} against height H as shown in Fig. 10.3. As is seen, Λ^{c} increases initially as H is increased from 1500 to $H^* \simeq 1541$, because the first critical point is a limit point, and the vertical resistance against snapthrough behavior is enhanced by increasing H. At $H = H^* \simeq 1541$, the truss has a hilltop branching point, at which a group-theoretic double bifurcation point collides with the limit point. For $H > H^*$, Λ^{c} decreases as H is further increased because horizontal stiffness resisting against bifurcation decreases.

Next choose A, as well as H, as a design variable and consider the optimization problem of minimizing V under constraint $\Lambda^{\mathrm{c}} = \overline{\Lambda}^{\mathrm{c}}$ of the critical load factor. Since Λ^{c} is proportional to A for a fixed value of H, the value of A and the corresponding V satisfying $\Lambda^{\mathrm{c}} = \overline{\Lambda}^{\mathrm{c}} = 1$ can be calculated for each value of H.

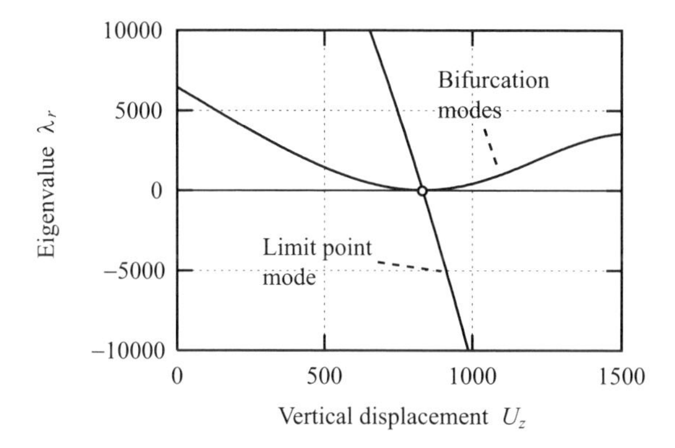

Fig. 10.5 Variation of eigenvalues λ_r ($r = 1, 2, 3$) of the optimal four-bar truss without spring plotted against vertical displacement U_z ($K = 0, \Lambda^c = \overline{\Lambda}^c = 1$). ○: degenerate hilltop point.

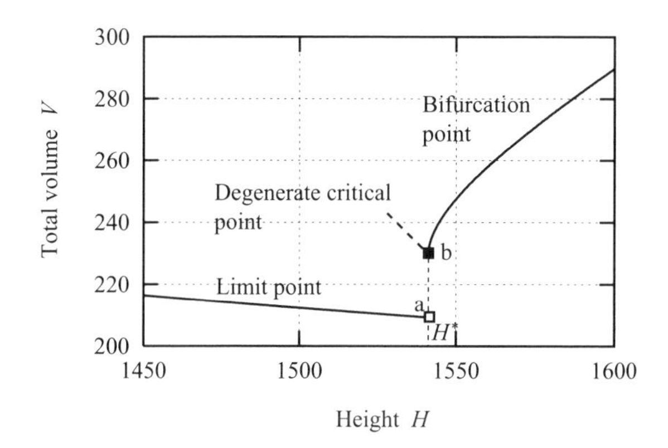

Fig. 10.6 Variation of structural volume V of the four-bar truss with a spring plotted against height H ($K = 20, \Lambda^c = \overline{\Lambda}^c = 1$).

Fig. 10.4 shows the relation between H and the volume V satisfying the constraint $\Lambda^c = \overline{\Lambda}^c = 1$. As is seen, $H = H^*$ is the optimal solution, for which V is minimized under this constraint. A hilltop branching point has thus been obtained as a result of optimization.

For the optimal truss, the eigenvalues of the tangent stiffness matrix are plotted in Fig. 10.5 against the z-directional displacement U_z of the center node. Note that the double eigenvalues corresponding to the bifurcation modes are tangential to the horizontal line $\lambda_r = 0$. The optimal solution has thus encountered a degenerate hilltop branching point [216, 217].

10.3.2 With a spring

Consider the four-bar truss with a spring ($K = 20$). Fig. 10.6 shows the relation between H and V satisfying the constraint $\Lambda^c = \overline{\Lambda}^c = 1$. The truss has a

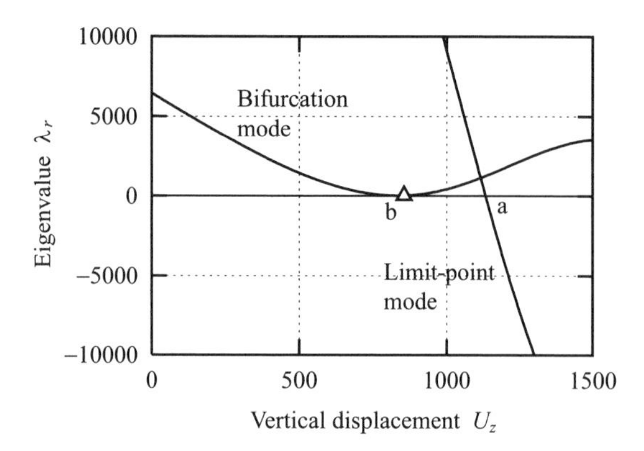

Fig. 10.7 Variation of eigenvalues λ_r $(r = 1, 2, 3)$ of the optimal four-bar truss $(H = H^*)$ with a spring plotted against the z-directional displacement U_z of the center node $(K = 20, \Lambda^{\mathrm{c}} = 1)$. $\triangle$: degenerate critical point.

degenerate critical point 'b' at $H = H^* \simeq 1541$, and $\square$ at 'a' indicates that this point is not included in the plot, i.e., V is a discontinuous function of H at $H = H^*$. Fig. 10.7 shows the variation of λ_r against U_z for the optimal truss with $H = H^*$. Note that the double eigenvalues corresponding to the bifurcation are tangential to the horizontal line $\lambda_r = 0$ at the degenerate critical point 'b' shown by $\triangle$. The degenerate point disappears if H is slightly decreased from H^* and the lowest critical point jumps to point 'a' in Fig. 10.7. Therefore, the sensitivity coefficient of Λ^{c} with respect to H is unbounded at $H = H^*$, and the value of H that is infinitesimally smaller than H^* serves as an optimal solution from an engineering standpoint[1].

The imperfect behavior of the truss with $H = H^*$ $(K = 20, A = 100)$ is investigated. The perfect system has a degenerate critical point at $\Lambda = \Lambda^{\mathrm{c}0} \simeq 32.0$. Imperfections of ± 1.0 are given in the x-direction at the center node of the truss. As can be seen from imperfect equilibrium paths in Fig. 10.8, the x-directional displacement U_x of the center node encounters a large variation around the degenerate critical point at $\Lambda = \Lambda^{\mathrm{c}0}$ for the perfect system. However, this variation decreases as Λ is increased beyond $\Lambda = \Lambda^{\mathrm{c}0}$. Therefore, the structure remains stable, and there is no bifurcation path or snapthrough behavior. The existence of the degenerate critical point is tolerable if the deformation along the equilibrium path is within a specified limit.

Since the design sensitivity is unbounded at the degenerate critical point even for a modification of H corresponding to a minor imperfection, it is desirable to formulate an optimization problem under constraints on the eigenvalues of the tangent stiffness matrix, as problem P3 presented in Section 4.4.

[1]In a strict sense, there is no optimal solution for this problem, because the feasible region is not a closed set.

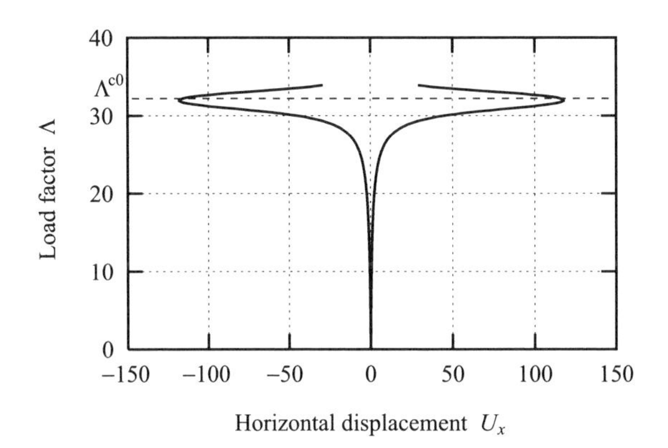

Fig. 10.8 Equilibrium paths of imperfect four-bar trusses with a spring ($K = 20$, $A = 100$). U_x: the x-directional displacement of the center node.

10.4 Symmetric Shallow Truss Dome

Consider the symmetric shallow truss dome as shown in Fig. 10.9, which has the same material properties and design variables as those in Section 2.7.2. Hilltop branching is produced as a result of an optimization of this dome.

The z-directional load Λp_i applied at node i is defined by

$$p_i = \begin{cases} -4.116: & \text{for} \quad i = 1 \\ -1.029: & \text{for} \quad i = 2, \ldots, 7 \end{cases} \tag{10.7}$$

An optimization problem to minimize the total structural volume V under the lower bound constraint $\overline{\Lambda}^{\mathrm{c}} = 10000$ on the critical load factor Λ^{c} is solved; this optimization problem is the same as that was employed in Section 4.5. At the

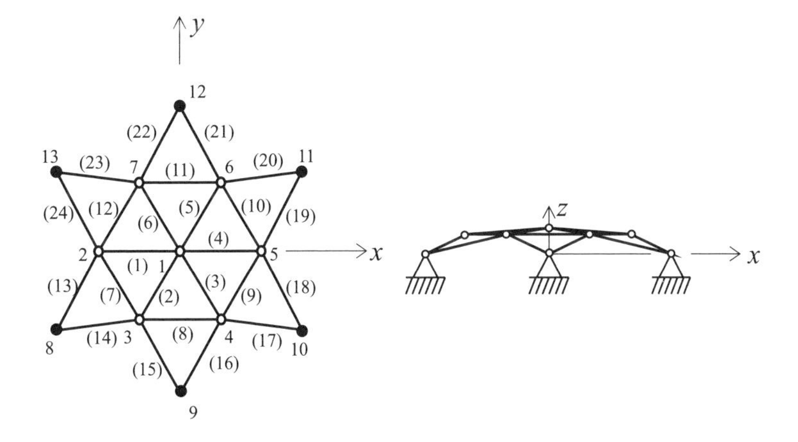

Fig. 10.9 Symmetric shallow truss dome. Nodal coordinates are listed in Table 2.1 in Section 2.7.2. $1, \ldots, 13$ denote node numbers; $(1), \ldots, (24)$ denote member numbers.

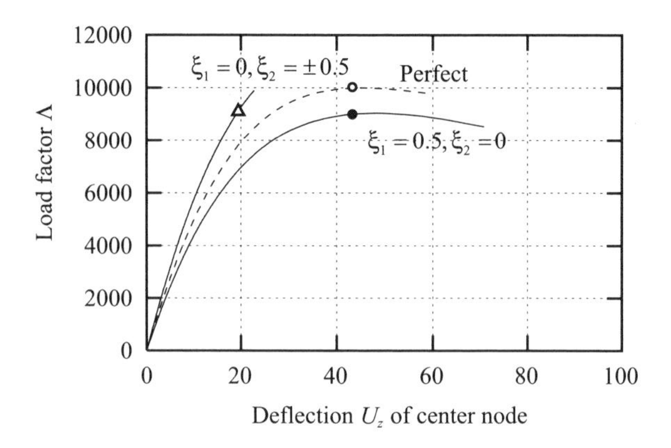

Fig. 10.10 Equilibrium paths of the perfect system and two imperfect systems with nodal imperfections. U_z: deflection of center node; ∘: degenerate hilltop point; △: bifurcation point; •: limit point.

optimal solution, the cross-sectional areas, the total structural volume, and so on, are

$$\begin{aligned} &A_1^* = 5056.0, \quad A_2^* = 4088.7, \quad A_3^* = 222.36, \quad V = 1.4585 \times 10^8 \\ &\lambda_1 = 0.0, \quad \lambda_2 = \lambda_3 = 0.38440 \simeq 0 \\ &V_{,1\Lambda} = 0.25300, \quad V_{,2\Lambda} = V_{,3\Lambda} = 0.0 \end{aligned} \tag{10.8}$$

Thus $\mathbf{\Phi}_1^{c0}$ is a limit point mode with $V_{,1\Lambda} \neq 0$; $\mathbf{\Phi}_i^{c0}$ $(i = 2, 3)$ are bifurcation modes with $V_{,i\Lambda} = 0$. This buckling behavior, accordingly, is approximated by a hilltop branching point with the multiplicity three, which occurs as a coincidence of a limit point and a double bifurcation point (cf., Section 9.2). The interpolation approach presented in Appendix A.2 is utilized to find the optimal truss with this hilltop branching point. The perfect equilibrium path of the optimal truss is plotted against the deflection U_z at the center node as shown by the dotted curve in Fig. 10.10.

Eigenvalues of the optimal truss are plotted against U_z in Fig. 10.11. These eigenvalues are not monotonically decreasing functions of U_z, and the curves of eigenvalues intersect with each other. The eigenvalue corresponding to a bifurcation mode is the lowest before reaching Λ^c (curve AB), but increases upon reaching the minimum at the point near B (curve BC). Thus this is a degenerate hilltop point. Note that the curve ABC is a duplicate of two coincident eigenvalues of a group-theoretic double bifurcation point.

Sensitivity to imperfections of nodal locations is investigated by considering two imperfections:

- minor imperfection $\xi_1 \mathbf{\Phi}_1^{c0}$ with a symmetric limit point mode, and
- major imperfection $\xi_2 \mathbf{\Phi}_2^{c0}$ with a geometrically antisymmetric mode corresponding to symmetric bifurcation.

A limit point shown by • is reached for the imperfect system with $(\xi_1, \xi_2) = (5.0, 0)$ as shown by the solid curve in Fig. 10.10. The equilibrium paths for

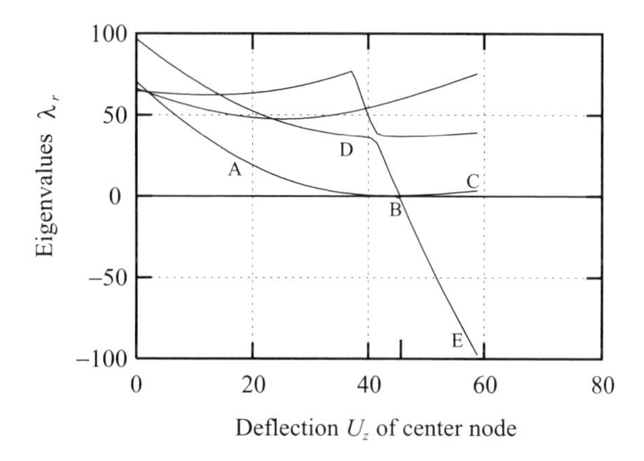

Fig. 10.11 Variation of eigenvalues λ_r of the optimal solution.

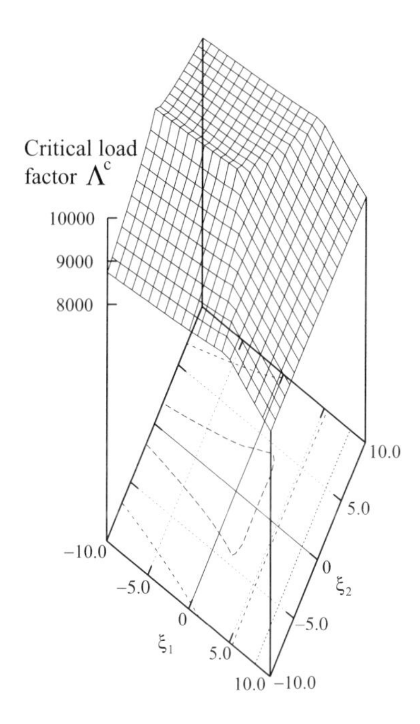

Fig. 10.12 Variation of critical load factor Λ^c of imperfect systems plotted against ξ_1 and ξ_2.

$(\xi_1, \xi_2) = (0, \pm 5.0)$, which are duplicate are shown by the solid curve with a bifurcation point indicated by $\triangle$.

The critical load factor Λ^c is plotted against the imperfection parameters ξ_1 and ξ_2 in Fig. 10.12. Note that the critical points along the line $\xi_1 < 0$, $\xi_2 = 0$ are associated with bifurcation points, but with limit points elsewhere. Hence, the critical point for an imperfect system corresponding to the minor imperfection ξ_1 is a limit point or a bifurcation point depending on the sign of ξ_1 (cf., Appendix A.9.3).

The critical load factors for $(\xi_1, \xi_2) = (-5.0, 0), (0, \pm 5.0), (5.0, 0)$ are $\Lambda^{\rm c} =$ 9895.4, 9386.4 and 9110.4, respectively. The three ridges in Fig. 10.12 intersect with each other at $(\xi_1, \xi_2) = (0, 0)$ to show complexity of the imperfect behaviors.

As an example of a non-optimal design, the cross-sectional areas are chosen to be $A_1^* = A_2^* = A_3^* = 2145.8$ so that the total volume is equal to the optimal value. The critical load factor of this uniform design is $\Lambda^{\rm c} = 4395.3$ at a simple limit point. The critical load $\Lambda^{\rm c}$ decreases to 3701.4 for $(\xi_1, \xi_2) = (5.0, 0)$ and increases to 5176.2 for $(\xi_1, \xi_2) = (-5.0, 0)$, as is typically the case for a limit point. The reduction of the critical load from 4395.3 is 15.787%, which is larger than 13.013% for the optimal truss. Thus the ratio of reduction of the critical load factor even decreases by virtue of optimization. This serves as a good counter example of "danger in naive optimization." The readers are encouraged to design optimal structures and to ensure their performances against imperfections.

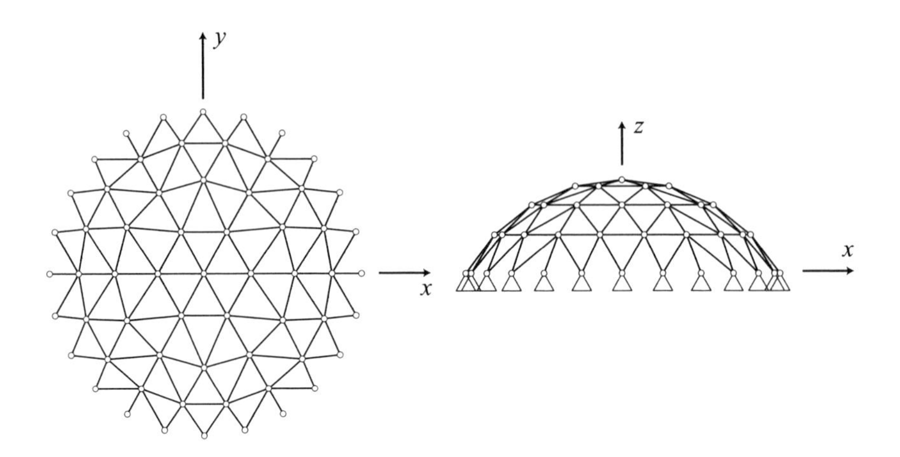

Fig. 10.13 Spherical truss.

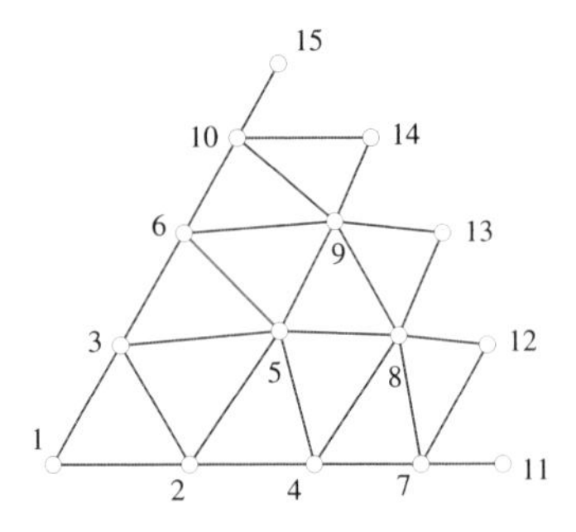

Fig. 10.14 Node numbers of the sixth of the spherical truss.

Table 10.1 Nodal coordinates of the sixth of the spherical truss.

Node number	x	y	z
1	0.0	0.0	4618.80
2	2390.87	0.0	4304.04
3	1195.43	2070.55	4304.04
4	4618.80	0.0	3381.20
5	4000.0	2309.40	3381.20
6	2309.40	4000.00	3381.20
7	6531.97	0.0	1913.17
8	6138.05	2234.07	1913.17
9	5003.78	4198.67	1913.17
10	3265.99	5656.85	1913.17
11	8000.00	0.0	0.0
12	7727.41	2070.55	0.0
13	6928.20	4000.0	0.0
14	5656.85	5656.85	0.0
15	4000.0	6928.20	0.0

10.5 Spherical Truss

Optimization of the spherical truss as shown in Fig. 10.13 is conducted [221]. Based on the symmetry, the members are divided into 14 groups; members belonging to the same group have the same cross-sectional area. Young's modulus is $E = 205.8\ \mathrm{kN/mm^2}$. Table 10.1 lists the coordinates of the nodes of the sixth of this truss as numbered in Fig. 10.14. The units of force kN and of length mm are suppressed in the following.

The optimization problem is formulated as P3 in (4.13a)–(4.13d). The total structural volume V is chosen to be the objective function. $A_i^{\mathrm{L}} = 100.0$ for all the members, and A_i^{U} is not specified.

10.5.1 Concentrated load

A concentrated nodal load Λp with $p = -9.8$ is applied in the z-direction of the center node of the spherical truss. The specified buckling load factor is $\overline{\Lambda}^{\mathrm{c}} = 100.0$. The optimal cross-sectional areas are as shown in Fig. 10.15(a), where the width of each member is proportional to its cross-sectional area. It is observed that the members near the center have large cross-sectional areas. The total volume and the first three critical load factors along the fundamental path are listed in the middle column of Table 10.2. Note that the three critical load factors Λ_1^{c}, Λ_2^{c} and Λ_3^{c} are nearly coincident as a consequence of optimization; Λ_2^{c} and Λ_3^{c} are slightly less than Λ_1^{c}, because the associated critical points are located slightly beyond the limit point.

In order to classify the three critical points by (3.25) in Section 3.3.1, the values of $V_{,r\Lambda}$ $(r = 1, 2, 3)$ have been computed to find that Λ_1^{c} with $V_{,1\Lambda} \neq 0$

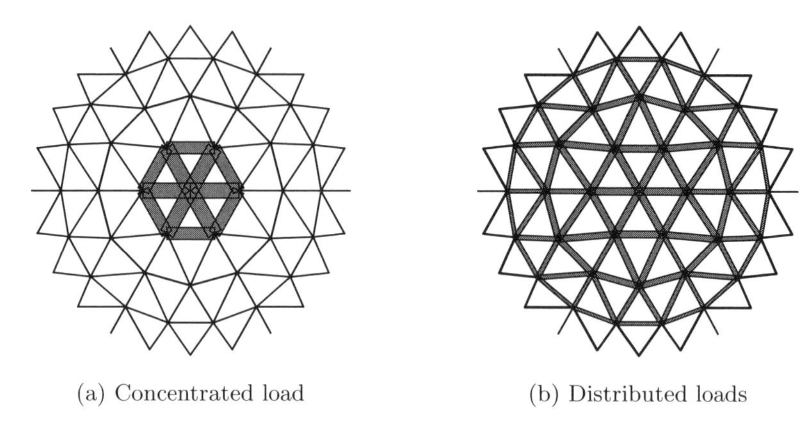

(a) Concentrated load (b) Distributed loads

Fig. 10.15 Optimum designs of spherical truss.

Table 10.2 Total volumes and critical load factors of the optimum designs. '–' denotes that the associated critical point is far apart from the hilltop point.

	Concentrated load	Distributed loads
Volume	1.2677×10^9	3.0704×10^9
Λ_1^c	99.961	99.998
Λ_2^c	99.952	100.409
Λ_3^c	99.728	100.409
Λ_4^c	–	101.148
Λ_5^c	–	101.148
Λ_6^c	–	102.340
Linear buckling load Λ_{L1}^c	228.153	135.230

corresponds to the limit point, and Λ_2^c and Λ_3^c with $V_{,2\Lambda} = V_{,3\Lambda} = 0$ correspond to a group-theoretic double bifurcation point. Hence the critical state of this optimum design can be approximated by a triple hilltop branching point.

Remark 10.5.1 The linear buckling load factor of the optimum design is $\Lambda_{L1}^c = 228.153$, which is more than twice of the specified nonlinear critical load factor $\Lambda^c = 100$; the deflection of the center node at the critical point is $U_z = 337.296$. Thus, prebuckling deformation is very large and, hence, should be incorporated in evaluating the buckling loads Λ^c for this case. □

The eigenvalues λ_r are plotted in Fig. 10.16 against the deflection of the center node U_z. Note that the curve 'a' in Fig. 10.16 corresponds to a limit point mode, and the curve 'b' corresponds to a double bifurcation point.

Imperfection sensitivity is investigated for two imperfections of nodal locations: the minor imperfection ξ_1 in the direction of the symmetric limit point mode $\mathbf{\Phi}_1^{c0}$, and the major imperfection ξ_2 in the direction of the bifurcation mode $\mathbf{\Phi}_2^{c0}$.

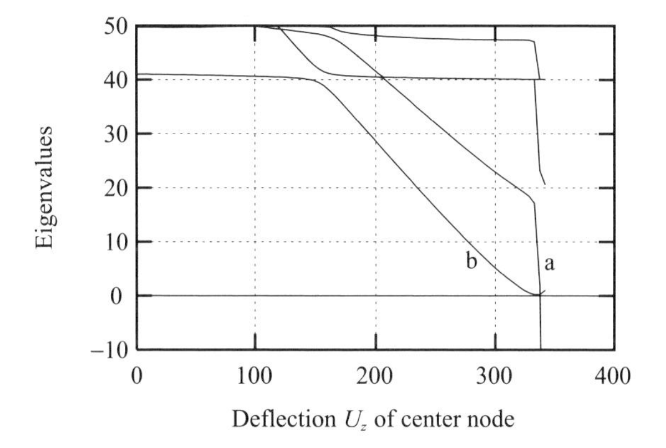

Fig. 10.16 Variation of eigenvalues λ_r plotted against the deflection U_z of the center node for the concentrated load.

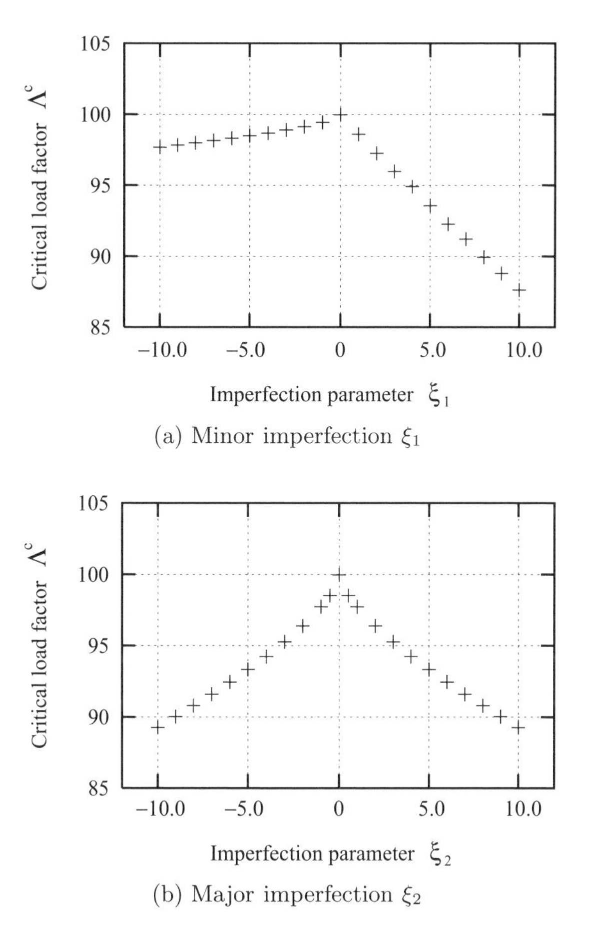

(a) Minor imperfection ξ_1

(b) Major imperfection ξ_2

Fig. 10.17 Imperfection sensitivity for the concentrated load. +: numerical path-tracing analysis.

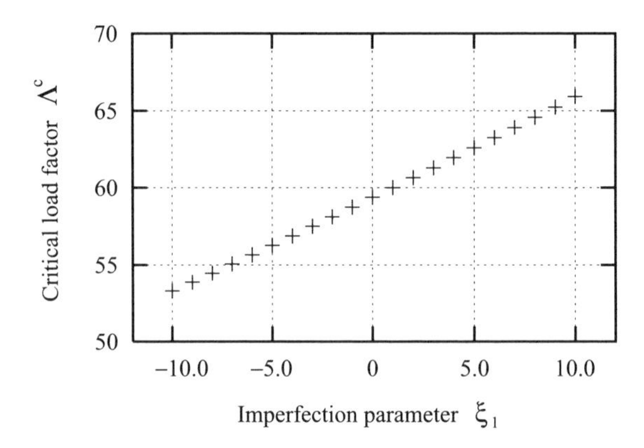

Fig. 10.18 Imperfection sensitivity of the uniform design. +: numerical path-tracing analysis.

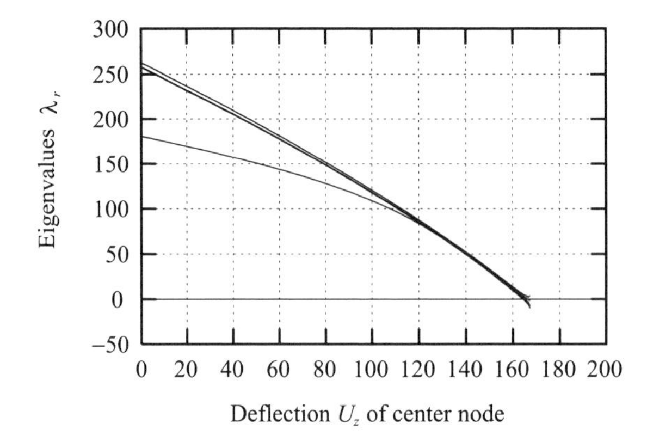

Fig. 10.19 Variation of eigenvalues λ_r $(r = 1, \ldots, 6)$ plotted against the deflection U_z of the center node for the distributed loads.

First, consider the minor imperfection ξ_1. For $\xi_1 > 0$, the first critical point is a limit point, as the imperfection of the center node is in the negative direction of the z-axis. For $\xi_1 < 0$, the first critical point is a bifurcation point. As shown in Fig. 10.17(a), the critical loads are bounded and are governed by the piecewise linear law with a kink at $\xi_1 = 0$, as is the case for a degenerate hilltop point with symmetric bifurcation (cf., (A.135) in Appendix A.9.3).

Next, we consider the major imperfection ξ_2. Fig. 10.17(b) shows the imperfection sensitivity. The critical loads of imperfect systems are symmetric with respect to $\xi \mapsto -\xi$ and follows the 2/3-power law (cf., (A.145) in Appendix A.9.4).

It should be noted, however, that the reduction of critical load factor Λ^c for this major imperfection ξ_2 is of the same order as that of the minor imperfection ξ_1 for a finite range of imperfection, e.g., $\xi_1 = \xi_2 = 10$. For this particular problem, minor imperfection ξ_1 should be properly considered in evaluating the critical load factors Λ^c of imperfect systems.

Fig. 10.18 shows the imperfection sensitivity in the direction of the limit point mode of the uniform design with $A_i = 2000.0$ for all the members, which exhibits almost linear relation with respect to the imperfection parameter ξ_1. The reduction ratio of the critical load, e.g., for $\xi_1 = -10$ is almost identical with that for the optimum design. Therefore, the imperfection sensitivity is not enhanced as a result of optimization.

10.5.2 Distributed loads

Distributed loads Λp with $p = -9.8$ are applied in the z-direction of all free nodes of the spherical truss. The optimal cross-sectional areas as shown in Fig. 10.15(b) are almost uniformly distributed except those for the members connected to the supports.

In this case, the six critical load factors have almost identical values as shown in the last column of Table 10.2, and the sixth critical point is a limit point. The relation between U_z and λ_r is as shown in Fig. 10.19, where some curves are duplicate due to the symmetry of the structure. It is observed that the six critical points are closely located along the fundamental path. It indeed is the inherent mechanism of optimization to produce coincident or nearly coincident critical points. Recall that hilltop branching with many bifurcation points was treated in Chapter 9.

The linear buckling load of the optimal truss is $\Lambda^{\rm c}_{\rm L1} = 135.230$, which is about 35% larger than the nonlinear buckling load $\Lambda^{\rm c} = 100$. Therefore, the effect of prebuckling deformation for this case is smaller than that for the case of the concentrated load.

10.6 Summary

In this chapter,

- the optimization algorithm in Section 4.4.2 has been shown to be applicable to a practical optimum design with coincident critical points,
- degenerate hilltop branching points have been obtained via optimization of structures, and
- imperfect behaviors at these points have been investigated.

The major findings of this chapter are as follows.

- A hilltop point with as many as five bifurcation points has been produced for the optimized spherical truss.
- Optimum designs with closely spaced critical points can be found without any difficulty by the optimization algorithm in Section 4.4.2.
- The reduction of the critical load by a minor imperfection is of the same order as that by a major imperfection. Hence, the effect of a minor imperfection should be appropriately considered for this kind of structure.

Part III: Worst and Random Imperfections

11
Worst Imperfection: Asymptotic Theory

11.1 Introduction

In Part II, optimization methodologies for stability design have been presented. In the stability design of structures, however, the performance of optimized structures against initial imperfections must be evaluated to ensure their safety.

The studies of imperfect structures in early days dealt with pre-specified imperfection modes. There arose a question, "What is the imperfection to be employed?" The geometrical imperfection in the shape of the relevant buckling mode was used initially [42, 112, 131, 132]. For example, classical axisymmetric buckling modes were used as initial shape imperfections for cylindrical shells [167], and checkerboard imperfections were used for spherical shells [130]. Underlying belief that such initial imperfections realize the worst-case scenario was addressed later [142, 299]. In addition, dimple imperfections were employed as realistic imperfections.

It is desirable to consider the *worst* imperfection, the worst-case scenario, that achieves a drastic simplification focusing on the lower bound for buckling strength. The study on the worst imperfection drew considerable attention [30, 63, 146, 211, 263, 265, 317]. The worst imperfection vector of an imperfect cubic potential system was proved to be in the direction of the perfect bifurcated path of the largest slope [123]. The worst imperfection shape of structures was studied in a more general setting [241, 242, 294], extended to imperfections other than structural shapes [137, 142, 205], and implemented into the framework of finite element analysis [136]. Optimization considering the reduction of the maximum load was conducted [253], where the buckling mode of the perfect system was given as the worst imperfection mode. The optimization problem for finding the worst-case scenario is called an *anti-optimization problem* [190].

In this chapter, two methodologies to obtain the worst imperfection are proposed:

- A closed asymptotic form is derived for the worst imperfection that minimizes the maximum load under a given norm of initial imperfections.
- As a numerical counterpart of this, the worst imperfection is formulated as an anti-optimization problem that is readily applicable to finite element analysis of large-scale structures [136, 137, 205].

The worst imperfection was studied in connection with the optimization of structures. As cited repeatedly throughout this book, optimization often produces an extremely imperfection-sensitive structure; therefore, it is practically important to investigate the worst imperfection of an optimized structure. The reduction of the maximum load against the worst imperfection can be formulated into an optimization problem so as to testify the performance of an optimized structure.

This chapter is organized as follows. Asymptotic theory of the worst imperfection is presented in Section 11.2 for simple critical points and a hilltop branching point with simple symmetric bifurcation. An optimum design considering the worst imperfection is presented in Section 11.3. The worst imperfection for the four-bar truss tent is studied in Section 11.4. Optimum design of trusses considering the worst imperfection is carried out in Section 11.5.

11.2 Asymptotic Theory of Worst Imperfection

Asymptotic theory of the worst initial imperfection is presented [142].

11.2.1 General formulation

We consider the first critical point $(\mathbf{U}^{c0}, \Lambda^{c0})$ on the fundamental path of the perfect system with $\mathbf{v} = \mathbf{v}^0$. The imperfection is expressed in terms of the increment $\xi\mathbf{d}$ of $\mathbf{v}$ from $\mathbf{v}^0$ as

$$\mathbf{v} = \mathbf{v}^0 + \xi\mathbf{d} \tag{11.1}$$

with an imperfection pattern vector $\mathbf{d}$ normalized as

$$\|\mathbf{d}\|_H^2 = \mathbf{d}^\top\mathbf{H}\mathbf{d} = 1 \tag{11.2}$$

where $\mathbf{H}$ is a positive-definite matrix to be specified in accordance with an engineering viewpoint. $\|\mathbf{d}\|_H^2$ is a convex function of $\mathbf{d}$.

We formulate the problem of finding the worst imperfection as that of finding the imperfection pattern vector $\mathbf{d}$ that minimizes the critical load Λ^c of an imperfect system under the normalization condition (11.2) for a specified value of ξ (> 0). An anti-optimization problem for the worst imperfection is formulated as

$$\text{WI1}: \ \text{minimize} \quad \Lambda^c \tag{11.3a}$$

$$\text{subject to} \quad \mathbf{d}^\top\mathbf{H}\mathbf{d} = 1 \tag{11.3b}$$

11.2.2 Simple critical points

In imperfection sensitivity laws, only major imperfections are considered in favor of theoretical simplicity, namely[1], (cf., (3.48) in Section 3.5)

$$\widetilde{\Lambda}^{\rm c}(\xi) = \begin{cases} C_0 V_{,1\xi}\xi : & \text{for limit point} \\ C_1 |V_{,1\xi}\xi|^{\frac{1}{2}} : & \text{for asymmetric bifurcation point } (V_{,111}V_{,1\xi}\xi > 0) \\ C_3 (V_{,1\xi}\xi)^{\frac{2}{3}} : & \text{for unstable-symmetric bifurcation point} \end{cases} \tag{11.4}$$

with $\widetilde{\Lambda}^{\rm c}(\xi) = \Lambda^{\rm c} - \Lambda^{\rm c0}$ and

$$C_0 = -\frac{1}{V_{,1\Lambda}}, \quad C_1 = \frac{|V_{,111}|^{\frac{1}{2}}}{V_{,11\Lambda}}, \quad C_3 = -\frac{3^{\frac{2}{3}}(V_{,1111})^{\frac{1}{3}}}{2V_{,11\Lambda}} \tag{11.5}$$

We set $\xi > 0$ to simplify the discussion and assume (cf., (3.29) and (3.50))

$$V_{,11\Lambda} < 0, \quad V_{,1111} < 0 \tag{11.6}$$

The worst imperfection is determined as follows. On the right-hand-sides of (11.4), $V_{,1\xi}$ alone is a function of $\mathbf{d}$. By (3.6) and (3.19a), we have

$$V_{,1\xi} = D_{,1\xi} = \sum_{1=1}^{n} S_{,i\xi}\phi_{1i}^{\rm c0} = \mathbf{\Phi}_1^{\rm c0\top}\mathbf{B}^{\rm c0}\mathbf{d} \tag{11.7}$$

for the perfect system, where

$$\mathbf{B}^{\rm c0} = \mathbf{B}(\mathbf{U}^{\rm c0}, \Lambda^{\rm c0}, \mathbf{v}^0) = [B_{ij}]^{\rm c0} = \left[\frac{\partial S_{,i}}{\partial v_j}\right]^{\rm c0}, \quad (i = 1, \ldots, n;\ j = 1, \ldots, \nu) \tag{11.8}$$

is a constant imperfection sensitivity matrix evaluated at $(\mathbf{U}^{\rm c0}, \Lambda^{\rm c0}, \mathbf{v}^0)$. We assume a major imperfection that satisfies $V_{,1\xi} \neq 0$ for some $\mathbf{d}$.

The minimum of $\widetilde{\Lambda}^{\rm c}$ in (11.4) for a specified value of ξ (> 0), therefore, is achieved by $\mathbf{d}$ that maximizes or minimizes $V_{,1\xi} = \mathbf{\Phi}_1^{\rm c0\top}\mathbf{B}^{\rm c0}\mathbf{d}$ in (11.7). The anti-optimization problem WI1 in (11.3a)–(11.3b) is reformulated as

$$\text{WI2}: \ \text{maximize or minimize} \quad \mathbf{\Phi}_1^{\rm c0\top}\mathbf{B}^{\rm c0}\mathbf{d} \tag{11.9a}$$

$$\text{subject to} \quad \mathbf{d}^\top\mathbf{H}\mathbf{d} = 1 \tag{11.9b}$$

where $\mathbf{d}$ is the only variable in this problem.

We set

$$\mathbf{d}^* = \frac{1}{|\mu|}\mathbf{H}^{-1}\mathbf{B}^{\rm c0\top}\mathbf{\Phi}_1^{\rm c0} \tag{11.10}$$

and call μ the *imperfection influence factor*. From Remark 11.2.1 below, the maximum and minimum of $\mathbf{\Phi}_1^{\rm c0\top}\mathbf{B}^{\rm c0}\mathbf{d}$ are attained by $\mathbf{d}^*$ and $-\mathbf{d}^*$, respectively.

[1] For a stable-symmetric bifurcation point, $\widetilde{\Lambda}^{\rm c}$ is non-existent for $\xi \neq 0$ and the minimization problem of $\widetilde{\Lambda}^{\rm c}$ becomes spurious.

We see from (11.5), (11.7) and $\xi > 0$ that $\widetilde{\Lambda}^{\rm c}(\xi)$ and, in turn, $\Lambda^{\rm c}$ in (11.3a) is minimized by the worst imperfection

$$\mathbf{d}^{\perp} = \begin{cases} \mathrm{sign}\, V_{,1\Lambda}\mathbf{d}^* : & \text{for limit point} \\ \mathrm{sign}\, V_{,111}\mathbf{d}^* : & \text{for asymmetric bifurcation point} \\ \pm\mathbf{d}^* : & \text{for unstable-symmetric bifurcation point} \end{cases} \tag{11.11}$$

Remark 11.2.1 Define the Lagrangian of problem WI2 in (11.9) by

$$L(\mathbf{d}, \mu) = \boldsymbol{\Phi}_1^{\rm c0\top}\mathbf{B}^{\rm c0}\mathbf{d} + \frac{\mu}{2}\left(1 - \mathbf{d}^\top\mathbf{H}\mathbf{d}\right) \tag{11.12}$$

where μ is the Lagrange multiplier. Stationary condition of L with respect to $\mathbf{d}$ leads to

$$\mathbf{B}^{\rm c0\top}\boldsymbol{\Phi}_1^{\rm c0} - \mu\mathbf{H}\mathbf{d} = \mathbf{0} \tag{11.13}$$

Hence from (11.13) and the positive definiteness of $\mathbf{H}$, we have the anti-optimal solution as

$$\mathbf{d} = \frac{1}{\mu}\mathbf{H}^{-1}\mathbf{B}^{\rm c0\top}\boldsymbol{\Phi}_1^{\rm c0} \tag{11.14}$$

The Lagrange multiplier μ is obtained by incorporation of (11.14) into the norm constraint (11.9b) as $((\mathbf{H}^{-1})^\top = \mathbf{H}^{-1})$

$$\mu = \pm(\boldsymbol{\Phi}_1^{\rm c0\top}\mathbf{B}^{\rm c0}\mathbf{H}^{-1}\mathbf{B}^{\rm c0\top}\boldsymbol{\Phi}_1^{\rm c0})^{\frac{1}{2}} \tag{11.15}$$

□

Remark 11.2.2 The worst imperfection pattern depends on the definition of the objective function and the constraints of the anti-optimization problem. Linear bound constraints are consistent with standard specifications in engineering practice as shown in Chapter 12. Other performance measures, such as the eigenvalues of the tangent stiffness matrix, can be used as an alternative of the objective function. □

11.2.3 Hilltop branching with simple bifurcation

In this section, the worst imperfection for a nondegenerate hilltop branching point with simple symmetric bifurcation is obtained [146], while a numerical example will be given in Section 11.4.

The imperfection sensitivity law for this point reads (cf., Sections 3.6.1 and 8.2)

$$\widetilde{\Lambda}^{\rm c}(\xi) = C_4\,|V_{,1\xi}|\xi + C_5 V_{,2\xi}\xi = C_4\xi\,|\boldsymbol{\Phi}_1^{\rm c0\top}\mathbf{B}^{\rm c0}\mathbf{d}| + C_5\xi\,\boldsymbol{\Phi}_2^{\rm c0\top}\mathbf{B}^{\rm c0}\mathbf{d} \tag{11.16}$$

where $\xi > 0$ and C_4 and C_5 are constants independent of $\mathbf{d}$ that are defined by

$$C_4 = \frac{1}{V_{,2\Lambda}}\left(\frac{V_{,222}}{V_{,112}}\right)^{\frac{1}{2}} < 0, \quad C_5 = -\frac{1}{V_{,2\Lambda}} > 0 \tag{11.17}$$

The mode $\boldsymbol{\Phi}_1^{\rm c}$ corresponds to a simple bifurcation mode, and $\boldsymbol{\Phi}_2^{\rm c}$ corresponds to a limit point mode (cf., (8.1), (8.20), (8.23) and (8.28)). Recall that we consider the most customary case of (8.20).

A candidate for the most influential imperfection takes the form of (cf., Remark 11.2.3)

$$\mathbf{d} = a_1 \mathbf{d}_1^* + a_2 \mathbf{d}_2^* \tag{11.18}$$

where

$$\mathbf{d}_i^* = \frac{1}{\mu_i} \mathbf{H}^{-1} \mathbf{B}^{\rm c0\top} \boldsymbol{\Phi}_i^{\rm c0}, \quad (i = 1, 2) \tag{11.19}$$

with

$$\mu_i = (\boldsymbol{\Phi}_i^{\rm c0\top} \mathbf{B}^{\rm c0} \mathbf{H}^{-1} \mathbf{B}^{\rm c0\top} \boldsymbol{\Phi}_i^{\rm c0})^{\frac{1}{2}}, \quad (i = 1, 2) \tag{11.20}$$

Substitution of (11.18) into (11.2) and (11.16) with (11.25) in Remark 11.2.3 leads to an anti-optimization problem

$$\text{minimize} \quad \widetilde{\Lambda}^{\rm c} = C_1^* \xi |a_1| + C_2^* \xi a_2 \tag{11.21a}$$

$$\text{subject to} \quad (a_1)^2 + (a_2)^2 = 1 \tag{11.21b}$$

with $C_1^* = C_4 \mu_1 < 0$, $C_2^* = C_5 \mu_2 > 0$.

As explained in Remark 11.2.4 below, the minimum value of $\widetilde{\Lambda}^{\rm c}$ is equal to $-[(C_1^*)^2 + (C_2^*)^2]^{\frac{1}{2}} \xi$ and is attained by (11.18) with (11.26), namely,

$$\mathbf{d}^{\perp} = \pm \cos \beta^* \, \mathbf{d}_1^* + \sin \beta^* \, \mathbf{d}_2^* \tag{11.22}$$

where

$$\beta^* = \tan^{-1}(C_2^* / C_1^*) \tag{11.23}$$

where $\tan^{-1}(\,\cdot\,)$ takes the principal value in the range $[-\pi/2, \pi/2]$.

The maximum of $\widetilde{\Lambda}^{\rm c}$ is equal to $C_2^* \xi$ and is attained by the optimal imperfection

$$\mathbf{d} = \mathbf{d}_2^* \tag{11.24}$$

Remark 11.2.3 Note that $\mathbf{H}^{-1} \mathbf{B}^{\rm c0\top} \boldsymbol{\Phi}_i^{\rm c0}$ $(i = 3, \ldots, n)$ are not influential on $\widetilde{\Lambda}^{\rm c}$ in an asymptotic sense and that the following pertinent orthogonality condition in general can be derived [142, 205]:

$$\boldsymbol{\Phi}_1^{\rm c0\top} \mathbf{B}^{\rm c0} \mathbf{H}^{-1} \mathbf{B}^{\rm c0\top} \boldsymbol{\Phi}_2^{\rm c0} = 0 \tag{11.25}$$

□

Remark 11.2.4 It is pertinent to set

$$a_1 = \pm \cos \beta, \quad a_2 = \sin \beta, \quad (-\pi/2 \le \beta \le \pi/2) \tag{11.26}$$

Then the constraint (11.21b) is always satisfied and the objective function (11.21a) becomes

$$\widetilde{\Lambda}^{\rm c} = C_1^* \xi \cos \beta + C_2^* \xi \sin \beta \tag{11.27}$$

The stationary condition of $\widetilde{\Lambda}^{\rm c}$ with respect to β yields

$$\frac{{\rm d}\widetilde{\Lambda}^{\rm c}}{{\rm d}\beta} = -C_1^*\xi \sin\beta + C_2^*\xi \cos\beta = 0 \tag{11.28}$$

which is satisfied by $\beta = \beta^*$ in (11.23). □

11.3 Optimization Incorporating Worst Imperfection

The worst imperfection is incorporated in the optimum design of structures under stability constraints. Numerical examples will be presented in Section 11.5.

11.3.1 Formulation of optimization problem

The problem considered here is to find the optimum symmetric system under the condition that the critical load factor against the worst asymmetric imperfection remains above the prescribed lower bound, instead of assigning the constraints on the critical load of a perfect system. The effect of initial imperfections is systematically incorporated in this manner.

Let $\mathbf{A} \in \mathbb{R}^{n^{\rm m}}$ denote the vector of design variables, such as the cross-sectional areas of truss members. The design variables are defined so that symmetry conditions are preserved for modified designs; i.e., only symmetric structures are considered to be feasible, so that the critical point under consideration is a symmetric bifurcation point.

The imperfection parameter vector is defined by $\mathbf{v} = \mathbf{v}^0 + \xi\mathbf{d}$ in (11.1). The worst imperfection pattern $\mathbf{d}^{\perp}$ is obtained by using the definitions in Section 11.2 with the normalization $\mathbf{d}^\top\mathbf{d} = 1$ by setting $\mathbf{H}$ to be the identity matrix $\mathbf{I}_\nu \in \mathbb{R}^{\nu\times\nu}$. The magnitude of the imperfection ξ is specified to be a positive constant and the worst critical load is denoted by $\Lambda^{\rm c\perp}$.

The upper and lower bounds for A_i are denoted by $A_i^{\rm U}$ and $A_i^{\rm L}$, respectively. The specified lower bound for $\Lambda^{\rm c\perp}$ is denoted by $\overline{\Lambda}^{\rm c}$. Then the optimum design problem for minimizing the objective function $V(\mathbf{A})$, such as the total structural volume, is stated as

$$\text{minimize} \quad V(\mathbf{A}) \tag{11.29a}$$

$$\text{subject to} \quad \Lambda^{\rm c\perp}(\mathbf{A}) \geq \overline{\Lambda}^{\rm c} \tag{11.29b}$$

$$A_i^{\rm L} \leq A_i \leq A_i^{\rm U}, \quad (i = 1, \dots, n^{\rm m}) \tag{11.29c}$$

11.3.2 Optimization algorithm

The optimization algorithm using nonlinear programming is summarized as

Step 1: Set the values of parameters ξ, $\overline{\Lambda}^{\rm c}$, $A_i^{\rm U}$ and $A_i^{\rm L}$, and assign the initial value of $\mathbf{A}$.

Step 2: Find the critical load factor $\Lambda^{\rm c0}$ of the perfect system by tracing the fundamental path, and compute the design sensitivity coefficients $\partial\Lambda^{\rm c0}/\partial A_i$ of $\Lambda^{\rm c0}$ by the interpolation approach in Section 2.5.

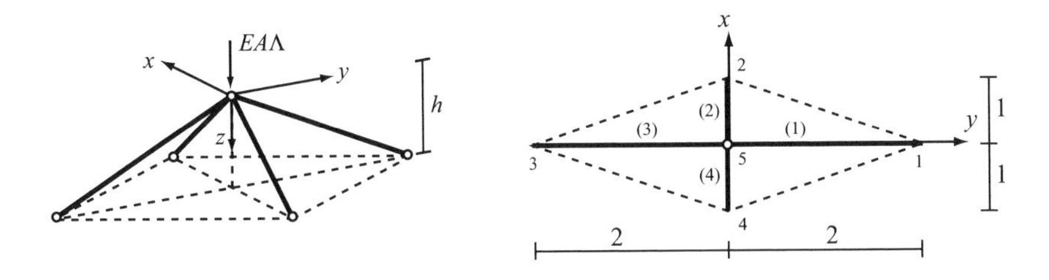

Fig. 11.1 Four-bar truss tent.

Step 3: Calculate the worst imperfection pattern from (11.11) with (11.10), and find the critical load factor $\Lambda^{c\perp}$ of the structure for the worst imperfection by path-tracing analysis.

Step 4: Approximate the sensitivity coefficient $\partial \Lambda^{c\perp}/\partial A_i$ by

$$\frac{\partial \Lambda^{c\perp}}{\partial A_i} \simeq \frac{\partial \Lambda^{c0}}{\partial A_i} \tag{11.30}$$

Step 5: Compute the sensitivity coefficients of the objective function $V(\mathbf{A})$.

Step 6: Update $\mathbf{A}$ using a nonlinear programming algorithm.

Step 7: Go to Step 2 if the stopping criteria are not satisfied.

11.4 Worst Imperfection for Four-Bar Truss: Hilltop Branching Point

The procedure to determine the worst imperfection at a hilltop branching point in Section 11.2 is applied to the four-bar truss tent in Fig. 11.1 (cf., Section 3.7). All truss members have the same cross-sectional area A and the same Young's modulus E.

As was shown in Fig. 3.6, the hilltop branching point 'A' is located at $(U_x^{c0}, U_y^{c0}, U_z^{c0}, \Lambda^{c0}) = (0, 0, 1.0356, 0.6624)$ on the fundamental path, and has two critical eigenvectors $\mathbf{\Phi}_1^{c0} = (1, 0, 0)^\top$ and $\mathbf{\Phi}_2^{c0} = (0, 0, 1)^\top$, which correspond to a bifurcation mode and a limit point mode, respectively.

We consider imperfections of member lengths L_i $(i = 1, \dots, 4)$. The vector $\mathbf{v}$ for representing the mechanical properties of the structure is chosen to be

$$\mathbf{v} = (L_1, L_2, L_3, L_4)^\top \tag{11.31}$$

which, for the perfect system ($h = h^0 = 1.8852$), is equal to

$$\mathbf{v}^0 = (\sqrt{4 + (h^0)^2}, \sqrt{1 + (h^0)^2}, \sqrt{4 + (h^0)^2}, \sqrt{1 + (h^0)^2}\,)^\top \tag{11.32}$$

We obtain the worst imperfection of the four-bar truss under constraint of $\mathbf{d}^\top \mathbf{d} = 1$ by putting $\mathbf{H} = \mathbf{I}_4$ in (11.2). The imperfection sensitivity matrix for

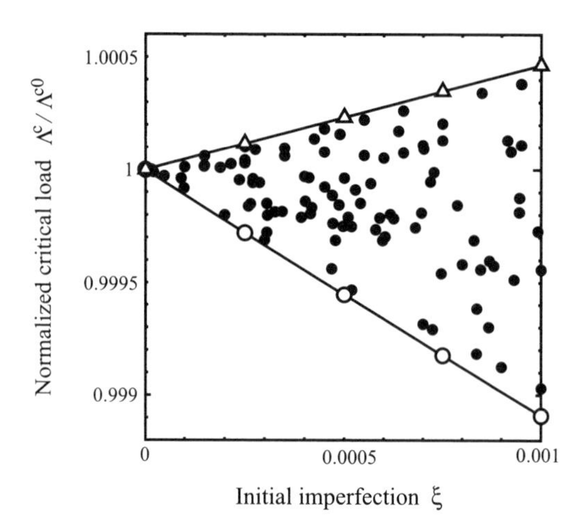

Fig. 11.2 Variation of Λ^{c}/Λ^{c0} plotted against ξ for the four-bar truss. •: random imperfection; ∘: the worst imperfection; △: the optimal imperfection.

this case becomes

$$\mathbf{B}^{c0} = \begin{pmatrix} 0 & -0.2196 & 0 & 0.2196 \\ -0.2648 & 0 & 0.2648 & 0 \\ 0.1125 & 0.1866 & 0.1125 & 0.1866 \end{pmatrix} \tag{11.33}$$

By (11.19) with $\mathbf{H} = \mathbf{I}_4$, $\boldsymbol{\Phi}_1^{c0} = (1,0,0)^\top$, $\boldsymbol{\Phi}_2^{c0} = (0,0,1)^\top$, we can evaluate

$$\mathbf{d}_1^* = \frac{1}{\mu_1}\mathbf{H}^{-1}\mathbf{B}^{c0\top}\boldsymbol{\Phi}_1^{c0} = (0, -0.7071, 0, 0.7071)^\top \tag{11.34a}$$

$$\mathbf{d}_2^* = \frac{1}{\mu_2}\mathbf{H}^{-1}\mathbf{B}^{c0\top}\boldsymbol{\Phi}_2^{c0} = (0.3651, 0.6056, 0.3651, 0.6056)^\top \tag{11.34b}$$

with $\mu_1 = 0.3744$, $\mu_2 = 0.3081$.

The optimal imperfection in (11.24) is given by $\mathbf{d} = \mathbf{d}_2^*$ in (11.34b), which strengthens the truss tent and preserves its symmetry.

The worst imperfection can be obtained as follows. By the path-tracing analyses for imperfection patterns $\mathbf{d}_1^*$ and $\mathbf{d}_2^*$ in (11.34a) and (11.34b) for $\xi = 10^{-6}$, we obtain $\widetilde{\Lambda}^c = -6.9068 \times 10^{-7}$, 3.0808×10^{-7}, respectively. Using (11.16), (11.19) and (11.20), we have

$$C_4\xi\,|\boldsymbol{\Phi}_1^{c0\top}\mathbf{B}^{c0}\mathbf{d}_1^*| = C_4\mu_1\xi = -6.9068 \times 10^{-7} \tag{11.35a}$$

$$C_5\,\xi\,\boldsymbol{\Phi}_2^{c0\top}\mathbf{B}^{c0}\mathbf{d}_2^* = C_5\,\mu_2\xi = 3.0808 \times 10^{-7} \tag{11.35b}$$

With (11.33)–(11.34) and $\xi = 10^{-6}$, (11.35a) and (11.35b) give

$$C_1^* = C_4\mu_1 = -0.69068, \quad C_2^* = C_5\mu_2 = 0.30808 \tag{11.36}$$

Then from (11.23), we can compute

$$\beta^* = -0.1336\pi \tag{11.37}$$

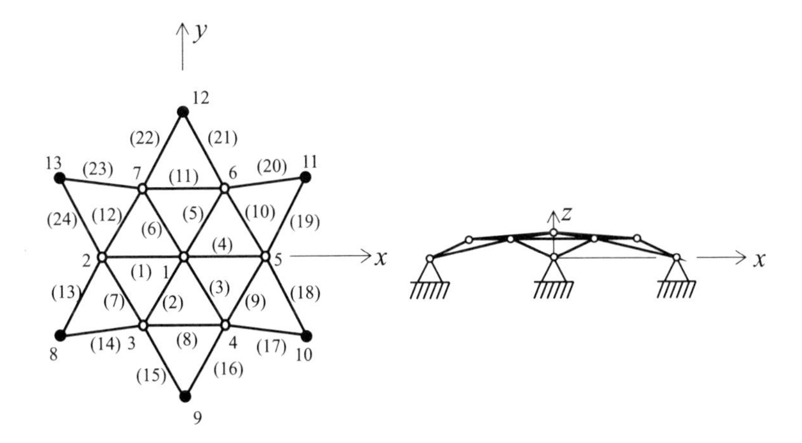

Fig. 11.3 Symmetric shallow truss dome. Nodal coordinates are listed in Table 2.1 in Section 2.7.2. $1, \dots, 13$ denote node numbers; $(1), \dots, (24)$ denote member numbers.

and by (11.22) with (11.34a) and (11.34b), we can compute the worst imperfection pattern

$$\mathbf{d}^{\perp} = (-0.7945, -0.2467, 0.4971, -0.2467)^{\top} \tag{11.38}$$

which breaks the symmetry.

Fig. 11.2 shows ξ versus Λ^{c}/Λ^{c0} relationship obtained by path-tracing for given initial imperfections. One can clearly see that the values of Λ^{c}/Λ^{c0} shown by • for white noise imperfections are bounded by those of the worst imperfection ○ from downward and by the optimal ones △ from upward. This assesses the validity of the present formulation.

11.5 Optimum Designs of Trusses with Worst Imperfection

Optimum designs are found for a symmetric shallow truss dome and a column-type plane truss by using the algorithm in Section 11.3.2. The first critical point is governed by unstable-symmetric bifurcation for both structures. The objective function is the total structural volume. Optimization is carried out by IDESIGN Ver. 3.5 [14], in which the SQP is used (cf., Section 4.3.2). Young's modulus is $E = 205.8$ kN/mm^2. The units of force kN and of length mm are suppressed in the following.

11.5.1 Symmetric shallow truss dome

Optimum designs are found for the symmetric shallow truss dome as shown in Fig. 11.3. Material properties and design variables are the same as those used in Section 2.7.2. The lower bound of cross-sectional area is $A_i^{\mathrm{L}} = 100$ for all the

Table 11.1 Sensitivity coefficients ($\times 10^{-7}$) of the optimal truss dome.

Numerical methods	A_1^*	A_2^*	A_3^*
Finite difference $\dfrac{\Delta\Lambda^{\mathrm{c}0}}{\Delta A_i^*}$	9.06652	9.03746	2.28457
Finite difference $\dfrac{\Delta\Lambda^{\mathrm{c}\perp}}{\Delta A_i^*}$	8.37854	8.89463	2.40644
Interpolation $\dfrac{\partial\Lambda^{\mathrm{c}0}}{\partial A_i^*}$	9.06618	9.03730	2.28455

members, and A_i^{U} is not specified. The load Λp_i is applied to the ith node in the z-direction, where p_i is given as

$$p_i = \begin{cases} -1.029 : & \text{for} \quad i = 1 \\ -4.116 : & \text{for} \quad i = 2, \dots, 7 \end{cases} \tag{11.39}$$

We set $\xi = 500$ and $\overline{\Lambda}^{\mathrm{c}} = 1.0$. The members are divided to three groups as (4.15) in Section 4.5, where the cross-sectional area of members in the ith group is denoted by A_i^*.

An optimum design is obtained by considering the worst imperfection of the cross-sectional areas. At the optimum, we have $A_1^* = 2370.0$, $A_2^* = 2929.0$, $A_3^* = 2644.0$. The maximum load of the optimum design with the worst imperfection is $\Lambda^{\mathrm{c}\perp} = 100.35$, which is sufficiently close to the specified value, $\overline{\Lambda}^{\mathrm{c}} = 100$.

The result of sensitivity analysis for the optimal design is as shown in Table 11.1, where $\Delta\Lambda^{\mathrm{c}0}/\Delta A_i^*$ is the sensitivity coefficient of the bifurcation load of the perfect system and $\Delta\Lambda^{\mathrm{c}\perp}/\Delta A_i^*$ is that of the limit point load of the imperfect system with the worst imperfection obtained by the central finite difference method with $\Delta A_i^* = 1.0 \times 10^{-3}$ (cf., (2.19) in Section 2.3.3). The value of $\Delta\Lambda^{\mathrm{c}0}/\Delta A_i^*$ agrees with $\partial\Lambda^{\mathrm{c}0}/\partial A_i^*$ found by the interpolation method within good accuracy. Although the difference between $\partial\Lambda^{\mathrm{c}0}/\partial A_i^*$ and $\Delta\Lambda^{\mathrm{c}\perp}/\Delta A_i^*$ is not negligibly small, the sensitivity coefficient $\partial\Lambda^{\mathrm{c}\perp}/\partial A_i^*$ for the imperfect system is successfully approximated by $\partial\Lambda^{\mathrm{c}0}/\partial A_i^*$. Thus the approximation (11.30) is sufficient for a practical purpose. The optimal solutions consequently can be found with good accuracy if the gradient-based approach with line search is used.

Remark 11.5.1 At each step of a nonlinear programming algorithm, such as SQP, the search direction is defined based on the sensitivity coefficients (gradients) of the objective function and the constraints. Since the best solution is searched in this direction, in the process of line search, errors in sensitivity coefficients are not amplified in the final optimal solution. However, approximate sensitivity coefficients cannot lead to the exact solution, because sensitivity coefficients appear in the optimality conditions that are used as the termination condition of the algorithm. □

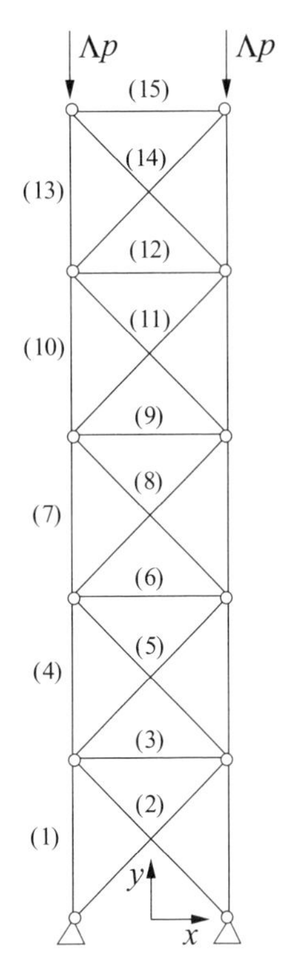

(a) Configuration and member numbers

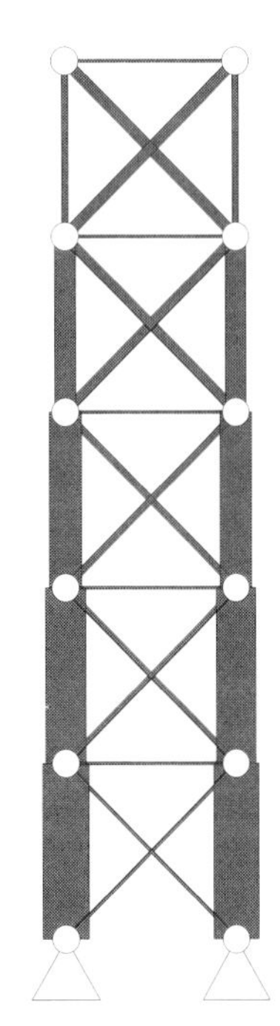

(b) Optimum cross-sectional areas

Fig. 11.4 Plane tower truss. (1), . . . , (15) represent group numbers of members assigned by considering the reflection symmetry with respect to the y-axis; the width of a member is proportional to its cross-sectional area.

11.5.2 Plane tower truss

An optimum design is found for the column-type plane tower truss as shown in Fig. 11.4(a). The lengths of all members in the x- and y-directions are equal to 1000. The lower bound of cross-sectional area is $A_i^{\mathrm{L}} = 600$ for all the members, and A^{U} is not specified. We set $p = 205.8$, $\xi = 1$, and $\overline{\Lambda}^{\mathrm{c}} = 100$.

The optimum cross-sectional areas considering the worst imperfection of the cross-sectional areas obtained by the approximation (11.30) are shown in Table 11.2 and are illustrated in Fig. 11.4(b). The optimal total structural volume is 6.109×10^5. Note that the cross-sectional areas of all horizontal members are equal to A_i^{L}. The curves of the critical load factor against the worst imperfection pattern with different imperfection magnitudes ξ for the optimal design are as shown in Fig. 11.5. The solid curve for the asymptotic formula of the 2/3-power law in (3.48) is in good accordance with the dashed curve that represents the limit point load factor Λ^{c} found by path-tracing of imperfect trusses.

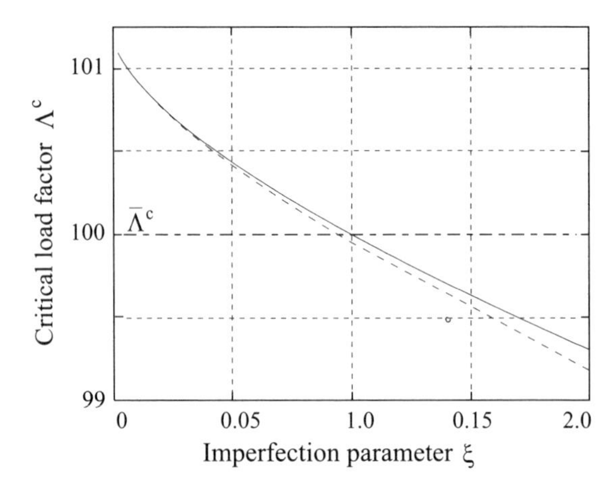

Fig. 11.5 Comparison of imperfection sensitivity curves. Solid curve: asymptotic equation (3.48); dashed curve: result of numerical path-tracing analysis.

Table 11.2 Optimal cross-sectional areas ($\times 10^3$) of the plane tower truss computed by the approximation (11.30) and by the finite difference method.

Group	1	2	3	4	5
Approximation (11.30)	6.738	0.6000	0.6000	5.970	0.7400
Finite difference	6.649	0.6000	0.6000	5.941	0.8000
	6	7	8	9	10
	0.6000	4.782	1.001	0.6000	3.120
	0.6000	4.737	1.022	0.6000	3.086
	11	12	13	14	15
	1.367	0.6000	0.9508	1.584	0.6000
	1.357	0.6000	1.059	1.555	0.6060

The optimal cross-sectional areas found by the direct computation of the sensitivity coefficient $\partial\Lambda^{c\perp}/\partial A_i^*$ by the central finite difference method, which are regarded as exact solution are also listed in Table 11.2. Note that the error due to the approximation (11.30) does not entail any significant difference in the results of optimization.

The critical load factor of a non-optimal design that has uniform cross-sectional areas with the same total structural volume as the optimal design is $\Lambda^c = 48.12$, which is 48.12% of $\Lambda^{c\perp} = 100$ of the optimal design with worst imperfection. Thus, the critical load has successfully been increased by optimization.

11.6 Summary

In this chapter,

- the asymptotic formulas for the worst imperfection have been presented for simple critical points and a hilltop branching point as the optimality conditions (stationary conditions) of an anti-optimization problem, and
- an optimization problem has been formulated for a symmetric structure for a specified lower bound of the critical load for the worst imperfection pattern, and the optimization of structures incorporating the worst imperfection has been conducted.

The major finding of this chapter is as follows.

- The optimization problem incorporating the worst imperfection is a powerful design methodology that is capable of achieving significant enhancement of the critical load.

12

Worst Imperfection: Anti-optimization by LP and QP

12.1 Introduction

In stability design of structures, it achieves a drastic simplification to consider the worst imperfection pattern that gives the lower bound of buckling loads among numerous possible imperfections. The theory of the worst imperfection introduced in Chapter 11, in principle, is a complete means to compute the worst imperfection for global buckling without mode interaction for a sufficiently small initial imperfection magnitude. However, in the application to practical stability design, the theory often encounters the following problems:

- This law is not applicable to a structure that undergoes an interaction between local and global buckling modes, as the eigenmode corresponding to the lowest critical load for the perfect system often becomes different from that of an imperfect system.

- Only major imperfections are incorporated in the theory. However, if an imperfection magnitude is not sufficiently small, minor imperfections cannot be ignored as shown in Chapter 13 [220].

- The imperfection sensitivity matrix $\mathbf{B}^{c0} = [\partial^2 \Pi / \partial U_i \partial v_j]^{c0}$ in (11.8) to be employed in this theory is not familiar to the communities of computational mechanics and structural optimization. A remedy of this will be given in Section 16.3.

In this chapter, we propose a numerical procedure to obtain the worst imperfection that can overcome these three problems [278]. This procedure is readily applicable to numerical analysis and mathematical programming by replacing mathematical details with numerical counterparts. The procedure consists of the following two steps of anti-optimization:

- The worst imperfection mode $\mathbf{d}_r^\perp$ is defined for each eigenvalue λ_r of the tangent stiffness matrix at a specified load level, such that the eigenvalue is most rapidly reduced for a given constraint on the imperfection mode $\mathbf{d}$. Then the critical loads for a series of worst imperfections $\mathbf{d}_r^\perp$ $(r = 1, 2, \dots)$ are estimated such that $\lambda_r = 0$ is satisfied.
- Among the worst imperfection modes $\mathbf{d}_r^\perp$, the *dominant worst imperfection* with the lowest critical load is to be depicted as the worst-case scenario.

According to constraints on the imperfection mode $\mathbf{d}$ to be employed, we propose two numerical anti-optimization problems for finding the worst imperfection:

- Linear Programming (LP) problem for linear inequality constraints is suitable for practical design, in which the upper and lower bounds on imperfection parameters, as well as additional linear constraints, are specified.
- Quadratic Programming (QP) problem for a weighted quadratic norm permits simple algebraic manipulation of the worst imperfection.

An appropriate problem has to be chosen in accordance with the purpose of stability design. The validity and applicability of the proposed methods are shown through the application to the design of laterally braced column structures, which serve as typical structures that undergo the mode interaction between the local and global buckling.

This chapter is organized as follows. In Section 12.2, a numerical procedure to obtain the worst imperfection is presented. In Section 12.3, the worst imperfection of the braced column structures is investigated.

12.2 Numerical Procedure to Obtain Worst Imperfection Mode

A numerical procedure to obtain the worst imperfection mode that consists of two steps of anti-optimization is proposed.

12.2.1 Minimization of eigenvalues

Consider imperfections of a finite-dimensional structure that are defined by (1.14) in Section 1.3 using imperfection parameter ξ and imperfection mode $\mathbf{d} \in \mathbb{R}^\nu$ as

$$\mathbf{v} = \mathbf{v}^0 + \xi \mathbf{d} \tag{12.1}$$

Recall that $\mathbf{v} \in \mathbb{R}^\nu$ is the vector defining the mechanical properties of the structure and that the superscript $(\,\cdot\,)^0$ denotes the value at the perfect system. In the formulation of the worst imperfection, we choose $\mathbf{d}$ as an independent variable vector, and set ξ to be constant. The value of ξ must be specified to define the constraint on $\mathbf{d}$, while it need not be specified in the asymptotic analysis in Chapter 11. Since the buckling mode may possibly change in association with the change of the imperfection magnitude ξ, the value of ξ is to be specified appropriately. We set $\xi = 1$ in the sequel for simplicity.

The rth eigenvalue λ_r of the tangent stiffness matrix $\mathbf{S}$ of an imperfect system for a specified $\mathbf{d}$ at a fixed load factor $\Lambda = \Lambda^*$ can be estimated by

$$\lambda_r = \lambda_r^0 + \delta\lambda_r \tag{12.2}$$

where $\delta\lambda_r$ is the increment of λ_r from its value λ_r^0 of the perfect system.

The worst imperfection mode $\mathbf{d}_r^\perp$ is defined as $\mathbf{d}$ that minimizes $\delta\lambda_r$ under a constraint on $\mathbf{d}$ at a fixed load factor $\Lambda = \Lambda^*$. The worst imperfection mode changes according to the constraint to be employed. The following two types of constraints are considered herein:

- Linear inequality constraints on $\mathbf{d}$ are specified to arrive at an LP problem (cf., Section 12.2.2).
- A quadratic norm of the imperfection mode $\mathbf{d}$ is specified to arrive at a QP problem (cf., Section 12.2.3).

There are several methods to evaluate $\delta\lambda_r$, e.g.:

- Nonlinear path-tracing analysis for the imperfect system is accurate, but is computationally inefficient for this purpose, as the analysis must be conducted repeatedly at each step of anti-optimization.
- It is far more computationally efficient to employ linearization

$$\delta\lambda_r = \sum_{i=1}^{\nu} \frac{\partial \lambda_r}{\partial d_i} d_i = \left(\frac{\partial \lambda_r}{\partial \mathbf{d}}\right)^\top \mathbf{d} = \mathbf{c}^\top \mathbf{d} \tag{12.3}$$

  Here sensitivity coefficients $\mathbf{c} = (c_i) = (\partial\lambda_r/\partial d_i)$ $(i = 1, \dots, \nu)$ can be obtained by replacing ξ with d_i in (1.35) in Section 1.5.2.

12.2.2 LP formulation

LP formulation of the worst imperfection is presented. Consider an imperfection pattern vector $\mathbf{d}$ of nodal locations, and specify the lower and upper bounds d_i^{L} and d_i^{U}, respectively, for $\mathbf{d} = (d_i)$ along with design codes for nodal location errors. The lower and upper bounds r_i^{L} and r_i^{U} are also specified for linear functions $r_i(\mathbf{d}) = \sum_{j=1}^{\nu} e_{ij} d_j$ of $\mathbf{d}$, such as inclination of members, where e_{ij} are constants.

The anti-optimization problem of the eigenvalue λ_r enjoys an LP formulation

$$\text{LP : minimize} \quad \delta\lambda_r = \mathbf{c}^\top \mathbf{d} \tag{12.4a}$$

$$\text{subject to} \quad d_i^{\mathrm{L}} \le d_i \le d_i^{\mathrm{U}}, \quad (i = 1, \dots, \nu) \tag{12.4b}$$

$$r_i^{\mathrm{L}} \le \sum_{j=1}^{\nu} e_{ij} d_j \le r_i^{\mathrm{U}}, \quad (i = 1, \dots, s) \tag{12.4c}$$

In this LP problem (12.4a)–(12.4c), $\mathbf{d}$ is the only independent variable vector. This problem can be easily solved by using a standard mathematical programming method, e.g., the simplex method and the interior-point method.

The LP formulation is consistent with design codes that specify the bounds on nodal dislocation and member inclination (cf., Fig. 12.4 in Section 12.3.2), and, accordingly, is more practical than the QP formulation presented below.

12.2.3 QP formulation

We consider a QP problem for which the imperfection is normalized by a quadratic norm with the positive-definite weight matrix $\mathbf{H} \in \mathbb{R}^{\nu \times \nu}$. This problem is formulated as

$$\text{QP : minimize} \quad \delta\lambda_r = \mathbf{c}^\top \mathbf{d} \tag{12.5a}$$

$$\text{subject to} \quad \mathbf{d}^\top \mathbf{H} \mathbf{d} = 1 \tag{12.5b}$$

The worst (or optimal) imperfection mode is evaluated from (12.5b) in the similar manner as Section 11.2.2 as

$$\mathbf{d}_r^\perp = \frac{1}{\mu} \mathbf{H}^{-1} \mathbf{c} \tag{12.6}$$

$$\mu = \pm (\mathbf{c}^\top \mathbf{H}^{-1} \mathbf{c})^{\frac{1}{2}} \tag{12.7}$$

Thus, the worst imperfection mode $\mathbf{d}_r^\perp$ by the QP formulation can be found simply by an algebraic manipulation, while the LP formulation does not permit such a simple manipulation.

12.2.4 Dominant worst imperfection

Based on the numerical procedure presented above, we can compute a series of worst imperfection modes $\mathbf{d}_r^\perp$ $(r = 1, \ldots, n^{\mathrm{e}})$ associated with eigenvalues λ_r. Here, the number n^{e} should be specified pertinently. It is in order to consider another anti-optimization problem to depict the *dominant worst imperfection* $\mathbf{d}_{\min}$ that gives the worst-case scenario among $\mathbf{d}_r^\perp$ $(r = 1, \ldots, n^{\mathrm{e}})$. To be concrete, this problem consists of two steps:

Step 1: Compute the critical load $\Lambda_r^{\mathrm{c}\perp}$ for an imperfect system for each of the worst imperfection modes $\mathbf{d}_r^\perp$ $(r = 1, \ldots, n^{\mathrm{e}})$.

Step 2: Depict, among the candidates $\mathbf{d}_r^\perp$ $(r = 1, \ldots, n^{\mathrm{e}})$, the dominant worst imperfection $\mathbf{d}_{\min} = \mathbf{d}_k^\perp$ (for some k) that achieves the minimum critical load $\Lambda_{\min}^{\mathrm{c}}$ among $\Lambda_r^{\mathrm{c}\perp}$ $(r = 1, \ldots, n^{\mathrm{e}})$.

In the following, the minimum critical load and the associated eigenmode are called *buckling load* and *buckling mode*, respectively. Upon the completion of Step 1, Step 2 can be carried out in a straightforward manner. Hence we focus on Step 1 in the sequel.

With the use of the worst imperfection mode $\mathbf{d}_r^\perp$ and corresponding increment $\delta\lambda_r$ of λ_r, the critical load of the associated imperfect system can be estimated as follows:

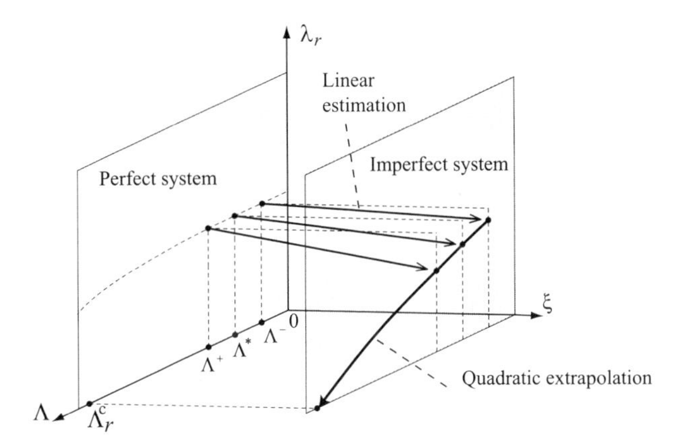

Fig. 12.1 Illustration of quadratic extrapolation to approximate critical load (after Takagi and Ohsaki, 2004; Copyright ©World Scientific Publishing Co., reprinted with permission).

Step 1-i: Linearly estimate the rth eigenvalue λ_r of an imperfect system at a fixed load factor $\Lambda = \Lambda^*$ by (12.2) and (12.3) with $\mathbf{d} = \mathbf{d}_r^{\perp}$.

Step 1-ii: Find the critical load factor $\Lambda_r^{c\perp}$ as the load factor $\Lambda = \Lambda^*$ that satisfies $\lambda_r = 0$.

In Step 1-ii, a trial-and-error search of $\Lambda_r^{c\perp}$ demands a substantial computational cost. In addition, the critical loads $\Lambda_r^{c\perp}$ for several eigenvalues λ_r have to be estimated (cf., Remark 12.2.1).

As a computationally efficient procedure, *quadratic extrapolation* is suggested to be employed. The eigenvalues of the imperfect system at three different load levels Λ^-, Λ^*, Λ^+ ($\Lambda^{\pm} = \Lambda^* \pm \Delta\Lambda$) are calculated, and the critical load with $\lambda_r = 0$ is estimated by extrapolating the relation between Λ and λ_r by a quadratic function as illustrated in Fig. 12.1. Details of Step 1-ii are presented in Appendix A.5.

Remark 12.2.1 It may appear redundant at a glance to consider a number of eigenmodes, as it is customary to choose the buckling mode as the worst mode of initial nodal imperfection [123]. Nevertheless, as demonstrated in the following example, the buckling mode of the imperfect system is not always the same as that of the perfect system, and it is necessary to calculate several possible or all eigenmodes to determine the dominant worst imperfection. □

12.3 Dominant Worst Imperfection of Braced Column Structures

The procedure to obtain the dominant worst imperfection presented in Section 12.2 is applied to braced column structures, which are widely used in civil engineering and architectural engineering.

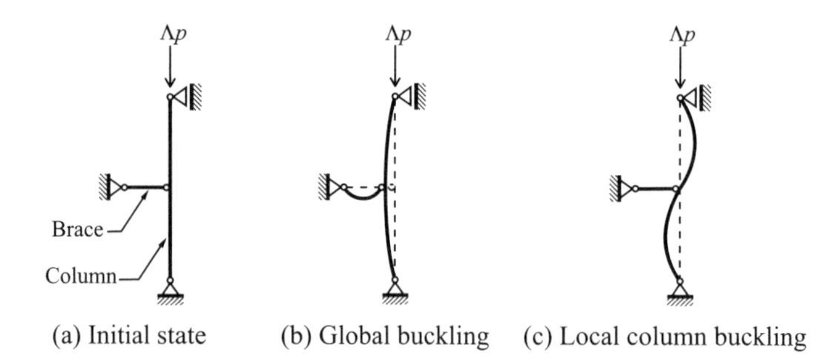

Fig. 12.2 Simple braced column structure (after Takagi and Ohsaki, 2004; Copyright ©World Scientific Publishing Co., reprinted with permission).

12.3.1 Buckling characteristics of braced column

Consider the simple braced column structure as shown in Fig. 12.2(a) subjected to a vertical load at the top of the column. The lateral brace is attached to prevent global buckling mode with a half sine wave of column deformation as shown in Fig. 12.2(b). The brace is assigned with enough stiffness so as to prevent premature brace buckling preceding column buckling.

The buckling mode of the perfect system corresponds to local buckling of the column with a full sine wave, which causes no deformation in the brace as shown in Fig. 12.2(c) [1]. A special feature of this structure is that the buckling strength is drastically reduced by an initial imperfection that triggers brace buckling causing column buckling with a half sine wave as shown in Fig. 12.2(b). Hence, the worst imperfection for this structure should contain brace buckling, and is different from the buckling mode of the perfect system. For this reason, the buckling mode of an imperfect system corresponds to a higher mode of the perfect system. Thus, the buckling behavior of the braced column structure suffers from mode interaction as stated in Section 12.1. The numerical procedure presented in Section 12.2 is employed to overcome this difficulty.

12.3.2 Numerical models

Models 1 and 2 of braced column structures as shown in Fig. 12.3 are used for numerical examples. The columns and the braces are modeled by beam elements; each member is divided into two elements; Green's strain is employed. We set $p = 1.0$.

For the LP formulation, the feasible regions of the nodes including the center node of the member are the squares as shown in Fig. 12.4, and constraints are also given for the inclination of each column element.

[1]The influence of stiffness of braces on the post-buckling behavior of a braced column was investigated [313]. In most existing studies, the braces are modeled by springs that do not buckle [198].

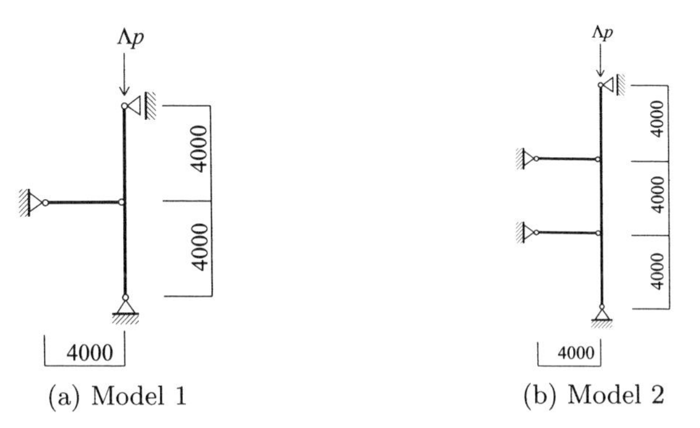

Fig. 12.3 Braced column structures (after Takagi and Ohsaki, 2004; Copyright ©World Scientific Publishing Co., reprinted with permission).

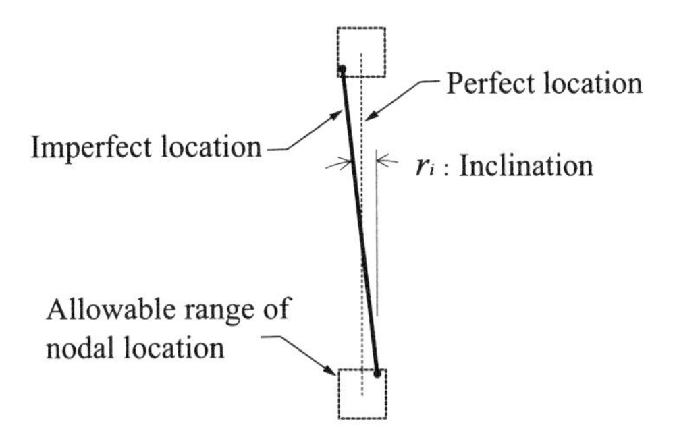

Fig. 12.4 Allowable range of nodal location and the inclination of column element owing to the imperfection of nodal location (after Takagi and Ohsaki, 2004; Copyright ©World Scientific Publishing Co., reprinted with permission).

Young's modulus is $E = 200$ kN/mm^2. The second moment of inertia is 3.60×10^6 mm^4 for the columns, 2.24×10^4 mm^4 and 2.72×10^4 mm^4 for the braces of Models 1 and 2, respectively. The cross-sectional area is 1.20×10^4 mm^2 for the columns, 518.0 mm^2 and 560.0 mm^2 for the braces of Models 1 and 2, respectively. The units of force kN and of length mm are suppressed in the following.

12.3.3 Eigenmodes and worst imperfection modes

In reference to the allowable errors for fabrication and construction, the bounds of imperfections of nodal locations and element inclination are $d^{\rm L} = -4.0$, $d^{\rm U} = 4.0$, $r^{\rm L} = -1/500$ and $r^{\rm U} = 1/500$, where imperfections of support locations are also considered. For the QP formulation, the weight matrix $\mathbf{H}$ for a quadratic norm of imperfection is chosen to be a diagonal matrix $2.25 \times \mathbf{I}_\nu$.

The eigenmodes of $\mathbf{S}$ and the worst imperfection modes by the LP and QP formulations are computed at $\Lambda = \Lambda^* = 98.0$. The perfect structure remains stable at $\Lambda = 98.0$, which is much smaller than the buckling load $\Lambda_1^{\rm c} = 448$ corresponding to the local column buckling mode. The second critical load $\Lambda_2^{\rm c} =$

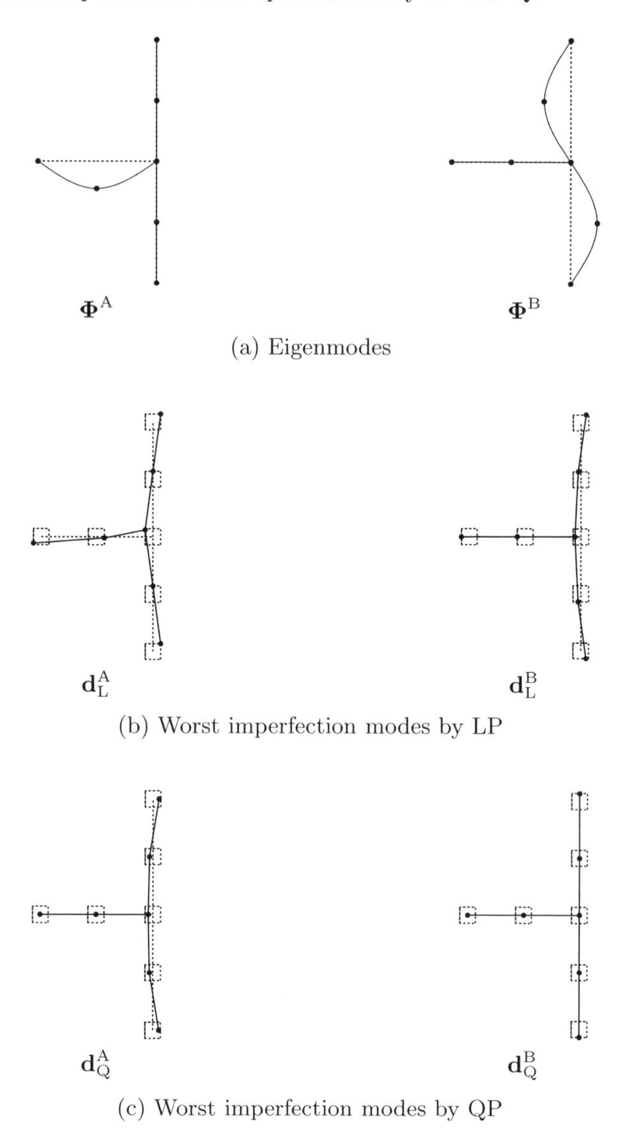

Fig. 12.5 Comparison of eigenmodes and worst imperfection modes for Model 1 (after Takagi and Ohsaki, 2004; Copyright ©World Scientific Publishing Co., reprinted with permission).

930 of the perfect structure corresponding to global buckling amounts to almost the twice of Λ_1^{c}.

For Model 1, eigenmodes $\mathbf{\Phi}^{\mathrm{A}}$ representing global buckling mode with brace buckling and $\mathbf{\Phi}^{\mathrm{B}}$ representing local column buckling mode as shown in Fig. 12.5(a) are considered. For $\mathbf{\Phi}^{\mathrm{A}}$, the brace buckles to cause its axial shortening; the column undergoes a very small half sine wave deformation, although it is not visible in Fig. 12.5(a). The worst imperfection modes $\mathbf{d}_{\mathrm{L}}^{\mathrm{A}}$ and $\mathbf{d}_{\mathrm{L}}^{\mathrm{B}}$

associated with the eigenvalues of $\mathbf{\Phi}^{\rm A}$ and $\mathbf{\Phi}^{\rm B}$, respectively, computed by the LP formulation are shown in Fig. 12.5(b). The dotted squares that are magnified for illustration are 8 mm $\times$ 8 mm boundaries of the nodal locations. The worst imperfection modes $\mathbf{d}_{\rm Q}^{\rm A}$ and $\mathbf{d}_{\rm Q}^{\rm B}$ computed by the QP formulation are shown in Fig. 12.5(c). The following properties for the worst imperfection modes are observed:

- The worst imperfection modes are distinctively different from the corresponding eigenmodes as those modes include minor imperfections. The buckling mode of the perfect structure is local column buckling, while that of an imperfect structure is global buckling. Therefore, the buckling mode of an imperfect structure corresponds to a higher critical mode of the perfect structure (cf., Remark 12.2.1). Thus it is necessary to analyze all or several possible eigenmodes to determine the dominant worst imperfection.

- The worst imperfection modes computed herein have physical necessity as follows. In the LP formulation, the worst mode $\mathbf{d}_{\rm L}^{\rm A}$ associated with $\mathbf{\Phi}^{\rm A}$ of global buckling with brace buckling exerts maximum compression to the brace by the inclination of the column. The worst mode $\mathbf{d}_{\rm L}^{\rm B}$ for $\mathbf{\Phi}^{\rm B}$ of local column buckling maximizes the length of the column as shown in Fig. 12.5(b), and, in turn, to minimize its buckling load. In the QP formulation, the inclination of the column that triggers brace buckling is maximized for $\mathbf{d}_{\rm Q}^{\rm A}$, and the length of the column is maximized for $\mathbf{d}_{\rm Q}^{\rm B}$ as shown in Fig. 12.5(c).

- For the LP formulation, each node is located on the boundary of a dotted square. Some nodes are not located at a corner of the square. It may invite a criticism that the solution of an LP problem, in general, should lie at a vertex of a convex polyhedron of the feasible region. Note that the uniqueness of the worst imperfection by LP is lost as the sensitivity of the eigenvalue with respect to the nodal location imperfection along the edge of the square almost diminishes.

- For the QP formulation, the nodes are smoothly distributed for $\mathbf{d}_{\rm Q}^{\rm A}$ and $\mathbf{d}_{\rm Q}^{\rm B}$ as shown in Fig. 12.5(c) because the nodal imperfections are proportional to the sensitivity coefficient vector $\mathbf{c}$ for the weight matrix $\mathbf{H} = 2.25 \times \mathbf{I}_\nu$ as observed in (12.6).

For Model 2 with two braces, eigenmodes $\mathbf{\Phi}^{\rm A}$, $\mathbf{\Phi}^{\rm B}$, $\mathbf{\Phi}^{\rm C}$ and $\mathbf{\Phi}^{\rm D}$ as shown in Fig. 12.6(a) are considered. The mode $\mathbf{\Phi}^{\rm A}$ is a global buckling mode, and the others are local column buckling modes, where $\mathbf{\Phi}^{\rm D}$ has the shortest wave length. The worst imperfection modes by LP and QP are shown in Figs. 12.6(b) and (c), respectively. One can observe also for Model 2 most of the general properties of the worst imperfections observed for Model 1.

12.3.4 Estimation of buckling loads of imperfect structures

The eigenvalues of $\mathbf{S}$ at $\Lambda = \Lambda^* = 98.0$ of the perfect and imperfect structures of Models 1 and 2 with the worst imperfection obtained by the LP formulation

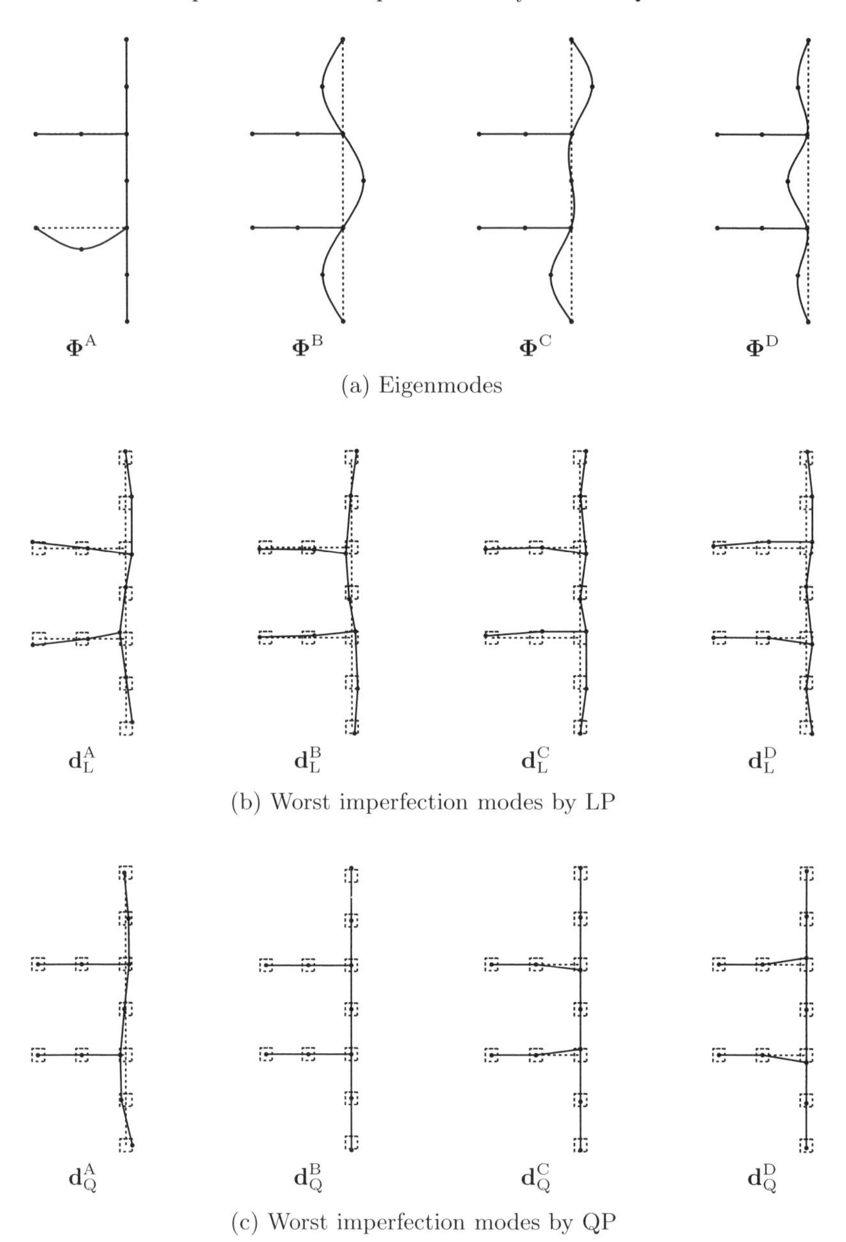

Fig. 12.6 Comparison of eigenmodes and worst imperfection modes for Model 2 (after Takagi and Ohsaki, 2004; Copyright ©World Scientific Publishing Co., reprinted with permission).

are compared in Table 12.1. The eigenvalues λ_r of imperfect structures in Table 12.1 are obtained by the path-tracing of the fundamental path, and, therefore, can be regarded as exact values within numerical errors. $\lambda_r^0 + \delta\lambda_r$ are the eigenvalues of the imperfect structures that are linearly estimated by the method in

Table 12.1 Eigenvalues of perfect and imperfect structures (after Takagi and Ohsaki, 2004; Copyright ©World Scientific Publishing Co., reprinted with permission).

	Model 1		Model 2	
	$\mathbf{\Phi}^{\mathrm{A}}$	$\mathbf{\Phi}^{\mathrm{B}}$	$\mathbf{\Phi}^{\mathrm{A}}$	$\mathbf{\Phi}^{\mathrm{B}}$
λ_r^0	3.4286×10^{-3}	4.3089×10^{-1}	4.0103×10^{-3}	4.3090×10^{-3}
λ_r	2.9488×10^{-3}	4.2936×10^{-1}	3.4906×10^{-3}	4.2988×10^{-3}
$\lambda_r^0 + \delta\lambda_r$	2.9487×10^{-3}	4.2937×10^{-1}	3.4894×10^{-3}	4.2989×10^{-3}

Table 12.2 Comparison of buckling load factors and associated buckling modes (after Takagi and Ohsaki, 2004; Copyright ©World Scientific Publishing Co., reprinted with permission).

		Model 1	Model 2
Perfect structure	Buckling mode	$\mathbf{\Phi}^{\mathrm{B}}$	$\mathbf{\Phi}^{\mathrm{B}}$
	$\Lambda_1^{\mathrm{c}0}$	448	448
Imperfect structure	Buckling mode	$\mathbf{\Phi}^{\mathrm{A}}$	$\mathbf{\Phi}^{\mathrm{A}}$
	$\widehat{\Lambda}_1^{\mathrm{c}}$	343	362
	$\Lambda_{\mathrm{min}}^{\mathrm{c}}$	348	361

Section 12.2.4 using the imperfection sensitivity coefficients for the perfect structures. The good agreement between λ_r and $\lambda_r^0 + \delta\lambda_r$ ensures the validity of the present method. It is also seen from Table 12.1 that the eigenvalue corresponding to the local buckling mode $\mathbf{\Phi}^{\mathrm{B}}$ is insensitive to the nodal imperfections because $\delta\lambda_r \simeq 0$.

In order to assess the validity of the quadratic extrapolation procedure of Λ_r^{c} in Section 12.2.4 and Appendix A.5, the worst imperfections are computed by the LP formulation with $d^{\mathrm{L}} = -8.0$, $d^{\mathrm{U}} = 8.0$, $r^{\mathrm{L}} = -1/250$, and $r^{\mathrm{U}} = 1/250$. Table 12.2 lists the values of the buckling load factors of perfect and imperfect structures as

- $\Lambda_1^{\mathrm{c}0}$: the buckling load of the perfect structure,
- $\widehat{\Lambda}_1^{\mathrm{c}}$: the buckling load of an imperfect structure found by path-tracing analysis, which is regarded herein as the exact value, and
- $\Lambda_{\mathrm{min}}^{\mathrm{c}}$: the buckling load of an imperfect structure computed by the present procedure.

The buckling load $\Lambda_{\mathrm{min}}^{\mathrm{c}}$ is estimated with $\Delta\Lambda = 1.0$ and $\Lambda^* = 98.0$; note that these values are not much influential on the results.

It is seen from Table 12.2 that the differences between $\widehat{\Lambda}_1^{\mathrm{c}}$ and $\Lambda_{\mathrm{min}}^{\mathrm{c}}$ are sufficiently small. Thus the buckling loads of the imperfect structures with the worst imperfection can be estimated accurately by the proposed method to demonstrate its usefulness.

12.4 Summary

In this chapter,

- LP and QP formulations of an anti-optimization problem have been presented to compute the worst imperfection for each eigenvalue of the tangent stiffness matrix for a fixed load factor,
- a computationally efficient procedure to estimate the lowest critical load associated with the worst imperfection has been presented, and
- the dominant worst imperfection corresponding to the lowest critical load of an imperfect braced column structure has been successfully computed.

The major findings of this chapter are as follows.

- The critical loads of imperfect systems with the worst imperfections can be efficiently and accurately calculated by extrapolating the linearly estimated eigenvalues of the tangent stiffness matrix at different load levels.
- It has been observed in the examples that by virtue of mode interaction, the second or higher eigenmode of the perfect system, in some cases, should be used for evaluating the worst imperfections. It is, therefore, insufficient to investigate only the lowest eigenmode of the perfect system to determine the buckling mode of the imperfect systems.

13 Worst Imperfection for Stable Bifurcation

13.1 Introduction

The procedures to determine the worst mode of imperfection presented in Chapters 11 and 12 are pertinent design methodologies for structures undergoing unstable bifurcation such as cylindrical shells and stiffened plates. However, these methodologies malfunction for a stable-symmetric bifurcation point, because this point on the fundamental path disappears in the presence of a major imperfection.

Let us observe the equilibrium paths for major and minor imperfections illustrated in Fig. 13.1. For a major imperfection, Λ increases above the bifurcation load factor $\Lambda^{\rm c}$ of the perfect system (cf., Fig. 13.1(a)). For a minor imperfection, an imperfect system retains a bifurcation point, and the bifurcation load factor may increase or decrease depending on the sign of the imperfection parameter (cf., Fig. 13.1(b)).

A question arises whether it is safe to allow loads above the bifurcation load $\Lambda^{\rm c}$. From a sole standpoint of stability, it seems possible to allow the loading along the bifurcation path exceeding the bifurcation load [246], and to determine the maximum load factor by constraints on displacements and/or stresses. For example, optimal design that has a stable-symmetric bifurcation point can be found by constraining the fourth-order differential coefficient $V_{,1111}$ to be nonnegative ($V_{,1111} \geq 0$) [31, 32, 97, 246].

Although a critical point does not exist for a structure with a major imperfection, the structure may undergo sudden dynamic large antisymmetric deformation near the stable bifurcation point as the load factor is further increased. The occurrence of such sudden deformation becomes abrupt if the major imperfection is extremely small.

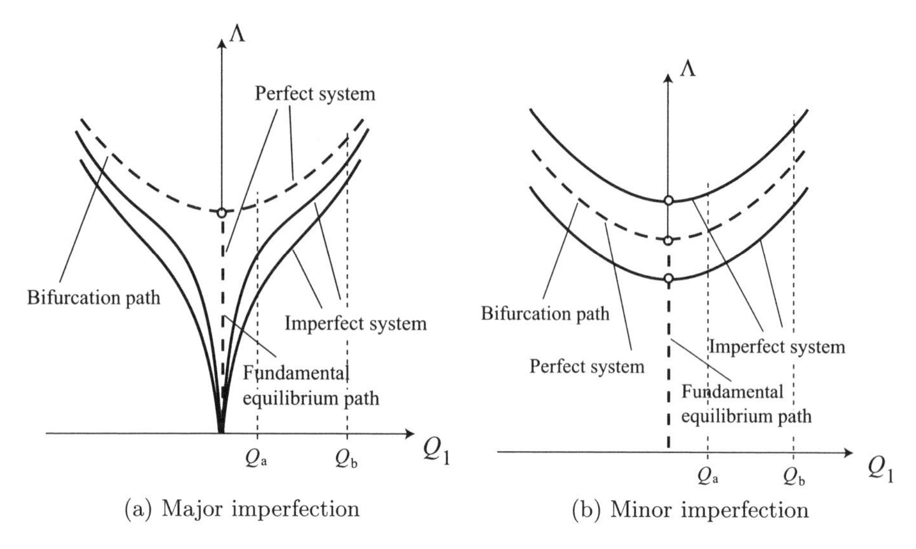

Fig. 13.1 Equilibrium paths of perfect and imperfect systems. ∘: bifurcation point; Q_1: generalized displacement in the direction of bifurcation mode; solid curve: perfect equilibrium path; dashed curve: imperfect equilibrium path.

In this chapter, a simple and numerically efficient procedure is presented for determining the maximum load factor of an imperfect elastic structure undergoing stable bifurcation for minor imperfections of nodal locations. We consider a flexible structure allowing moderately large deformation. An anti-optimization problem is formulated so as to minimize the bifurcation load factor within the convex bounds on imperfection parameters. The method called *simultaneous analysis and design* (SAND) [104, 315] is used, for which imperfections of nodal loads are introduced, and the displacements are also considered as independent variables to avoid costly nonlinear path-tracing analysis. It is shown for a plane column-type truss that the worst mode of imperfection, which turns out to be a minor imperfection, can be successfully obtained by the present approach.

This chapter is organized as follows. A procedure for determining the maximum load factor is introduced in Section 13.2. An anti-optimization problem is formulated in Section 13.3. The worst imperfections of column-type trusses are studied in Section 13.4.

13.2 Maximum Load Factor for Stable Bifurcation

The maximum load factor of a structure exhibiting stable bifurcation may be defined by either bifurcation load factor Λ^{c}, or the load factor Λ^{M} to be defined in accordance with the specified bounds on stresses and/or displacements.

The worst imperfection to be obtained changes according to whether Λ^{c} or Λ^{M} is employed. First, if the bifurcation load factor Λ^{c} is used, the worst imperfection should be a minor imperfection because the bifurcation point exists for a minor imperfection but disappears for a major imperfection.

Next we use the maximum load factor $\Lambda^{\rm M}$ that is defined by the upper-bound constraint $Q_1 \leq Q_1^{\rm U}$ of the generalized displacement Q_1 in the direction of the bifurcation mode. As illustrated in Fig. 13.1, the reduction of $\Lambda^{\rm M}$ due to a major imperfection is very large for a relatively strict constraint $Q_1^{\rm U} = Q_{\rm a}$, but becomes smaller as $Q_1^{\rm U}$ is relaxed to, e.g., $Q_1^{\rm U} = Q_{\rm b}$.

For a minor imperfection, as can be observed from Fig. 13.1(b), the amount of reduction is not sensitive to $Q_1^{\rm U}$, and the sensitivity of the maximum load to a minor imperfection is almost equivalent to that of the bifurcation load. The reduction is larger than that for a major imperfection if $Q_1^{\rm U}$ is moderately large, e.g., $Q_1^{\rm U} = Q_{\rm b}$. Therefore, the worst mode of imperfection for $\Lambda^{\rm M}$ corresponds to the major imperfection for small $Q_1^{\rm U}$, and to the minor imperfection for large $Q_1^{\rm U}$.

Thus, for a flexible structure allowing moderately large deformation, a minor imperfection plays a key role in defining both load factors $\Lambda^{\rm c}$ and $\Lambda^{\rm M}$, and the maximum load defined by deformation constraints may be dramatically reduced by minor imperfections rather than by major imperfections. For this reason, we consider worst minor imperfections of the bifurcation load in the remainder of this chapter.

13.3 Anti-Optimization Problem

An anti-optimization problem for minimizing the bifurcation load factor $\Lambda^{\rm c}$ against a minor imperfection is formulated.

Consider an imperfection pattern vector $\mathbf{d} \in \mathbb{R}^{\nu}$ for, e.g., nodal locations and cross-sectional areas. The norm of $\mathbf{d}$ is denoted by $\|\mathbf{d}\|_H^2 = \mathbf{d}^\top \mathbf{H} \mathbf{d}$ for a positive-definite weight matrix $\mathbf{H}$, and $\|\mathbf{d}\|_H^2$ is a convex function of $\mathbf{d}$.

13.3.1 Direct formulation

Specify an upper bound $\overline{\|\mathbf{d}\|_H^2}$ for $\|\mathbf{d}\|_H^2$. Decompose the imperfection mode $\mathbf{d}$ into major imperfection $\mathbf{d}^+$ and minor imperfection $\mathbf{d}^-$. The maximum load of an imperfect system considering reduction by the worst mode of a minor imperfection is defined as the solution of the following anti-optimization problem:

$$\text{AP1: minimize} \quad \Lambda^{\rm c}(\mathbf{d}^-) \tag{13.1a}$$

$$\text{subject to} \quad \|\mathbf{d}^-\|_H^2 \leq \overline{\|\mathbf{d}^-\|_H^2} \tag{13.1b}$$

Problem AP1 may be solved by using an appropriate gradient-based optimization algorithm if sensitivity coefficients of $\Lambda^{\rm c}$ can be computed. However, AP1 is computationally expensive, because $\Lambda^{\rm c}$ for a given $\mathbf{d}^-$ should be determined by path-tracing analysis at each iterative step of optimization.

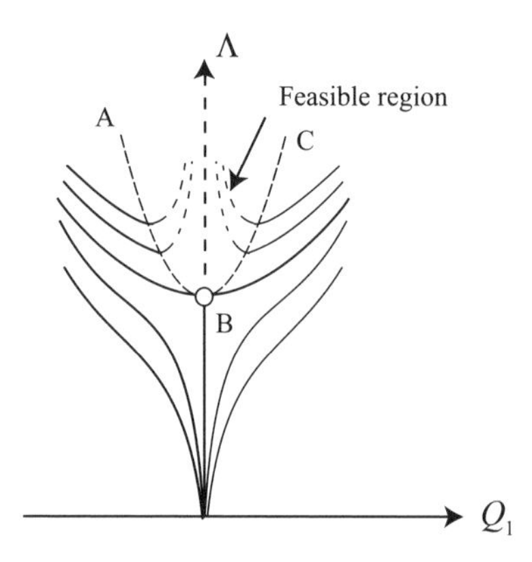

Fig. 13.2 Feasible region for the eigenvalue constraint $\lambda_1 \leq 0$. Solid curve: stable; dashed curve: unstable; ∘: stable bifurcation point.

13.3.2 Numerically efficient formulation

A numerically efficient formulation to find $\Lambda^{\rm c}$ as the minimum value of Λ under constraint $\lambda_1 \leq 0$ is presented.

If we fix $\mathbf{d}^-$ and only consider a major imperfection $\mathbf{d}^+$, the feasible region in the (Q_1, Λ)-space satisfying $\lambda_1 \leq 0$ lies above the curve ABC in Fig. 13.2. For stable bifurcation under consideration, the feasible region in the vicinity of the bifurcation point is convex in the (Q_1, Λ)-space. Hence, the bifurcation load of the imperfect system for a minor imperfection is found by the two steps of minimization:

Step 1: Minimize Λ with respect to $\mathbf{d}^+$ under constraint of $\lambda_1 \leq 0$ to find the bifurcation load $\Lambda^{\rm c}$.

Step 2: Minimize $\Lambda = \Lambda^{\rm c}$ with respect to $\mathbf{d}^-$ to obtain the worst imperfection.

Since both steps correspond to minimization of Λ, they can be carried out by considering major and minor imperfections simultaneously as variables.

The method called *simultaneous analysis and design*, which is abbreviated as SAND, is very efficient for reducing the required number of path-tracing analyses that are to be carried out at each step of optimization or anti-optimization of geometrically nonlinear structures. The state variable vector $\mathbf{U}$, as well as the design variables, are considered as independent variables [104, 315].

To utilize SAND, we consider the nodal displacement vector $\mathbf{U}$ as independent variables in addition to $\mathbf{d}$. In the conventional formulation of SAND, $\mathbf{U}$ is modified to satisfy the equilibrium equations, or to minimize the total potential energy. In order to make it easier to solve the optimization problem, AP1 in (13.1a) and (13.1b) is relaxed by permitting imperfections in nodal loads, because in our problem for obtaining the worst imperfection it is not very important to employ exactly *perfect* nodal loads. The ranges of the nodal loads Λp_i are given as

$$\Lambda p_i^{\rm L} \leq \Lambda p_i \leq \Lambda p_i^{\rm U}, \quad (i = 1, \dots, n) \tag{13.2}$$

where $p_i^{\rm L}$ and $p_i^{\rm U}$ are the specified lower and upper bounds, respectively, and n is the number of degrees of freedom.

Suppose the case where the errors in the loads are bounded by $\pm\Lambda\Delta p$, and rewrite (13.2) into

$$\Lambda(p_i^0 - \Delta p) \leq \Lambda p_i \leq \Lambda(p_i^0 + \Delta p), \quad (i = 1, \dots, n) \tag{13.3}$$

where p_i^0 is the perfect value of p_i.

The equivalent nodal load $R_i(\mathbf{U}, \mathbf{d})$ $(i = 1, \dots, n)$ in the direction of U_i of an imperfect system is then calculated for current values of $\mathbf{U}$ and $\mathbf{d}$ during anti-optimization. Recall that R_i is the derivative of the strain energy for the proportional loading (cf., (1.11) with $R_i = H_{,i}$ in Section 1.2).

The anti-optimization problem for finding the minimum $\Lambda^{\rm min}$ of Λ under constraints on the norms of imperfections and the sign of the lowest eigenvalue $\lambda_1(\mathbf{U}, \mathbf{d})$ of the tangent stiffness matrix is formulated as

$$\text{AP2: minimize} \quad \Lambda \tag{13.4a}$$

$$\text{subject to} \quad \|\mathbf{d}\|_H^2 \leq \overline{\|\mathbf{d}\|_H^2} \tag{13.4b}$$

$$\Lambda(p_i^0 - \Delta p) \leq R_i(\mathbf{U}, \mathbf{d}) \leq \Lambda(p_i^0 + \Delta p), \quad (i = 1, \dots, n) \tag{13.4c}$$

$$\lambda_1(\mathbf{U}, \mathbf{d}) \leq 0 \tag{13.4d}$$

$$U_i^{\rm L} \leq U_i \leq U_i^{\rm U}, \quad (i = 1, \dots, n) \tag{13.4e}$$

The variables of this problem are $\mathbf{U}, \mathbf{d}$ and Λ, and the upper- and lower-bound constraints (13.4e) are assigned for $\mathbf{U}$ to improve convergence. Only $R_i(\mathbf{U}, \mathbf{d})$ $(i = 1, \dots, n)$ and $\lambda_1(\mathbf{U}, \mathbf{d})$ are to be computed for the current values of $\mathbf{U}$ and $\mathbf{d}$ at each iterative step of optimization without resort to costly path-tracing analysis.

13.4 Worst Imperfection of Column-Type Trusses

Consider the plane column-type truss as shown in Fig. 13.3. The two springs attached at nodes 7 and 8 have the same extensional stiffness K. We consider two cases

- *column-type truss* with $K = 0$, and
- *laterally supported truss* with $K \neq 0$.

The lengths of x- and y-directional members are 1000 mm and 2000 mm, respectively. All the truss members have the same cross-sectional area of 200.0 mm^2. The proportional loads Λp in the negative y-direction are applied at nodes 7 and 8, where $p = 98$ kN. Young's modulus is $E = 205.8$ kN/mm^2. The axial strain is defined by Green's strain. The units of force kN and of length mm are suppressed in the following.

Optimization problem AP2 is solved by IDESIGN Ver. 3.5 [14], in which the SQP (cf., Section 4.3.2) is used, and the gradients of the objective and constraint functions are computed by the finite difference approach.

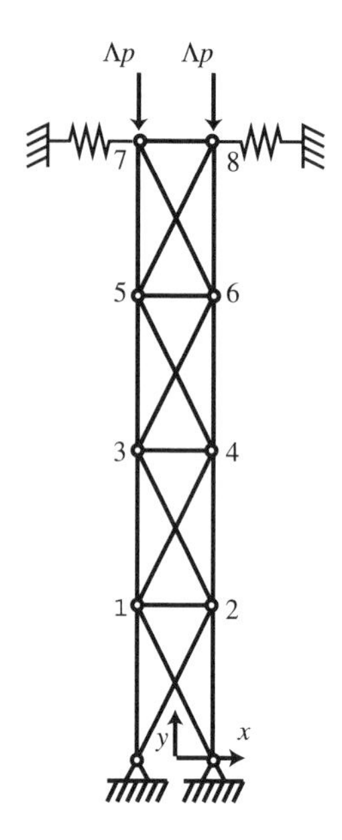

Fig. 13.3 Column-type plane truss.

The components of the imperfection pattern vector consist of the coordinates of all the nodes except for the two supports. Therefore, the size of $\mathbf{d}$ is equal to n (= 16). The weight matrix is given as $\mathbf{H} = (1/n^2)\mathbf{I}_n$, where $\mathbf{I}_n \in \mathbb{R}^{n\times n}$ is the identity matrix. The upper bound for the error in the nodal loads is $\Delta p = 0.98$. The total number of variables $\mathbf{d}, \mathbf{U}$ and Λ in AP2 is 33. The upper and lower bounds, which are given as $U_i^{\mathrm{L}} = -3000$ and $U_i^{\mathrm{U}} = 3000$ for U_i, are inactive for the following anti-optimal solutions.

Let $\mathbf{\Phi}^+$ and $\mathbf{\Phi}^-$ denote the lowest antisymmetric and symmetric linear buckling modes of the perfect system, respectively. Imperfection sensitivity is investigated for the major imperfections in the directions of $\mathbf{\Phi}^+$ and the minor ones in the direction of $\mathbf{\Phi}^-$, in comparison with the worst imperfection. Since the prebuckling deformation is not very large for the perfect column-type trusses, the eigenmode associated with the null eigenvalue of the tangent stiffness matrix at the critical point can be approximated by a linear buckling mode.

13.4.1 Column-type truss

Consider the column-type truss ($K = 0$). The critical load factor of the perfect system is $\Lambda^{c0} = 3.9366$ at a stable-symmetric bifurcation point with a bifurcation mode that is antisymmetric with respect to the y-axis.

We first investigate imperfection sensitivity of the maximum load factor in the directions of $\mathbf{\Phi}^+$ and $\mathbf{\Phi}^-$ shown in Figs. 13.4(a) and (b), respectively. Let U_x denote the x-directional displacement of node 8, and Fig. 13.5 shows equilibrium paths for the perfect and imperfect systems with major imperfections in the direction of $\mathbf{\Phi}^+$ with $\|\mathbf{d}\|_H^2 = 10^2$ and 50^2. As is seen, Λ increases very slightly

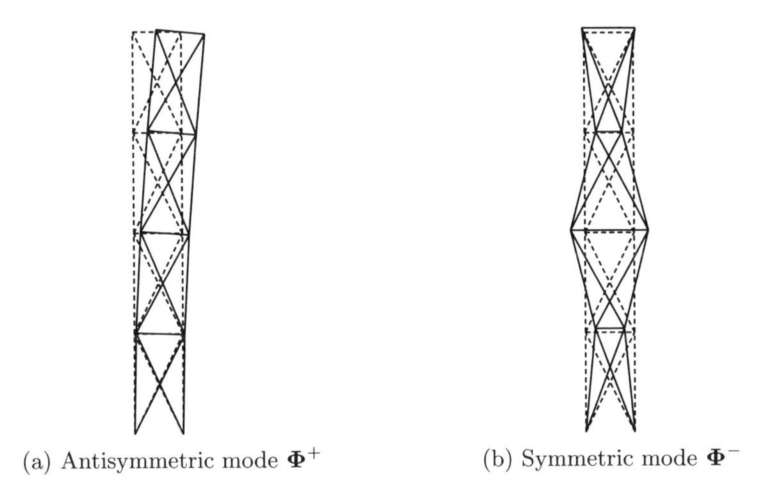

(a) Antisymmetric mode $\mathbf{\Phi}^+$ (b) Symmetric mode $\mathbf{\Phi}^-$

Fig. 13.4 Lowest symmetric and antisymmetric linear buckling modes of the column-type truss.

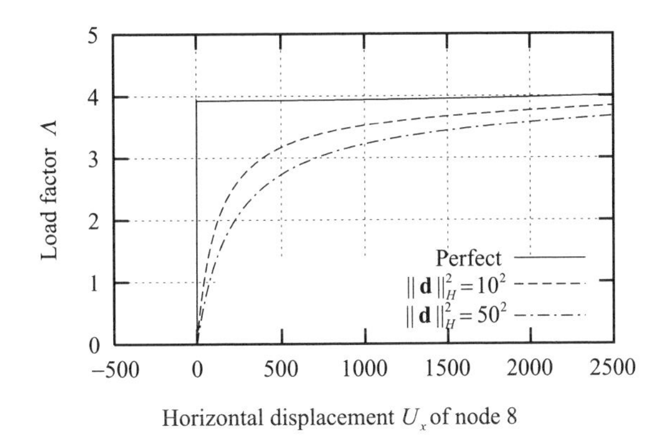

Fig. 13.5 Equilibrium paths for perfect and imperfect column-type trusses with major imperfections in the direction of $\mathbf{\Phi}^+$.

along the bifurcation path for the perfect system. Fig. 13.6 shows the paths for minor imperfections corresponding to $\mathbf{\Phi}^-$ with $\|\mathbf{d}\|_H^2 = 10^2$ and 50^2.

Define the maximum load factor Λ^{M} by the displacement constraint $U_x \leq \overline{U}_x$. It may be observed from Figs. 13.5 and 13.6 that the reduction of Λ^{M} due to a major imperfection is larger than that to a minor imperfection for small $\overline{U}_x$, but a minor imperfection is more influential than a major imperfection for sufficiently large $\overline{U}_x$. For instance, for $\|\mathbf{d}\|_H^2 = 50^2$, the reduction for a minor imperfection is larger than that for a major imperfection in the range $U_x > 1790$. It is to be emphasized, for minor imperfections, that the amount of reduction of Λ^{M} does not strongly depend on the value of $\overline{U}_x$. Therefore, the worst minor imperfection mode for Λ^{M} can be successfully obtained by solving AP2 that minimizes the bifurcation load.

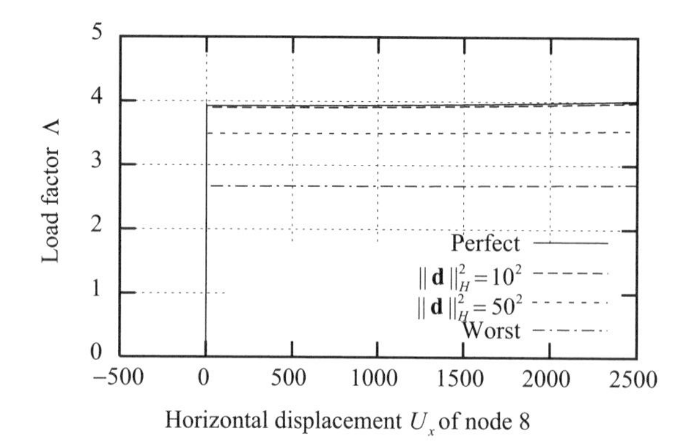

Fig. 13.6 Equilibrium paths for perfect and imperfect column-type trusses with minor imperfections in the direction of $\mathbf{\Phi}^-$.

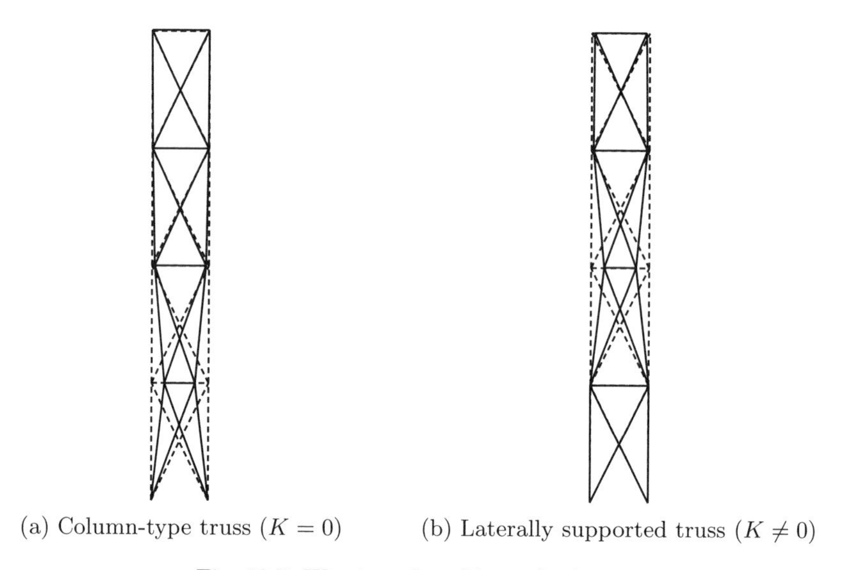

(a) Column-type truss ($K = 0$) (b) Laterally supported truss ($K \neq 0$)

Fig. 13.7 Worst modes of imperfection.

The minimum of Λ of AP2 for $\|\mathbf{d}\|_H^2 \leq \overline{\|\mathbf{d}\|_H^2} = 50^2$ is $\Lambda^{\min} = 2.6693$, which is about 68% of $\Lambda^{c0} = 3.9366$ of the perfect system. The worst mode of nodal imperfection $\mathbf{d}^\perp$ is as shown in Fig. 13.7(a), which is symmetric with respect to the y-axis and corresponds to a minor imperfection. The worst imperfections of nodal locations and nodal loads are also listed in Table 13.1; all the components of $\Delta\mathbf{p} = \mathbf{p} - \mathbf{p}^0$ are equal to the upper or lower bound. The equilibrium path for the worst imperfection is plotted in Fig. 13.6 by the dotted-dashed line.

Note that Λ^c of the imperfect system corresponding to $\|\mathbf{d}\|_H^2 = 50^2$ in the direction of $\mathbf{\Phi}^-$ is 3.4747, which is much larger than that for the worst imperfection $\mathbf{d}^\perp$. Therefore, $\mathbf{\Phi}^-$ cannot be used to approximate $\mathbf{d}^\perp$.

Table 13.1 Worst imperfections of nodal locations and nodal loads.

Node	Direction	Location	Load ($\Delta p_i/\lvert p\rvert$)
1	x	137.81	0.01
	y	−2.2715	−0.01
2	x	−137.83	−0.01
	y	−2.2766	−0.01
3	x	29.410	0.01
	y	2.2275	−0.01
4	x	−29.375	−0.01
	y	2.2211	−0.01
5	x	6.8235	0.01
	y	0.13910	−0.01
6	x	−6.8769	−0.01
	y	0.11772	−0.01
7	x	−0.49643	0.01
	y	9.1936	−0.01
8	x	0.49153	−0.01
	y	9.1907	−0.01

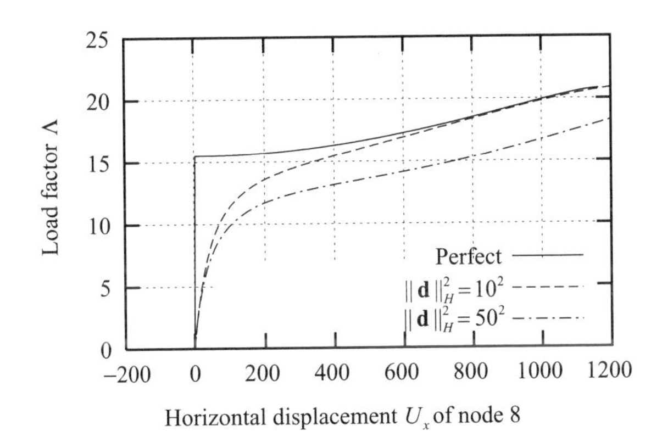

Fig. 13.8 Equilibrium paths for perfect and imperfect laterally supported trusses with major imperfections in the direction of $\mathbf{\Phi}^+$.

13.4.2 Laterally supported truss

Consider next a laterally supported truss with $K = 0.1029$. The ratio of the extensional stiffness of the spring to that of the horizontal truss member is 0.005. The critical load factor of the perfect system is $\Lambda^{c0} = 15.497$.

Fig. 13.8 shows equilibrium paths for major imperfections corresponding to $\mathbf{\Phi}^+$ with $\|\mathbf{d}\|_H^2 = 10^2$ and 50^2. As is seen, the critical point of the perfect system is a stable-symmetric bifurcation point, and the critical loads Λ^c of imperfect systems are far above the bifurcation load. Note that, under displacement constraint of a moderately large upper bound, the reduction of the maximum load Λ^M of the

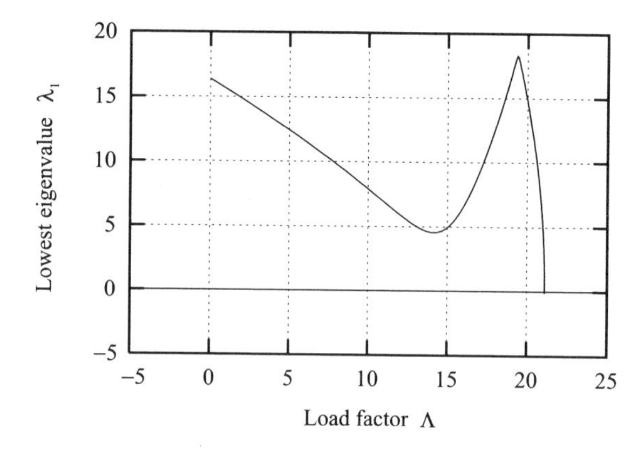

Fig. 13.9 Relation between load factor Λ and the lowest eigenvalue λ_1 with $\|\mathbf{d}\|_H^2 = 10^2$.

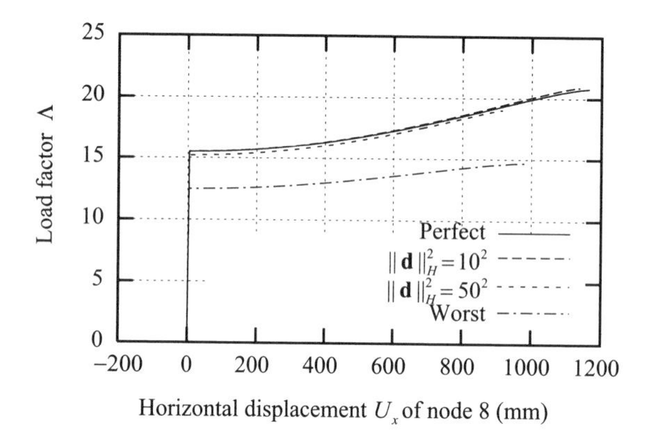

Fig. 13.10 Equilibrium paths for perfect and imperfect laterally supported trusses with minor imperfections in the direction of $\mathbf{\Phi}^-$.

imperfect system from that of the perfect system is very small for $\|\mathbf{d}\|_H^2 = 10^2$. Variation of λ_1 with respect to Λ for $\|\mathbf{d}\|_H^2 = 10^2$ is as shown in Fig. 13.9. The lowest eigenvalue has a local minimum at $\Lambda \simeq 14.5$ near the bifurcation point of the perfect system, but does not reach 0.

Fig. 13.10 shows equilibrium paths for minor imperfections in the direction of $\mathbf{\Phi}^-$ with $\|\mathbf{d}\|_H^2 = 10^2$ and 50^2. Note that the bifurcation load factor for $\|\mathbf{d}\|_H^2 = 50^2$ is 14.107, where imperfections in nodal loads in the direction of $\mathbf{\Phi}^-$ are also considered.

The load factor of the worst imperfection obtained by solving AP2 for $\|\mathbf{d}\|_H^2 \leq \overline{\|\mathbf{d}\|_H^2} = 50^2$ is $\Lambda^{\min} = 12.483$ which is about 81% of $\Lambda^{\mathrm{c}} = 15.497$ of the perfect system. The equilibrium path for the worst imperfection $\mathbf{d}^{\perp}$, which has reflection symmetry as shown in Fig. 13.7(b), is also plotted in Fig. 13.10. In this case the reduction of Λ^{M} due to the worst imperfection is much larger than that to the imperfection with the same norm in the direction of $\mathbf{\Phi}^-$. Note from Fig. 13.7 that $\mathbf{d}^\perp$ strongly depends on the presence of the extensional stiffness of the spring.

13.5 Summary

In this chapter,

- the maximum load factor for stable bifurcation based on displacement constraint has been defined,
- an anti-optimization problem for minimizing the critical load factor has been formulated, and
- the worst imperfection of a column-type plane truss has been studied.

The major findings of this chapter are as follows.

- The critical point of an imperfect system disappears if the perfect system has a stable-symmetric bifurcation point. In this case, the maximum load may be defined in reference to displacements and stresses of imperfect systems with a specified norm of the worst imperfection.
- The worst minor imperfection can be successfully obtained as a solution of an anti-optimization problem under a constraint on the lowest eigenvalue of the tangent stiffness matrix. The problem is relaxed incorporating imperfections of nodal loads and is solved by employing nodal displacements as independent variables without resort to costly path-tracing analysis at each step of anti-optimization.
- For a flexible structure allowing moderately large displacements, the antisymmetric buckling mode is not always the worst mode of imperfection and that a minor imperfection is very important for estimating the reduction of the maximum load factor defined by displacement constraints.

14

Random Imperfections: Theory

14.1 Introduction

A series of pertinent and powerful design methodologies that are introduced up to this chapter enable us to evaluate structural performances in a strategic manner. These methodologies, however, are all deterministic and limited to a given mode of initial imperfection chosen among infinite number of possible ones. To make the study of initial imperfections realistic, it is desirable to deal with the probabilistic variations of imperfections.

Extensive studies have been devoted to describing the probabilistic variation of critical loads, as summarized in the historical development in Section 14.5. In the description of the probabilistic variation of critical loads, it is robust but costly to obtain directly buckling loads of a large number of imperfect structures with initial imperfections subject to probabilistic variations.

In this chapter, an efficient methodology to describe the probabilistic variation of critical loads is presented. In order to realize drastic simplification, critical loads are expressed asymptotically by the imperfection sensitivity laws. For Gaussian initial imperfections, the closed forms of the probability density functions of buckling loads are derived to achieve a drastic simplification.

This chapter is organized as follows. Probability density functions of critical loads are derived based on an asymptotic theory in Section 14.2. The numerical procedure to obtain the probability density function is presented in Section 14.3. The probabilistic strength variations of a double-layer hexagonal truss roof and a spherical truss dome are investigated in Section 14.4. The historical development is reviewed in Section 14.5.

14.2 Probability Density Functions of Critical Loads

An asymptotic theory for random initial imperfections can be developed, as a natural extension of the results in the previous chapters, under the assumption that the initial imperfections are normally distributed [139–141, 206].

The imperfection sensitivity laws for simple critical points for major imperfections are expressed as (cf., (3.48) and (11.4))

$$\widetilde{\Lambda}^{\mathrm{c}}(\xi) = \begin{cases} C_0 V_{,1\xi}\xi : & \text{for limit point} \\ C_1 |V_{,1\xi}\xi|^{\frac{1}{2}} : & \text{for asymmetric bifurcation point} \\ & \qquad (V_{,111}V_{,1\xi}\xi > 0) \\ C_3 (V_{,1\xi}\xi)^{\frac{2}{3}} : & \text{for unstable-symmetric bifurcation point} \end{cases} \tag{14.1}$$

where $\widetilde{\Lambda}^{\mathrm{c}}(\xi) = \Lambda^{\mathrm{c}}(\xi) - \Lambda^{\mathrm{c}0}$. The variable $V_{,1\xi}$ in (14.1) depends on $\mathbf{d}$ through one variable

$$V_{,1\xi} = \mathbf{\Phi}_1^{\mathrm{c}0\top}\mathbf{B}^{\mathrm{c}0}\mathbf{d} = \sum_{i=1}^{\nu} a_i d_i = \mathbf{a}^\top \mathbf{d} \tag{14.2}$$

by (3.26) and (11.8), where $\mathbf{\Phi}_1^{\mathrm{c}0}$ is the critical mode and

$$\mathbf{a} = (a_1, \ldots, a_\nu)^\top = (\mathbf{\Phi}_1^{\mathrm{c}0\top}\mathbf{B}^{\mathrm{c}0})^\top \tag{14.3}$$

is called *major imperfection influence vector.*

Given the joint probability density function of d_i ($i = 1, \ldots, \nu$), the probability density function of $V_{,1\xi}$ can be calculated from (14.2). Then a simple transformation from $V_{,1\xi}$ to the critical load Λ^{c}, through (14.1), yields the probability density function of Λ^{c}, as we will see below.

We shall investigate the behavior of Λ^{c} when the initial imperfection $\mathbf{v} - \mathbf{v}^0 = \xi\mathbf{d}$ is subject to a normal distribution[1] $\mathrm{N}(\mathbf{0}, \xi^2\mathbf{W})$ with mean $\mathbf{0}$ and variance–covariance matrix $\xi^2\mathbf{W}$, where $\mathbf{W}$ is a given symmetric positive-definite matrix. Then the variable $V_{,1\xi}$ is subject to a normal distribution $\mathrm{N}(0, \widetilde{\sigma}^2)$ with mean 0 and variance

$$\widetilde{\sigma}^2 = \mathbf{\Phi}_1^{\mathrm{c}0\top}\mathbf{B}^{\mathrm{c}0}\mathbf{W}\mathbf{B}^{\mathrm{c}0\top}\mathbf{\Phi}_1^{\mathrm{c}0} \tag{14.4}$$

By transforming the probability density function $\phi_{V_{,1\xi}}$ of $V_{,1\xi}$, which is a normal distribution, we obtain the probability density functions $\phi_{\widetilde{\Lambda}^{\mathrm{c}}}$ of the critical load $\widetilde{\Lambda}^{\mathrm{c}}$:

[1] In this book, we focus on the case of normal distribution, while the case of uniform distribution is to be consulted with Ikeda and Murota [139].

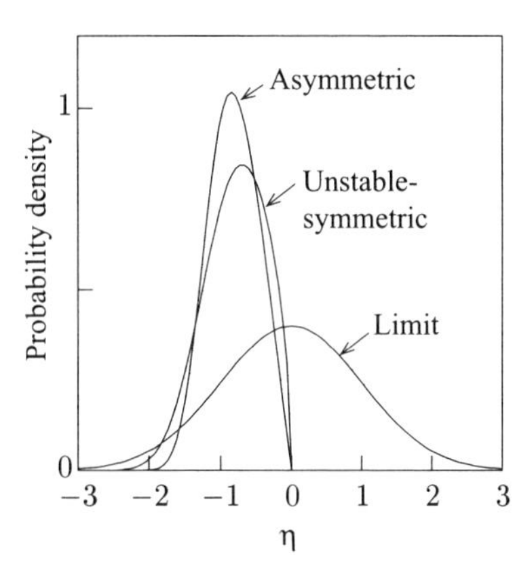

Fig. 14.1 Probability density function $\phi_\eta(\eta)$ for normalized critical load increment $\eta = \widetilde{\Lambda}^{\rm c}/\widehat{C}$.

$$
\phi_{\widetilde{\Lambda}^{\rm c}} = \phi_{V_{,1\xi}} \frac{{\rm d}V_{,1\xi}}{{\rm d}\widetilde{\Lambda}^{\rm c}}
= \begin{cases}
\dfrac{1}{\sqrt{2\pi}\widehat{C}} \exp\Big[-\dfrac{1}{2}\Big(\dfrac{\widetilde{\Lambda}^{\rm c}}{\widehat{C}}\Big)^2\Big], \quad (-\infty < \widetilde{\Lambda}^{\rm c} < \infty): \\
\qquad\qquad\qquad\qquad \text{for limit point} \\
\dfrac{4|\widetilde{\Lambda}^{\rm c}|}{\sqrt{2\pi}\widehat{C}^2} \exp\Big[-\dfrac{1}{2}\Big(\dfrac{\widetilde{\Lambda}^{\rm c}}{\widehat{C}}\Big)^4\Big], \quad (-\infty < \widetilde{\Lambda}^{\rm c} < 0): \\
\qquad\qquad\qquad\qquad \text{for asymmetric bifurcation point} \\
\dfrac{3|\widetilde{\Lambda}^{\rm c}|^{\frac{1}{2}}}{\sqrt{2\pi}\widehat{C}^{\frac{3}{2}}} \exp\Big[-\dfrac{1}{2}\Big|\dfrac{\widetilde{\Lambda}^{\rm c}}{\widehat{C}}\Big|^3\Big], \quad (-\infty < \widetilde{\Lambda}^{\rm c} < 0): \\
\qquad\qquad\qquad\qquad \text{for unstable symmetric bifurcation point}
\end{cases}
\tag{14.5}
$$

where

$$
\widehat{C} = \begin{cases}
|C_0\widetilde{\sigma}\xi| : & \text{for limit point} \\
|C_1||\widetilde{\sigma}\xi|^{\frac{1}{2}} : & \text{for asymmetric bifurcation point} \\
|C_3|(\widetilde{\sigma}\xi)^{\frac{2}{3}} : & \text{for unstable-symmetric bifurcation point}
\end{cases}
\tag{14.6}
$$

The probability density functions $\phi_\eta(\eta)$ of the three simple critical points are illustrated in Fig. 14.1, where $\eta = \widetilde{\Lambda}^{\rm c}/\widehat{C}$ is a normalized critical load increment.

14.3 Numerical Procedure

As we have seen in Section 14.2, the probability density function of critical loads permits a simple formulation via the imperfection sensitivity law (14.1). A numerical procedure to compute the parameters for the sensitivity law (14.1) and the probability density function (14.5) for particular structures are as follows:

Step 1: Conduct the path-tracing analysis of the perfect system to obtain the first critical point and the associated critical load $\Lambda^{\rm c0}$ and eigenvector $\boldsymbol{\Phi}_1^{\rm c0}$.

Step 2: Compute $\mathbf{B}^{c0}$ by (11.8) at the critical point.

Step 3: Compute the major imperfection influence vector $\mathbf{a}$ in (14.3) with the use of $\mathbf{B}^{c0}$ and $\mathbf{\Phi}_1^{c0}$.

Step 4: Compute the coefficients C_i ($i = 0, 1, 3$) in the sensitivity law (14.1) based on the path-tracing analyses for imperfections $\xi\mathbf{d}$ with a given $\mathbf{d}$ and several values of ξ.

Step 5: Compute $\widetilde{\sigma}^2$ in (14.4) and $\widehat{C}$ in (14.6).

In this procedure, the computation of the imperfection sensitivity matrix $\mathbf{B}^{c0}$ is the most important addition to the customary nonlinear finite element buckling analysis of structures. There are two methods to compute $\mathbf{B}^{c0}$:

- A general and numerically efficient procedure based on numerical differentiation will be presented in Section 16.3.
- For truss structures, $\mathbf{B}^{c0}$ can be computed compatibly with the framework of the finite element analysis [136].

It is possible to avoid computation of $\mathbf{B}^{c0}$ in the evaluation of $\widehat{C}$ when a set of critical loads for random imperfections are available by equating the sample variance with its theoretical value to be obtained from (14.5).

14.4 Probabilistic Variation of Strength of Truss Domes

Probabilistic variations of critical loads of truss domes are investigated. We consider imperfections of member lengths that produce initial stresses of the members. The imperfection $\xi\mathbf{d}$ is subjected to a normal distribution $\mathrm{N}(\mathbf{0}, \xi^2\mathbf{W})$ with $\xi^2\mathbf{W} = 1.5 \times 10^{-8}\mathbf{I}_\nu$, where $\mathbf{I}_\nu \in \mathbb{R}^{\nu\times\nu}$ is the identity matrix. All the truss members have the same cross-sectional area $A = 8.366 \times 10^{-4}$ m^2 and the same Young's modulus $E = 2.1 \times 10^5$ MN/m^2. The units of the force MN and of length m are suppressed in the sequel.

14.4.1 Double-layer hexagonal truss roof: limit point

Consider the double-layer hexagonal truss roof with 1165 members as shown in Fig. 14.2 subjected to the proportional z-directional load $-\Lambda$ at the hexagonal nodes shown by $\triangle$ and -2Λ at the center node shown by $\circ$ both in the lower layer. The fundamental path of the truss roof has the limit point at $\Lambda^{c0} = 0.7674$ as shown in Fig. 14.3.

At the limit point, we computed by path-tracing analysis $\mathbf{B}^{c0}$ and $\mathbf{\Phi}_1^{c0}$, $\widetilde{\sigma} = 1.051 \times 10^{-3}$ (cf., (14.4)), $C_0 = 0.2967$ and $\widehat{C} = |C_0\widetilde{\sigma}\xi| = 3.118 \times 10^{-4}$. With the use of the values of Λ^{c0} and $\widehat{C}$ thus obtained in (14.5), the curve of probability density function as shown in Fig. 14.4 is obtained. The histogram computed by the Monte Carlo simulation for an ensemble of 100 sets of random initial imperfections based on path-tracing analyses are also shown in Fig. 14.4. The theoretical curve can simulate the histogram fairly well.

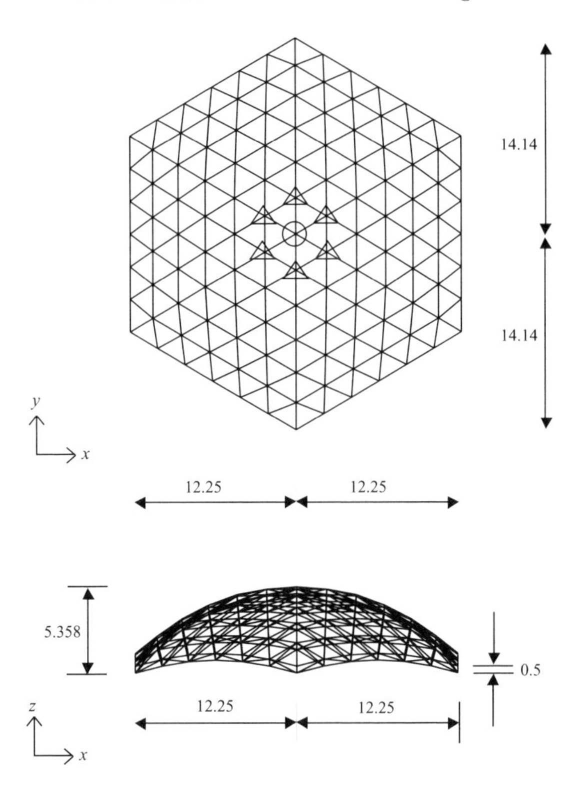

Fig. 14.2 Double-layer hexagonal truss roof.

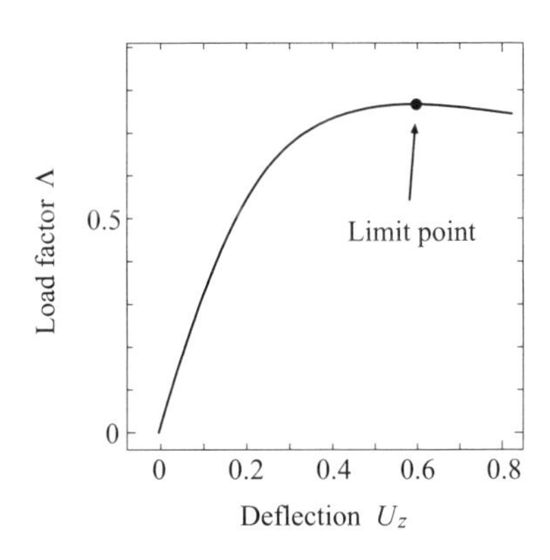

Fig. 14.3 Equilibrium path of the truss roof. U_z: deflection of the center node.

14.4.2 Spherical truss dome: unstable-symmetric bifurcation

Consider the spherical truss dome with 756 members as shown in Fig. 14.5 subjected to the proportional z-directional load $-\Lambda$ at the hexagonal nodes shown by $\triangle$ and -2Λ at the center node shown by $\circ$. The equilibrium paths of the spherical truss are as shown in Fig. 14.6, where U_z is the deflection of the center node.

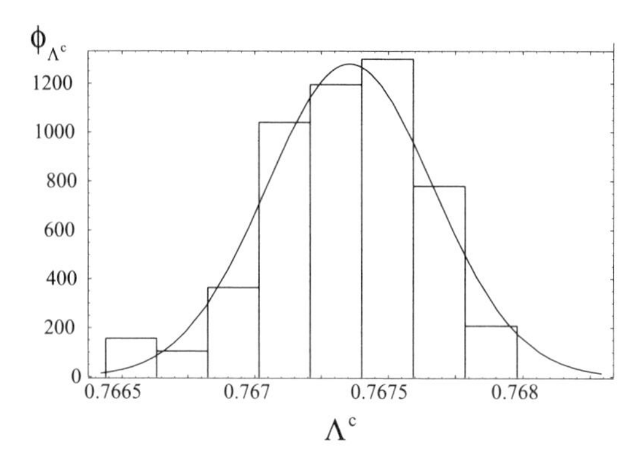

Fig. 14.4 Probability density of the limit point load Λ^c of the truss roof.

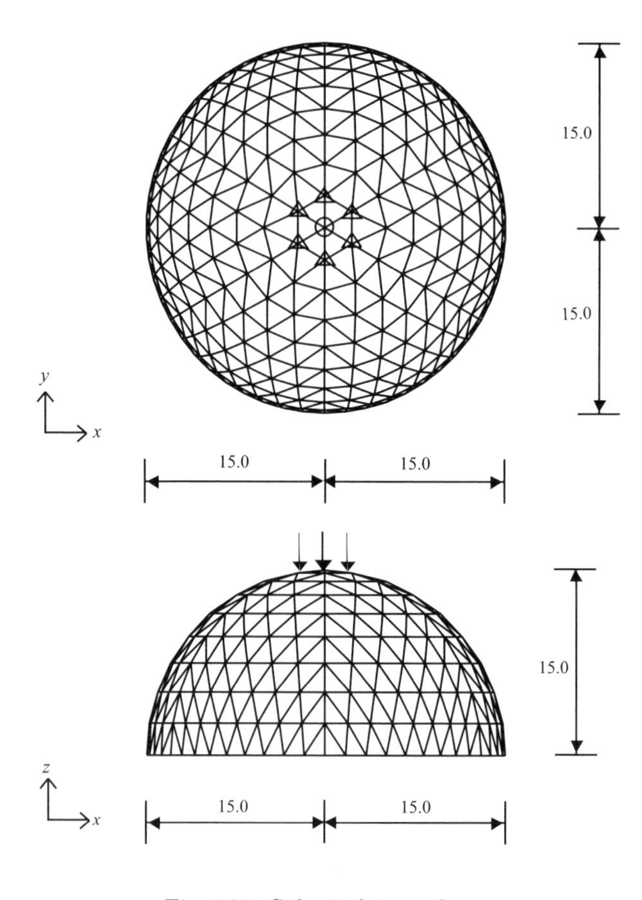

Fig. 14.5 Spherical truss dome.

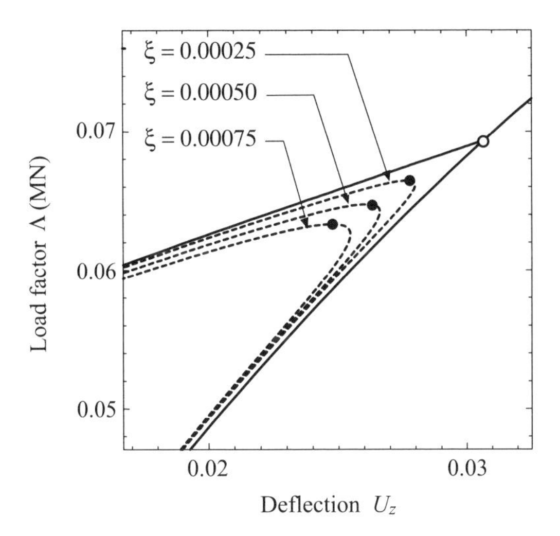

Fig. 14.6 Equilibrium paths of the spherical truss dome. Solid curve: perfect equilibrium path; dashed curve: imperfect equilibrium path; ∘: bifurcation point; •: limit point.

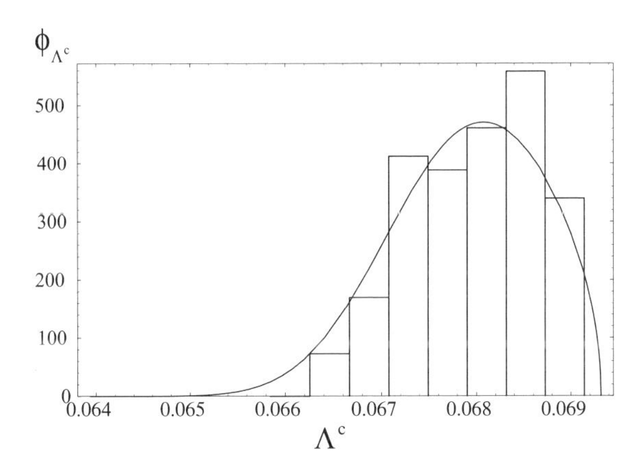

Fig. 14.7 Probability density of the buckling load Λ^c of the spherical truss dome.

The perfect equilibrium paths have an unstable-symmetric bifurcation point at $\Lambda^{c0} = 6.931 \times 10^{-2}$ shown by ∘.

The worst imperfection pattern $\mathbf{d}^{\perp}$ in (11.11) was computed and associated equilibrium paths as shown by dotted curves in Fig. 14.6 were obtained. The reductions of the critical loads are 2.873×10^{-3}, 4.573×10^{-3}, 6.006×10^{-3} for $\xi = 0.00025, 0.00050, 0.00075$, respectively. We have $C_3 = -0.1819$ and $\widehat{C} = |C_3|(\widetilde{\sigma}\xi)^{2/3} = 1.792 \times 10^{-3}$ with $\widetilde{\sigma} = 9.777 \times 10^{-4}$.

With the use of the values of Λ^{c0} and $\widehat{C}$ in (14.5), the curve of probability density function as shown in Fig. 14.7 is obtained. This curve is in good accordance with the associated Monte Carlo simulation shown by the histogram, and, hence, to demonstrate the validity of the present method.

14.5 Historical Development

Detailed knowledge of geometric imperfections of shells was gathered in association with the development of techniques for measuring initial imperfections [12, 27, 54, 199, 248, 304]. Compilation and extensive analysis of the international initial data banks were pursued actively at the Delft University of Technology [8, 9, 58, 62, 164, 165] and at the Israel Institute of Technology [2, 275]. These data banks are useful in deriving characteristic initial imperfection distributions that a given fabrication process is likely to produce. At Imperial College London, information on measured imperfections was gathered to produce characteristic imperfection shapes [52, 53, 269].

Through the search for the prototype imperfections, it came to be acknowledged that initial imperfections are subjected to probabilistic scatter and that the study of initial imperfections has to be combined with probabilistic treatment to make them practical.

Reliability function of the buckling load of a column supported by springs was obtained considering random imperfections and deflection criteria. For columns and bars, random imperfections were employed [5, 28, 36, 60, 88, 89, 200, 232, 306]. For shells, random axisymmetric imperfections [4, 6, 261, 280, 300] and general (non-symmetric) random initial imperfections [10, 29, 73, 76, 86, 111, 192, 267, 300] were employed. Initial imperfections were often assumed to be Gaussian random variables [4, 70, 111, 112]. In addition to the initial imperfection of structural shapes, various kinds of imperfections, such as loadings [47, 136, 174, 180, 182, 260], material properties (elastic moduli) [41, 136, 154, 272], and thickness variation [30, 103, 168, 181, 297] were considered.

The imperfection sensitivity law was used as a transfer function from an initial imperfection to a deterministic critical load and, in turn, to obtain the probabilistic variation of critical load for an imperfection with a known probabilistic property [33, 259, 261, 282]. Such use of imperfection sensitivity, however, was limited to a certain prototype imperfection. As a remedy of this, an initial imperfection was represented as a random process and a stochastic differential equation was used to obtain an asymptotic relationship between the critical load and initial imperfection [4–6, 89, 300].

The probability of failure was employed to express the influence of the randomness of experimentally measured imperfections [70, 112, 149] and paved the way for the introduction of statistical methods [7, 16, 23, 34, 35, 72, 108, 281]. A series of studies based on measured data were conducted [10, 11, 68, 71, 143, 185, 247, 279, 295] often with resort to the international initial imperfection data banks, from which stochastic imperfections with known average and autocorrelation can be produced. The Monte Carlo simulation came to be conducted to compute numerically the reliability of the buckling strength for measured or random initial imperfections [69, 70, 235, 309]. An asymptotic approach was combined with statistical analysis [234, 235, 293]. Koiter's special theory [167] for axisymmetric imperfections was combined with the Monte Carlo method and, in turn, to introduce imperfection-sensitivity concept into design. This method was replaced by the first-order second-moment method to reduce computational costs considerably [10, 81, 153]. Palassopoulos [233] presented a stochastic method of optimization of an imperfection sensitive column on elastic foundation, where randomly generated imperfection modes are used.

In order to overcome possible limitations of probabilistic methods [75], several attempts to arrive at a lower bound of strength were conducted:

- *Knockdown factor* based on the so-called *lower-bound design philosophy* is the most primitive but most robust way [310]. Engineers are reluctant to use the concept of imperfection sensitivity and prefer to rely on the *knockdown factor*, as was pointed out repeatedly [78].
- The *reduced stiffness* method finds a lower bound of design strength of a shell through the identification of the components of the membrane energy that are eroded by imperfections and mode interactions [57].
- Convex modeling of uncertainty, robust reliability, and anti-optimization approach were developed to estimate the lower bounds based on the worst-case-scenario of problems where scarce knowledge is present and the use of a probabilistic method cannot be justified [22–24, 74, 77–79, 186, 190, 237].

Finite element methods, such as STAGS [3, 11], were employed to deal with realistic imperfections of structures and to investigate the stochastic properties and reliability of their strength. Koiter's asymptotic approach was combined with the finite element method to be consistent with computer aided engineering environments [46, 107].

Stochastic finite element method (SFEM) [15, 25, 64, 65, 80, 93, 109, 120, 121, 124, 163, 187, 208, 238, 252, 273, 302, 318] was employed to numerically tackle the probabilistic properties of structures. The perturbation method was employed for most cases to deal with random quantities involved, and the second-moment analysis was often employed to compute the mean and the variance of the displacement or stress. The elastic modulus was often modeled as a random field in SFEM. See the exhaustive review by Schuëller [268] for more account of SFEM.

The response surface approach was used to evaluate the reliability of structures [40, 51, 59, 84, 85, 92, 101, 102, 161, 188, 207, 251, 256, 303]; nonlinear finite element analyses, for example, were conducted for the parameter sweep of a few initial imperfections and/or design parameters to evaluate the reliability.

14.6 Summary

In this chapter,

- the explicit forms of probability density functions of critical loads for simple critical points have been obtained, and

- the usefulness of the procedure presented in this chapter has been demonstrated through its application to structural models.

The major finding of this chapter is as follows.

- The present method is promising in that it permits the estimation of the probability density function of critical loads with much smaller cost than by the conventional method of random imperfections based on the Monte Carlo simulation.

15

Random Imperfections of Elasto-Plastic Solids

15.1 Introduction

A considerable number of numerical studies have been made on limit behavior of a tensile steel block (specimen) undergoing plastic instability (cf., [244, 292, 298] for comprehensive review). It is observed that a bifurcation point exists just after a limit point on an equilibrium path. For a long steel block, the limit point and the bifurcation point tend to coincide [43, 119, 134, 210], and such a pair of nearly coincident critical points are approximated by a hilltop branching point, as was done for elastic structures in Chapters 8–10.

The imperfection sensitivity law at the hilltop branching point of an elastic system was extended to arrive at an asymptotic formula for stochastic scatter of critical loads [146]. In this chapter, we search for applicability of this formula for description of imperfection sensitivity and probabilistic variation of tensile strengths of steel blocks. Application of elastic stability theory to a completely different problem, elasto-plastic instability behavior, may invite criticism on the possibility of unloading, which in mathematics means a lack of differentiability of the governing equation. Nonetheless, a numerical study demonstrates that such a lack does not significantly influence imperfection sensitivities of steel blocks.

This chapter is organized as follows. In Section 15.2, the explicit form of probability density function of critical loads is derived for a hilltop branching point. In Section 15.3, numerical analyses are conducted to investigate perfect and imperfect behaviors of steel blocks undergoing plastic deformation and the histograms of the maximum loads are compared with these theoretical distributions to ensure their validity.

15.2 Probability Density Function of Critical Loads

Probabilistic variation of $\Lambda^{\rm c}$ is investigated for a hilltop branching point when initial imperfection parameter vector $\mathbf{v} - \mathbf{v}^0 = \xi \mathbf{d}$ is subject to a normal distribution $\mathrm{N}(\mathbf{0}, \xi^2 \mathbf{W})$ with mean $\mathbf{0}$ and variance–covariance matrix $\xi^2 \mathbf{W}$, where $\mathbf{W}$ is a positive-definite matrix.

Let $\mathbf{\Phi}_1^{\rm c0}$ and $\mathbf{\Phi}_2^{\rm c0}$ correspond to symmetric bifurcation point and a limit point, respectively. Recall the imperfection sensitivity law (11.16) for a hilltop point

$$\widetilde{\Lambda}^{\rm c}(\xi) = C_4 |V_{,1\xi} \xi| + C_5 V_{,2\xi} \xi = C_4 |\xi| |\mathbf{\Phi}_1^{\rm c0\top} \mathbf{B}^{\rm c0} \mathbf{d}| + C_5 \xi \mathbf{\Phi}_2^{\rm c0\top} \mathbf{B}^{\rm c0} \mathbf{d} \tag{15.1}$$

with $C_4 < 0$ and $C_5 > 0$ (cf., (8.23) and (8.28)).

Since $\mathbf{d}$ is subject to normal distribution $\mathrm{N}(\mathbf{0}, \mathbf{W})$, it is easy to show that variables $\mathbf{\Phi}_1^{\rm c0\top} \mathbf{B}^{\rm c0} \mathbf{d}$ and $\mathbf{\Phi}_2^{\rm c0\top} \mathbf{B}^{\rm c0} \mathbf{d}$ are statistically independent and respectively are subject to normal distributions $\mathrm{N}(0, (\sigma_1)^2)$ and $\mathrm{N}(0, (\sigma_2)^2)$ with mean 0 and variances [146]

$$(\sigma_i)^2 = \mathbf{\Phi}_i^{\rm c0\top} \mathbf{B}^{\rm c0} \mathbf{W} \mathbf{B}^{\rm c0\top} \mathbf{\Phi}_i^{\rm c0}, \quad (i = 1, 2) \tag{15.2}$$

Then the probability density function of $\widetilde{\Lambda}^{\rm c}$ in (15.1) is evaluated to

$$\begin{aligned} \phi_{\widetilde{\Lambda}^{\rm c}} &= \int_{-\infty}^{\infty} \frac{1}{\sqrt{2\pi}\sigma_1} \exp\Big(-\frac{x^2}{2(\sigma_1)^2}\Big) \frac{1}{C_5|\xi|} \frac{1}{\sqrt{2\pi}\sigma_2} \exp\Big(-\frac{(|C_4 \xi x| + \widetilde{\Lambda}^{\rm c})^2}{2(\sigma_2^* \xi)^2}\Big) \mathrm{d}x \\ &= \frac{1}{2\pi \sigma_1 \sigma_2^* |\xi|} \exp\Big(-\frac{(\widetilde{\Lambda}^{\rm c})^2}{2\widehat{\sigma}^2}\Big) \int_{-\infty}^{\infty} \exp\Big(-\frac{k}{2}(|x| + m\widetilde{\Lambda}^{\rm c})^2\Big) \mathrm{d}x \\ &= \frac{1}{2\pi \sigma_1 \sigma_2^* |\xi|} \exp\Big(-\frac{(\widetilde{\Lambda}^{\rm c})^2}{2\widehat{\sigma}^2}\Big) 2\sqrt{\frac{2\pi}{k}} \int_{-\infty}^{-\sqrt{k}m\widetilde{\Lambda}^{\rm c}} \frac{1}{\sqrt{2\pi}} \exp(-x^2/2) \mathrm{d}x \\ &= \frac{2}{\sqrt{2\pi}\widehat{\sigma}} \exp\Big(-\frac{1}{2}\frac{(\widetilde{\Lambda}^{\rm c})^2}{\widehat{\sigma}^2}\Big) \Phi_{\rm N}\Big(-\frac{r}{\widehat{\sigma}}\widetilde{\Lambda}^{\rm c}\Big) \end{aligned} \tag{15.3}$$

where

$$\begin{aligned} &\sigma_1^* = |C_4| \sigma_1, \quad \sigma_2^* = C_5 \sigma_2 \\ &\widehat{\sigma} = \sqrt{(\sigma_1^*)^2 + (\sigma_2^*)^2} |\xi|, \quad r = \sigma_1^* / \sigma_2^* \\ &k = \frac{\widehat{\sigma}^2}{(\sigma_1 \sigma_2^* \xi)^2}, \quad m = |\xi| \frac{\sigma_1^* \sigma_1}{\widehat{\sigma}^2} \end{aligned} \tag{15.4}$$

and

$$\Phi_{\rm N}(\zeta) = \int_{-\infty}^{\zeta} \frac{1}{\sqrt{2\pi}} \exp(-\zeta^2/2) \mathrm{d}\zeta \tag{15.5}$$

is the cumulative distribution function of the standard normal distribution $\mathrm{N}(0, 1)$.

The expected value of $\widetilde{\Lambda}^{\rm c}$ is expressed as

$$\mathrm{E}[\widetilde{\Lambda}^{\rm c}] = -\sqrt{\frac{2}{\pi}} \sigma_1^* |\xi| \tag{15.6}$$

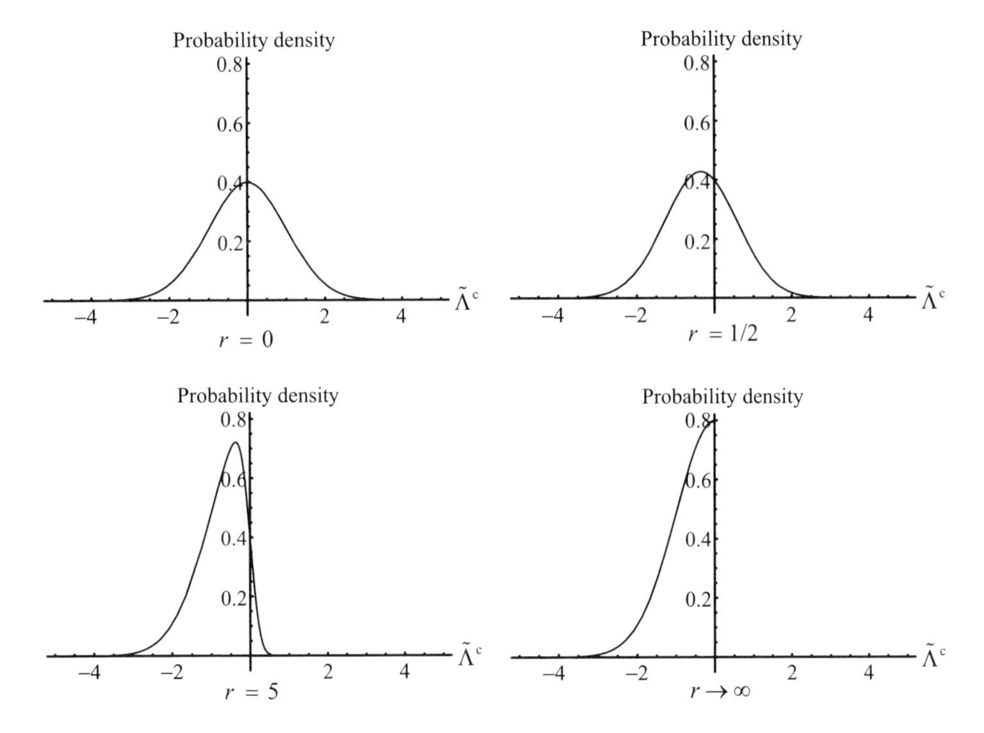

Fig. 15.1 Probability density function of $\widetilde{\Lambda}^{\rm c}$ plotted for several values of $r = \sigma_1^*/\sigma_2^*$ ($\widehat{\sigma} = 1$).

and its variance is

$$\mathrm{Var}[\widetilde{\Lambda}^{\rm c}] = \mathrm{E}[(\widetilde{\Lambda}^{\rm c})^2] - (\mathrm{E}[\widetilde{\Lambda}^{\rm c}])^2 = \widehat{\sigma}^2 - \frac{2}{\pi}(\sigma_1^*\xi)^2 \tag{15.7}$$

Note that the probability density function $\phi(\widetilde{\Lambda}^{\rm c})$ in (15.3) has two parameters $\widehat{\sigma}$ and r. Parameter $\widehat{\sigma}$ characterizes deviation of $\widetilde{\Lambda}^{\rm c}$ ($\widehat{\sigma}^2$ is equal to the average of $(\widetilde{\Lambda}^{\rm c})^2$, to be precise), whereas r characterizes the shape of the function. Curves of the probability density function for several values of r are shown in Fig. 15.1. For $r = 0$ ($C_4 = 0$), in which only the variation of coefficient $V_{,2\xi}$ for the limit point is influential, $\phi(\widetilde{\Lambda}^{\rm c})$ reduces to normal distribution of $\mathrm{N}(0, (\sigma_2^*\xi)^2)$. For another extreme case of $r \to +\infty$ ($C_5 \to 0$), in which only the variation of coefficient $V_{,1\xi}$ for the symmetric bifurcation point is influential, $\phi(\widetilde{\Lambda}^{\rm c})$ reduces to

$$\begin{cases} 2\mathrm{N}(0, (\sigma_1^*\xi)^2): & \text{for } \widetilde{\Lambda}^{\rm c} < 0 \\ 0: & \text{for } \widetilde{\Lambda}^{\rm c} > 0 \end{cases} \tag{15.8}$$

We can employ one of the following two procedures to compute parameters $\widehat{\sigma}$ and r in (15.4) of the probability density function $\phi(\widetilde{\Lambda}^{\rm c})$ in (15.3):

- Compute $\mathbf{B}^{\rm c0}$ in (11.8) and, in turn, compute σ_i ($i = 1, 2$) by (15.2) and C_i ($i = 4, 5$) by sensitivity law (15.1). Then compute $\widehat{\sigma}$ and r by (15.4).

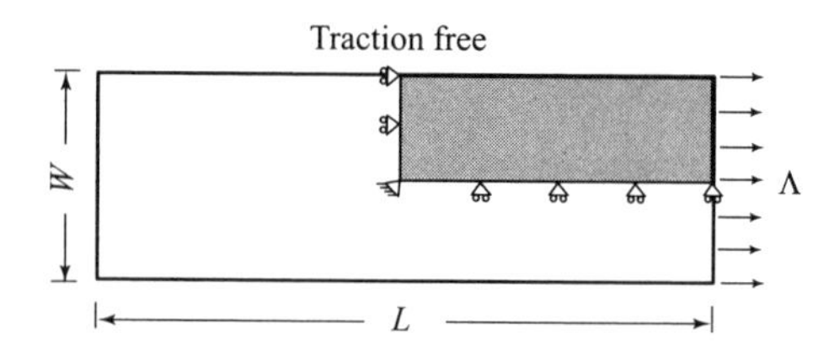

Fig. 15.2 Rectangular steel block.

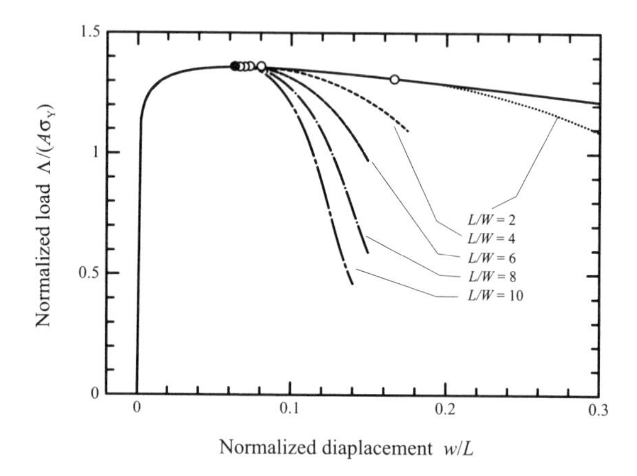

Fig. 15.3 Equilibrium paths of steel blocks with $L/W = 2, 4, 6, 8, 10$. A $(= W)$: initial cross section; w: axial displacement; •: limit point; ○: bifurcation point.

- When a histogram or a set of critical loads $\Lambda^{\rm c}$ is given for a particular case, we can employ (15.6) and (15.7) to approximate $\sigma_1^* \xi$ and $\widehat{\sigma}$ and to compute r by (15.4).

For the second procedure, which is based on the Monte Carlo simulation, we need not to compute $\mathbf{B}^{\rm c0}$, as seen in the next section. This indeed is a merit of the formalism of this procedure.

15.3 Probabilistic Strength Variation of Steel Blocks

Probabilistic strength variation of steel blocks subjected to uniform tensile load Λ is investigated based on the probabilistic theory presented in Section 15.2. Numerical analyses[1] are conducted on a rectangular steel block under plane strain condition illustrated in Fig. 15.2. Young's modulus is $E = 200$ kN/mm^2, Poisson's ratio $\nu = 0.333$ for elasticity, and yield stress $\sigma_{\rm Y} = 0.4$ kN/mm^2. The units of force kN and of length mm are suppressed in the following.

[1] See [229] for details of numerical analysis.

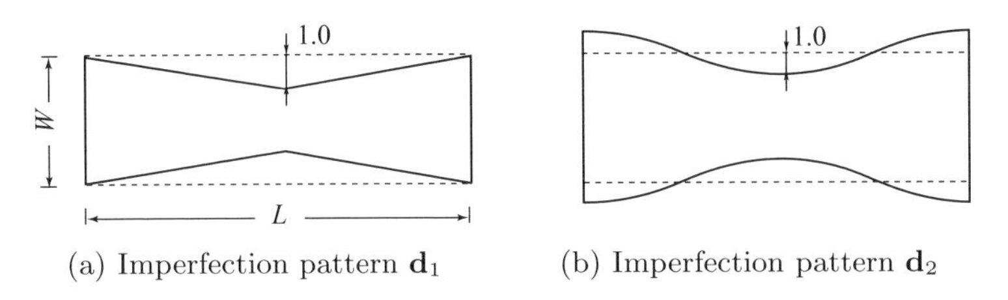

(a) Imperfection pattern $\mathbf{d}_1$ (b) Imperfection pattern $\mathbf{d}_2$

Fig. 15.4 Non-harmonic and harmonic imperfection patterns of steel blocks.

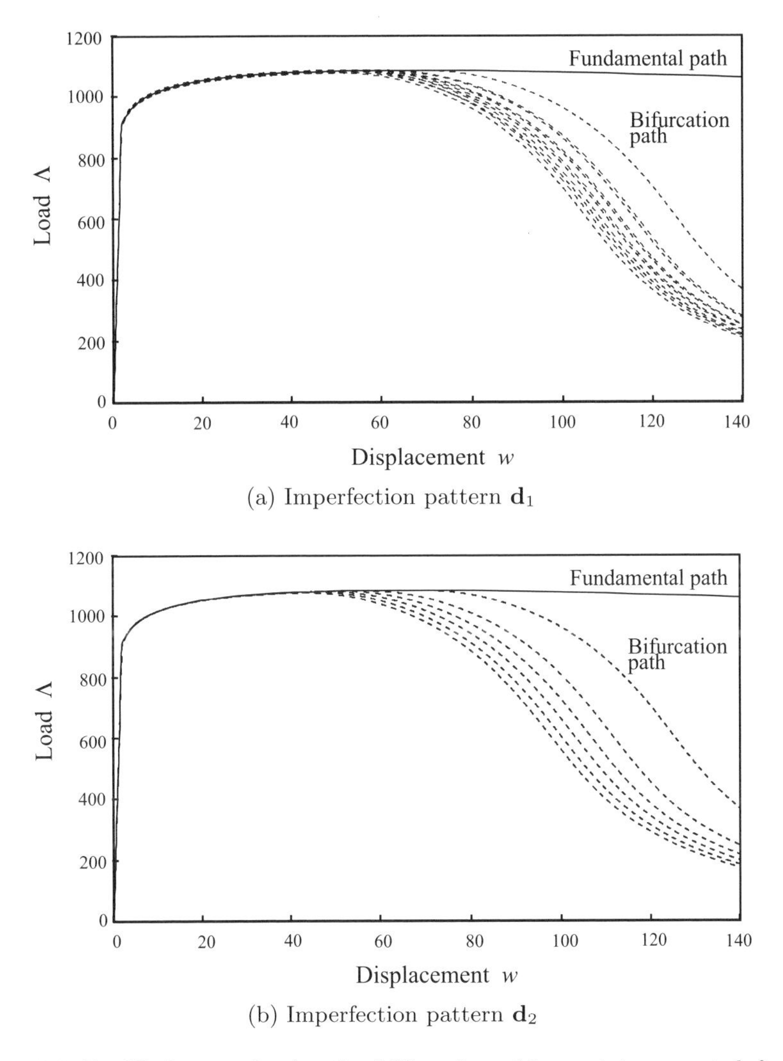

(a) Imperfection pattern $\mathbf{d}_1$

(b) Imperfection pattern $\mathbf{d}_2$

Fig. 15.5 Equilibrium paths for the hilltop branching point computed for a number of imperfection magnitudes.

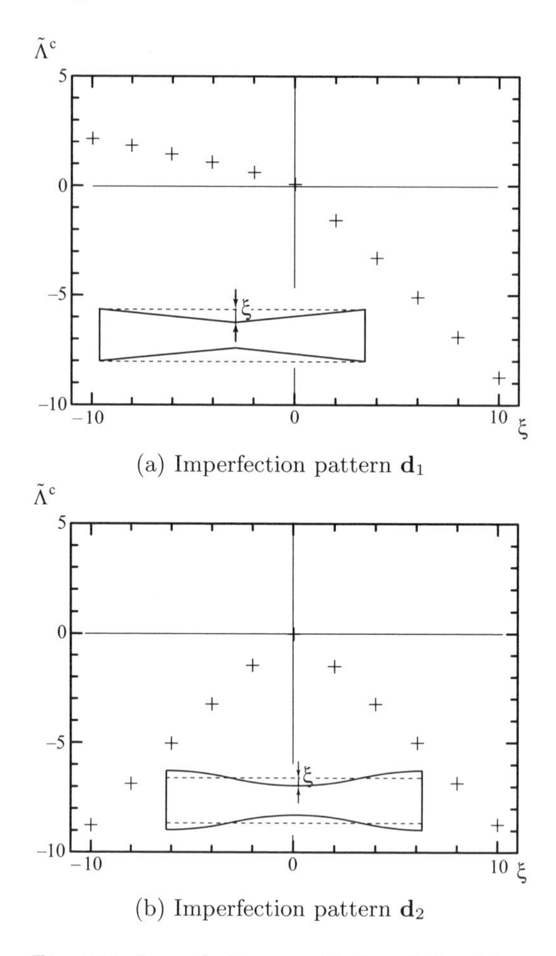

Fig. 15.6 Imperfection sensitivity relationships.

Plastic bifurcation analyses of the rectangular domain for aspect ratios of $L/W = 2, 4, 6, 8, 10$ are conducted to obtain the equilibrium paths as shown in Fig. 15.3. The locations of limit points denoted by • are identical for all aspect ratios. The first bifurcation point, denoted by ∘, approaches the limit point as the specimen becomes slender [43, 210]. The bifurcation load $\Lambda^{c0} = 1086$ for the aspect ratio of $L/W = 10$ is only 0.2% smaller than the limit point load. Hence, $L/W = 10$ is used herein to approximate the hilltop branching point.

15.3.1 Imperfection sensitivity

Imperfect behaviors are investigated by employing two imperfection patterns $\mathbf{d} = \mathbf{d}_1$ and $\mathbf{d}_2$ as shown in Fig. 15.4 with several imperfection magnitudes in the range $-10.0 \leq \xi \leq 10.0$. The pattern $\mathbf{d}_1$ has both volumetric[2] and harmonic

[2]A volumetric mode causes a volumetric change of the specimen, while the harmonic mode does not.

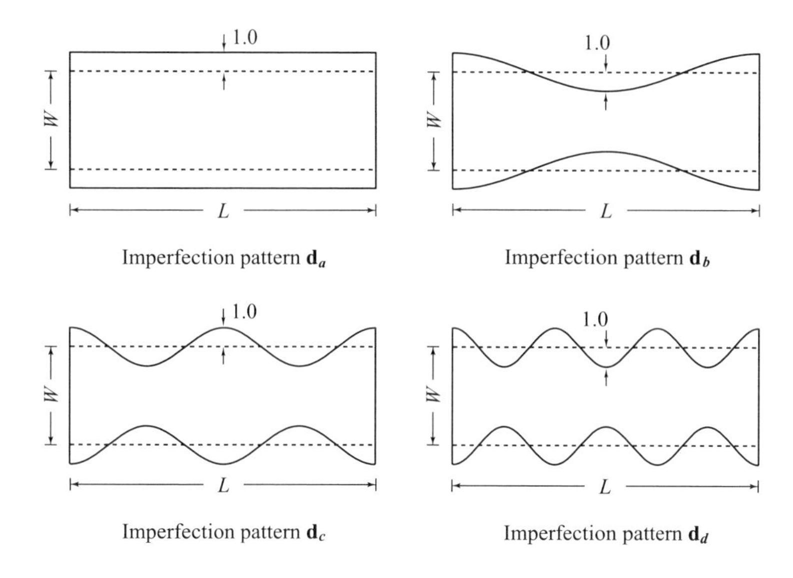

Fig. 15.7 Uniform and harmonic imperfection patterns of steel blocks.

modes. Note that the critical eigenvector $\mathbf{\Phi}_1^{c0}$ of the tangent stiffness matrix $\mathbf{S}$, which is a pure harmonic mode, is chosen for the imperfection pattern $\mathbf{d}_2$.

Fig. 15.5 shows equilibrium paths for several imperfection magnitudes ξ. For $\mathbf{d} = \mathbf{d}_1$ with $V_{,1\xi} \neq 0$ and $V_{,2\xi} \neq 0$, as shown in Fig. 15.6(a), imperfection magnitude ξ and critical load increment $\widetilde{\Lambda}^c$ display a piecewise linear relationship with a kink at $\xi = 0$; the relationship for $\xi > 0$ and that for $\xi < 0$ have different slopes in agreement with the law (15.1).

For $\mathbf{d} = \mathbf{d}_2$ with $V_{,1\xi} \neq 0$ and $V_{,2\xi} \simeq 0$ as shown in Fig. 15.6(b), imperfection magnitude ξ and critical load $\widetilde{\Lambda}^c$ display a piecewise linear relationship with a kink at $\xi = 0$ that is symmetric with respect to the reflection with respect to the $\widetilde{\Lambda}^c$-axis, in agreement with the law (15.1) with $V_{,2\xi} = 0$.

15.3.2 Probabilistic variation of critical loads

We carry out the Monte Carlo simulation on imperfect steel blocks to build a data bank of their buckling strengths.

The imperfection parameter vector is defined as

$$\mathbf{v} = \xi_a \mathbf{d}_a + \xi_b \mathbf{d}_b + \xi_c \mathbf{d}_c + \xi_d \mathbf{d}_d \tag{15.9}$$

where $\mathbf{v}^0 = \mathbf{0}$, and $\mathbf{d}_a$, $\mathbf{d}_b$, $\mathbf{d}_c$ and $\mathbf{d}_d$ are uniform and harmonic modes shown in Fig. 15.7. We choose an ensemble of 100 imperfection patterns $\xi\mathbf{d}$ that is subject to a multivariate normal distribution $\xi\mathbf{d} \sim \mathrm{N}(\mathbf{0}, 0.01^2\mathbf{I}_4)$, where $\mathbf{I}_4$ is the identity matrix.

For a limit point with $L/W = 2$ and a hilltop point with $L/W = 10$, we computed maximum loads for the 100 imperfection patterns. Fig. 15.8 shows histograms obtained in this manner and curves for the theoretical probability density function, which is a normal distribution (14.5) for (a) the limit point,

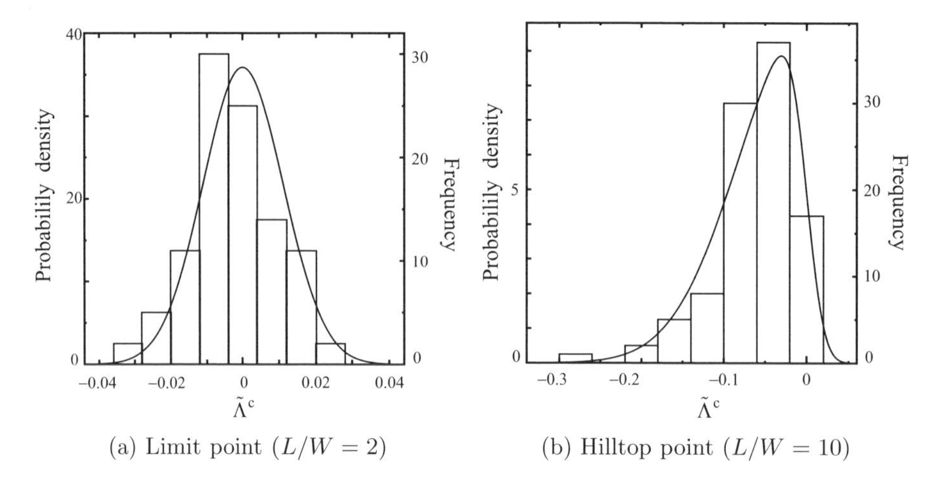

(a) Limit point ($L/W = 2$) (b) Hilltop point ($L/W = 10$)

Fig. 15.8 Comparison of histograms and theoretical probability density functions.

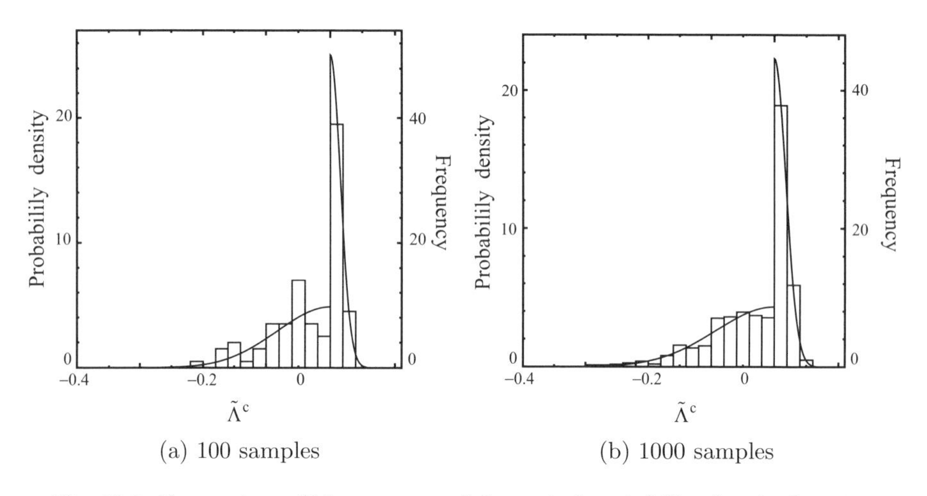

(a) 100 samples (b) 1000 samples

Fig. 15.9 Comparison of histograms and theoretical probability density functions for the hilltop point ($L/W = 10$).

and (15.3) for (b) the hilltop point, respectively. The theoretical curves have passed the χ^2 test at a significance level of 0.05 or less, and are fairly accurate. In particular, for the hilltop point, the Weibull-like histogram is represented well by the theoretical curve.

Remark 15.3.1 Consider a case where $\mathbf{d}$ is kept fixed and ξ is subject to a normal distribution $\mathrm{N}(0, \sigma^2)$. The probability density function of $\widetilde{\Lambda}^{\rm c}$ is dependent on cases. For $|C_4 V_{,1\xi}| < |C_5 V_{,2\xi}|$, $\widetilde{\Lambda}^{\rm c}$ is subject to

$$\left\{ \begin{array}{ll} \mathrm{N}(0, (|C_4 V_{,1\xi}| + |C_5 V_{,2\xi}|)^2 \sigma^2) : & \text{for } \widetilde{\Lambda}^{\rm c} < 0 \\ \mathrm{N}(0, (|C_4 V_{,1\xi}| - |C_5 V_{,2\xi}|)^2 \sigma^2) : & \text{for } \widetilde{\Lambda}^{\rm c} > 0 \end{array} \right. \tag{15.10}$$

and for $|C_4 V_{,1\xi}| > |C_5 V_{,2\xi}|$, $\widetilde{\Lambda}^{\rm c}$ is subject to

$$\begin{cases} \mathrm{N}(0, (|C_4 V_{,1\xi}| + |C_5 V_{,2\xi}|)^2 \sigma^2) + \mathrm{N}(0, (|C_4 V_{,1\xi}| - |C_5 V_{,2\xi}|)^2 \sigma^2) : & \text{for } \widetilde{\Lambda}^{\rm c} < 0 \\ 0 : & \text{for } \widetilde{\Lambda}^{\rm c} > 0 \end{cases} \tag{15.11}$$

For a fixed pattern $\mathbf{d} = \mathbf{d}_1$ corresponding to $|C_4 V_{,1\xi}| < |C_5 V_{,2\xi}|$, choose an ensemble of 1000 random imperfection magnitudes ξ subject to normal distribution $\mathrm{N}(0, 0.1013^2)$. As shown in Fig. 15.9, the curves of the theoretical probability density function (15.10) agree fairly well with histograms. Compatibility between the histograms and the theoretical curves is improved with increased sample size. Yet the simple case treated here, in favor of its simplicity, is not as realistic as the standard case where $\mathbf{d}$ is subject to random variations. □

15.4 Summary

In this chapter,

- the probability density function of the critical load for a hilltop branching point has been derived, and
- strength variation of steel specimens has been investigated.

The major findings of this chapter are as follows.

- The applicability of elastic stability theory to description of tensile strength variation of steel blocks has been assessed. In fact, the theoretical imperfection sensitivity laws agree well with empirical imperfection sensitivities of these blocks obtained by numerical simulations, and the theoretical formulas for the probability of critical loads are shown to be useful to describe tensile strength of steel blocks.
- The limit point and hilltop branching point have thoroughly different imperfection sensitivity laws and probabilistic variations of strengths. It is, therefore, vital in successful description of strengths of steel blocks to identify the type of critical point that governs the critical load.

16

Random Imperfections: Higher-Order Analysis

16.1 Introduction

The imperfection sensitivity law was extended in Chapters 11 and 14 to a number of imperfection parameters to develop methodologies to deal with the worst imperfection and the probability density function of critical loads for random initial imperfections [136, 137, 140, 142]. These methodologies were successfully applied to an elasto-plastic problem in Chapter 15. Yet they have the following problems in practical application:

1. They capture the influence of a major imperfection, and are valid asymptotically for a small imperfection magnitude, but are inaccurate for large imperfection magnitudes. In fact, a minor imperfection is sometimes more influential than a major imperfection [220].

2. In the asymptotic method [33, 259, 282], it is necessary to compute the imperfection sensitivity matrix $\mathbf{B}^{c0}$ in (11.8).

In this chapter, in order to address the first problem, the imperfection sensitivity laws are generalized to be applicable to a large number of imperfection variables, and the methodology to obtain the probability density function of structures is extended up to minor imperfections. The second problem is resolved by replacing the computation of the imperfection sensitivity matrix by numerical differentiation. This serves as a combination of the theoretical framework in Chapter 14 with the concept of the first-order and the second-order imperfections [106, 202, 258] and the finite difference method (cf., Remark 3.2.2).

This chapter is organized as follows. The asymptotic probabilistic theory is upgraded to be applicable up to the minor imperfection in Section 16.2. A numerical procedure for this theory is presented in Section 16.3. The proposed methodology is applied to truss structures in Sections 16.4 and 16.5.

16.2 Higher-Order Asymptotic Theory

Higher-order asymptotic theory is introduced [142].

16.2.1 Generalized sensitivity law

Recall the sensitivity law (3.48) for a simple unstable-symmetric bifurcation point

$$\widetilde{\Lambda}^{\mathrm{c}}(\xi) = \Lambda^{\mathrm{c}} - \Lambda^{\mathrm{c}0} = C_3(V_{,1\xi}\xi)^{\frac{2}{3}} + C_2 V_{,11\xi}\xi \tag{16.1}$$

where by (3.49) and (3.50)

$$C_2 = -\frac{1}{V_{,11\Lambda}} > 0, \quad C_3 = -\frac{3^{\frac{2}{3}}(V_{,1111})^{\frac{1}{3}}}{2V_{,11\Lambda}} < 0 \tag{16.2}$$

We define the incremental initial imperfection vector as

$$\widetilde{\mathbf{v}} = \xi\mathbf{d} = \mathbf{v} - \mathbf{v}^0 \tag{16.3}$$

with $\widetilde{\mathbf{v}} = (\widetilde{v}_1, \ldots, \widetilde{v}_\nu)^\top$ (cf., (1.12)). Then with the use of (11.7), we have

$$V_{,1\xi}\xi = \mathbf{\Phi}_1^{\mathrm{c}0\top}\mathbf{B}^{\mathrm{c}0}\widetilde{\mathbf{v}}, \quad V_{,11\xi}\xi = \mathbf{\Phi}_1^{\mathrm{c}0\top}(\partial\mathbf{B}/\partial q_1)^{\mathrm{c}0}\widetilde{\mathbf{v}} \tag{16.4}$$

where $\mathbf{B}$ is the imperfection sensitivity matrix defined in (11.8).

With reference to (16.4), we define a pair of vectors expressing the influence of initial imperfections as

$$\mathbf{a} = (a_1, \ldots, a_\nu)^\top = (\mathbf{\Phi}_1^{\mathrm{c}0\top}\mathbf{B}^{\mathrm{c}0})^\top \tag{16.5a}$$

$$\mathbf{b} = (b_1, \ldots, b_\nu)^\top = (\mathbf{\Phi}_1^{\mathrm{c}0\top}(\partial\mathbf{B}/\partial q_1)^{\mathrm{c}0})^\top \tag{16.5b}$$

which are called the *major imperfection influence vector* and the *minor imperfection influence vector*, respectively. Then the imperfection sensitivity law (16.1) is generalized into

$$\widetilde{\Lambda}^{\mathrm{c}}(\widetilde{\mathbf{v}}) = C_3(\mathbf{a}^\top\widetilde{\mathbf{v}})^{\frac{2}{3}} + C_2\mathbf{b}^\top\widetilde{\mathbf{v}} = -[(-C_3)^{\frac{3}{2}}\mathbf{a}^\top\widetilde{\mathbf{v}}]^{\frac{2}{3}} + C_2\mathbf{b}^\top\widetilde{\mathbf{v}} \tag{16.6}$$

16.2.2 Probability density functions of critical loads

The probability density function of critical loads is derived using the generalized sensitivity law (16.6) under the assumption that the incremental initial imperfection vector $\widetilde{\mathbf{v}} = \mathbf{v} - \mathbf{v}^0$ is subject to a normal distribution $\mathrm{N}(\mathbf{0}, \mathbf{W}^2)$ with mean $\mathbf{0}$ and variance–covariance matrix $\mathbf{W}$.

We further rewrite (16.6) as

$$\widetilde{\Lambda}^{\mathrm{c}}(\widetilde{\mathbf{v}}) = a + b \tag{16.7}$$

with

$$a = -[(-C_3)^{\frac{3}{2}}\mathbf{a}^\top\widetilde{\mathbf{v}}]^{\frac{2}{3}} < 0, \qquad b = C_2\mathbf{b}^\top\widetilde{\mathbf{v}} \tag{16.8}$$

Since $\widetilde{\mathbf{v}}$ is subject to the normal distribution $\mathrm{N}(\mathbf{0}, \mathbf{W}^2)$, it is easy to show

$$\begin{cases} (-C_3)^{\frac{3}{2}}\mathbf{a}^\top \widetilde{\mathbf{v}} \text{ is subject to } \mathrm{N}(0, (\sigma_a)^2) \\ C_2\mathbf{b}^\top \widetilde{\mathbf{v}} \text{ is subject to } \mathrm{N}(0, (\sigma_b)^2) \end{cases} \tag{16.9}$$

with

$$(\sigma_a)^2 = -(C_3)^3\mathbf{a}^\top\mathbf{W}\mathbf{a} = -(C_3)^3\mathbf{\Phi}_1^{\mathrm{c}0\top}\mathbf{B}^{\mathrm{c}0}\mathbf{W}\mathbf{B}^{\mathrm{c}0\top}\mathbf{\Phi}_1^{\mathrm{c}0} \tag{16.10}$$

$$(\sigma_b)^2 = (C_2)^2\mathbf{b}^\top\mathbf{W}\mathbf{b} = (C_2)^2\mathbf{\Phi}_1^{\mathrm{c}0\top}(\partial\mathbf{B}/\partial q_1)^{\mathrm{c}0}\mathbf{W}(\partial\mathbf{B}/\partial q_1)^{\mathrm{c}0\top}\mathbf{\Phi}_1^{\mathrm{c}0} \tag{16.11}$$

The probability density functions of a and b respectively are (cf., (14.5))

$$\phi_a = \frac{3|a|^{\frac{1}{2}}}{\sqrt{2\pi}\sigma_a}\exp\Big(-\frac{|a|^3}{2(\sigma_a)^2}\Big), \quad (-\infty < a < 0) \tag{16.12}$$

$$\phi_b = \frac{1}{\sqrt{2\pi}\sigma_b}\exp\Big(-\frac{b^2}{2(\sigma_b)^2}\Big), \quad (-\infty < b < \infty) \tag{16.13}$$

The probability density function of $\widetilde{\Lambda}^{\mathrm{c}}$ in (16.7) is obtained as

$$\phi_{\widetilde{\Lambda}^{\mathrm{c}}} = \int_{-\infty}^{0} \phi_a(a)\phi_b(\widetilde{\Lambda}^{\mathrm{c}} - a)\mathrm{d}a, \quad (-\infty < \widetilde{\Lambda}^{\mathrm{c}} < \infty) \tag{16.14}$$

The probability density function (16.12) gives the first-order reliability, and the probability density function (16.14) gives more accurate second-order reliability.

16.3 Numerical Procedure

The probability density function (16.14) of critical loads permits a simple formulation via the generalized sensitivity law (16.6). However, the differential coefficients $V_{,11\xi}$, $V_{,11\Lambda}$, and so on, cannot be easily computed for large-scale structures, because these coefficients suffer from the contamination of passive coordinates (cf., (3.20a) and (3.20b)). In this section, a numerical procedure is proposed to rewrite the formulas in Section 16.2 to be readily applicable to a practical numerical analysis.

The numerical procedure is summarized as follows, in which the imperfection parameter ξ is to be replaced by $\widetilde{v}_i$ for general imperfections:

Step 1: Conduct path-tracing analysis of the perfect system to obtain the unstable simple-symmetric bifurcation point $(\mathbf{U}^{\mathrm{c}0}, \Lambda^{\mathrm{c}0})$ in question and the associated eigenvectors $\mathbf{\Phi}_r^{\mathrm{c}0}$.

Step 2: Compute $\mathbf{a}$ in (16.5a) from (16.4) with $\xi = v_i$. Here $V_{,1v_i}$ is computed by a finite difference method by (3.7a) as (cf., Remark 3.2.2)

$$\begin{aligned} V_{,1v_i} \simeq \frac{1}{\widetilde{v}_i}\sum_{j=1}^{n}\phi_{1j}^{\mathrm{c}0}[&S_j(\mathbf{U}^{\mathrm{c}0}, \Lambda^{\mathrm{c}0}, v_1^0, \ldots, v_i^0 + \widetilde{v}_i, \ldots, v_\nu^0) \\ &- S_j(\mathbf{U}^{\mathrm{c}0}, \Lambda^{\mathrm{c}0}, v_1^0, \ldots, v_i^0, \ldots, v_\nu^0)] \end{aligned} \tag{16.15}$$

for some pertinent finite difference $\widetilde{v}_i$ of v_i.

Step 3: Compute C_3 from (16.2) with (3.20a) and (3.21).

Step 4: Compute $\mathbf{b}$ in (16.5b) from (16.4) with $\xi = v_i$ and (3.20a). Here $D_{,11v_i}$ is computed by a finite difference method using (3.7b) as (cf., Remark 3.2.2)

$$D_{,11v_i} \simeq \frac{1}{\widetilde{v}_i}[\boldsymbol{\Phi}_1^{c0\top}\mathbf{S}(\mathbf{U}^{c0}, \Lambda^{c0}, v_1^0, \ldots, v_i^0 + \widetilde{v}_i, \ldots, v_\nu^0)\boldsymbol{\Phi}_1^{c0} - \boldsymbol{\Phi}_1^{c0\top}\mathbf{S}(\mathbf{U}^{c0}, \Lambda^{c0}, v_1^0, \ldots, v_i^0, \ldots, v_\nu^0)\boldsymbol{\Phi}_1^{c0}] \tag{16.16}$$

Step 5: Compute C_2 from (16.2) with (3.20a).

Step 6: Compute the variances $(\sigma_a)^2$ and $(\sigma_b)^2$ in (16.10) and (16.11).

Note the computation of C_2 and C_3 can be alternatively achieved by the nonlinear path-tracing analysis for a given imperfection vector $\mathbf{v}$.

16.4 Four-Bar Truss Tent

The four-bar truss tent as shown in Fig. 16.1 subjected to the z-directional load Λ is considered. For the perfect system, all members have the same cross-sectional area A and the same elastic modulus E; $EA = 1.0$ for normalization. In the following, the units of length and force are omitted for brevity.

As initial imperfections of the four-bar truss, we employ the initial location (x^*, y^*, z^*) of the crown node and the cross sections A_1, A_2, A_3, and A_4 of the truss members. Recall $\widetilde{\mathbf{v}} = \mathbf{v} - \mathbf{v}^0$ in (16.3) defining initial imperfections, and set

$$\mathbf{v} = \begin{pmatrix} x^* \\ y^* \\ z^* \\ A_1 \\ A_2 \\ A_3 \\ A_4 \end{pmatrix}, \quad \mathbf{v}^0 = \begin{pmatrix} 0 \\ 0 \\ 0 \\ A \\ A \\ A \\ A \end{pmatrix}, \quad \widetilde{\mathbf{v}} = \begin{pmatrix} \widetilde{v}_1 \\ \widetilde{v}_2 \\ \widetilde{v}_3 \\ \widetilde{v}_{A_1} \\ \widetilde{v}_{A_2} \\ \widetilde{v}_{A_3} \\ \widetilde{v}_{A_4} \end{pmatrix} \tag{16.17}$$

16.4.1 Perfect system

To achieve Step 1 in Section 16.3, the equilibrium paths for the perfect system of the truss tent obtained by path-tracing analysis are shown in Fig. 16.2. The critical load is governed by the simple unstable-symmetric bifurcation point $(U_x^{c0}, U_y^{c0}, U_z^{c0}, \Lambda^{c0}) = (0, 0, 1.0288, 0.30292)$ on the fundamental path shown by $\circ$; the associated bifurcation mode $\boldsymbol{\Phi}_1^{c0} = (1, 0, 0)^\top$ denotes x-directional deflection.

By Steps 2–3, we computed

$$(-C_3)^{\frac{3}{2}}\mathbf{a} = (-C_3)^{\frac{3}{2}}(\boldsymbol{\Phi}_1^{c0\top}\mathbf{B}^{c0})^\top = 1.0200^{\frac{3}{2}} \times (0.0256131, 0, 0, 0.0358361, 0, -0.0358361, 0)^\top \tag{16.18}$$

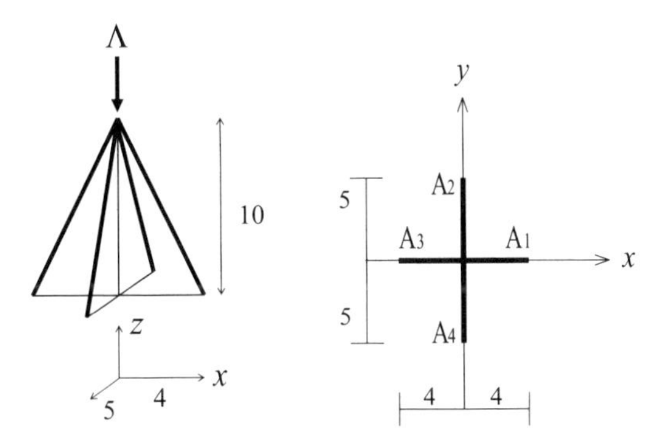

Fig. 16.1 Four-bar truss tent.

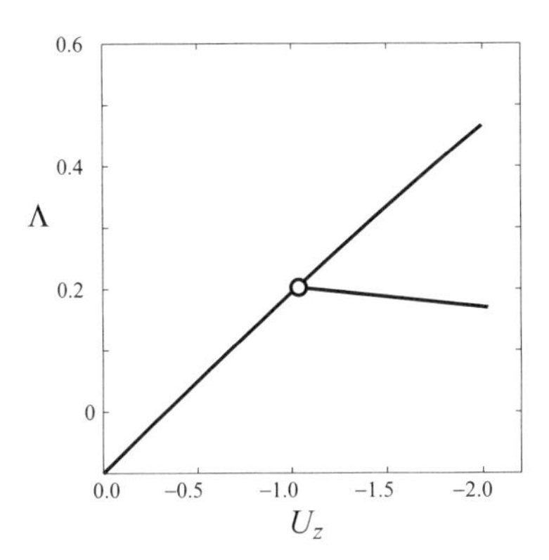

Fig. 16.2 Equilibrium paths for the four-bar truss tent. U_z: z-directional displacement of the center node; ∘: bifurcation point.

By Steps 4–5, we computed

$$\begin{aligned} C_2\mathbf{b} &= C_2(\mathbf{\Phi}_1^{\mathrm{c}0\top}(\partial\mathbf{B}/\partial q_1)^{\mathrm{c}0})^\top \\ &= (0, 0, -5.8976\times 10^{-2}, 0.16688, -1.5416\times 10^{-2}, 0.16688, -1.5416\times 10^{-2})^\top \end{aligned} \tag{16.19}$$

16.4.2 Generalized imperfection sensitivity law

With the use of $(-C_3)^{\frac{3}{2}}\mathbf{a}$ and $C_2\mathbf{b}$ in (16.18) and (16.19), the generalized imperfection sensitivity law considering up to second order influence of imperfections is obtained as

$$\begin{aligned} \widetilde{\Lambda}^{\mathrm{c}}(\widetilde{\mathbf{v}}) &= C_3(\mathbf{a}^\top\widetilde{\mathbf{v}})^{\frac{2}{3}} + C_2\mathbf{b}^\top\widetilde{\mathbf{v}} \\ &= -\left[2.6385\times 10^{-2}\widetilde{v}_1 + 3.6916\times 10^{-2}(\widetilde{v}_{A_1} - \widetilde{v}_{A_3})\right]^{\frac{2}{3}} \\ &\quad +0\times\widetilde{v}_2 - 5.8976\times 10^{-2}\widetilde{v}_3 + 0.16688(\widetilde{v}_{A_1} + \widetilde{v}_{A_3}) \\ &\quad -1.5416\times 10^{-2}(\widetilde{v}_{A_2} + \widetilde{v}_{A_4}) \end{aligned} \tag{16.20}$$

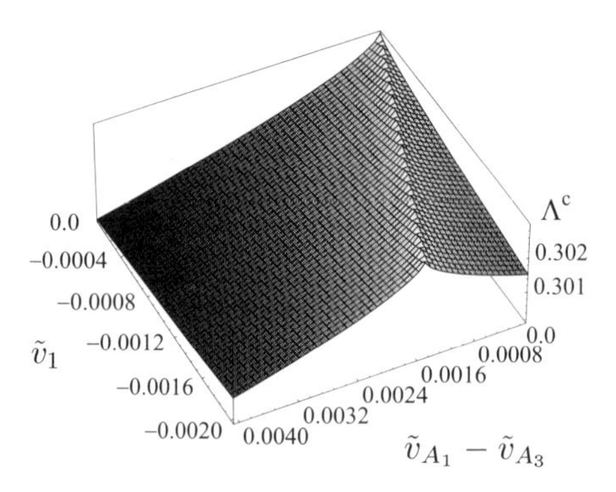

Fig. 16.3 Imperfection sensitivity for major imperfections $\widetilde{v}_1$ and $\widetilde{v}_{A_1} - \widetilde{v}_{A_3}$.

In (16.20), we focus only on major imperfections to arrive at the first-order formula

$$\widetilde{\Lambda}^{\rm c}(\widetilde{\mathbf{v}}) \simeq - \left[2.6385 \times 10^{-2}\widetilde{v}_1 + 3.6916 \times 10^{-2}(\widetilde{v}_{A_1} - \widetilde{v}_{A_3})\right]^{\frac{2}{3}} \quad (16.21)$$

which is plotted in Fig. 16.3; the critical load $\Lambda^{\rm c}$ has a peak on the straight line

$$2.6385 \times 10^{-2}\widetilde{v}_1 + 3.6916 \times 10^{-2}(\widetilde{v}_{A_1} - \widetilde{v}_{A_3}) = 0 \quad (16.22)$$

and decreases sharply at both sides of this straight line.

The accuracy of the generalized laws is investigated on the basis of the following steps:

Step 1: An ensemble of 100 sets of initial imperfections distributing uniformly in the range $[-10^{-4}, 10^{-4}]$ are produced.

Step 2: The associated critical loads $\Lambda^{\rm c}$ are computed by path-tracing analyses.

Step 3: The relative errors

$$\left|[(\text{Theoretical } \Lambda^{\rm c}) - (\text{Computed } \Lambda^{\rm c})]/\Lambda^{\rm c0}\right| \times 100 \quad (16.23)$$

are computed for the theoretical $\Lambda^{\rm c}$ computed by the second-order law (16.20) or the first-order law (16.21).

The histograms of an ensemble of 100 relative errors as shown in Fig. 16.4 were obtained by the above steps. The maximum relative error for the second-order law (16.20) in Fig. 16.4(a) is only 0.0090% to show the accuracy of this formula. The error is 0.0155% for the law (16.21) in Fig. 16.4(b) to show the adequacy of the first-order formula when imperfections are small.

16.4.3 Probability density function of critical loads

The probabilistic variation of critical loads of the four-bar truss tent is investigated. The imperfection variables $\widetilde{v}_i$ $(i = 1, 2, 3)$ and $\widetilde{v}_{A_i}$ $(i = 1, \ldots, 4)$ are assumed to respectively follow the normal distribution with the mean 0 and the

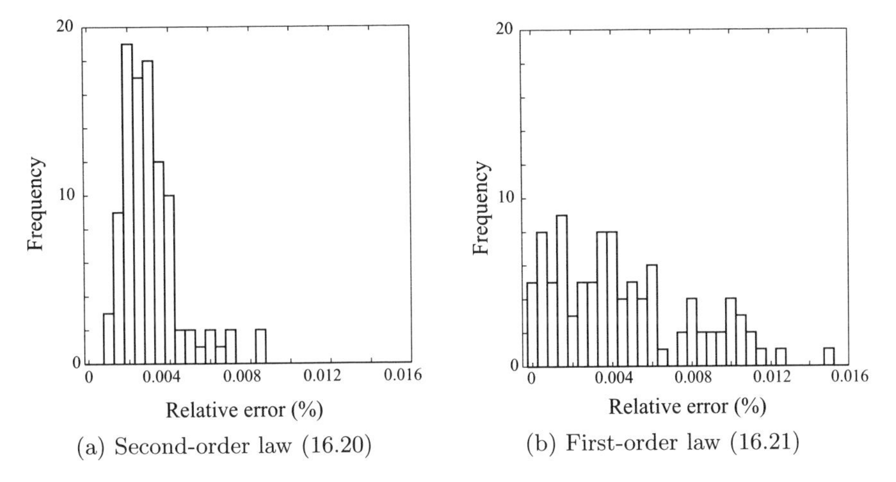

(a) Second-order law (16.20) (b) First-order law (16.21)

Fig. 16.4 Probabilistic distribution of errors of $\Lambda^{\rm c}$ (100 buckled tents).

variance $(\sigma_i)^2$ and $(\sigma_{A_i})^2$. Namely, the variance–covariance matrix $\mathbf{W}$ reads

$$\mathbf{W} = \begin{pmatrix} (\sigma_1)^2 & 0 & 0 & 0 & 0 & 0 & 0 \\ 0 & (\sigma_2)^2 & 0 & 0 & 0 & 0 & 0 \\ 0 & 0 & (\sigma_3)^2 & 0 & 0 & 0 & 0 \\ 0 & 0 & 0 & (\sigma_{A_1})^2 & 0 & 0 & 0 \\ 0 & 0 & 0 & 0 & (\sigma_{A_2})^2 & 0 & 0 \\ 0 & 0 & 0 & 0 & 0 & (\sigma_{A_3})^2 & 0 \\ 0 & 0 & 0 & 0 & 0 & 0 & (\sigma_{A_4})^2 \end{pmatrix} \tag{16.24}$$

We further set $\sigma_i = \sigma_{A_i} = 10^{-3}$. Then with the use of $(-C_3)^{\frac{3}{2}}\mathbf{a}$ and $C_2\mathbf{b}$ in (16.18) and (16.19), and $\mathbf{W}$ in (16.24), we can evaluate the variances in (16.10) and (16.11) as

$$\begin{aligned} (\sigma_a)^2 &= \{(2.6385 \times 10^{-2})^2(\sigma_1)^2 + (3.6916 \times 10^{-2})^2[(\sigma_{A_1})^2 + (\sigma_{A_3})^2]\} \\ &= 3.6312 \times 10^{-9} \end{aligned} \tag{16.25}$$

$$\begin{aligned} (\sigma_b)^2 &= \{(5.8976 \times 10^{-2})^2(\sigma_3)^2 + 0.16688^2[(\sigma_{A_1})^2 + (\sigma_{A_3})^2] \\ &\quad +(1.5416 \times 10^{-2})^2[(\sigma_{A_2})^2 + (\sigma_{A_4})^2]\} \\ &= 5.9651 \times 10^{-8} \end{aligned} \tag{16.26}$$

Then we obtain the probability density functions in (16.12)–(16.14)

$$\phi_a = 2.0460 \times 10^4 |a|^{1/2} \exp(-1.4613 \times 10^8 |a|^3) \tag{16.27}$$

$$\phi_b = 1.6335 \times 10^3 \exp(-8.3823 \times 10^6 b^2) \tag{16.28}$$

$$\phi_{\widetilde{\Lambda}^{\rm c}} = \int_{-\infty}^{0} \phi_a(a)\phi_b(\widetilde{\Lambda}^{\rm c} - a){\rm d}a \tag{16.29}$$

The path-tracing analyses of the four-bar truss tent were conducted for an ensemble of 1000 sets of initial imperfections following the normal distribution prescribed above to arrive at the histogram of critical loads $\widetilde{\Lambda}^{\rm c}$ as shown in Fig. 16.5. The first-order asymptotic approximation by (16.27) shown by the

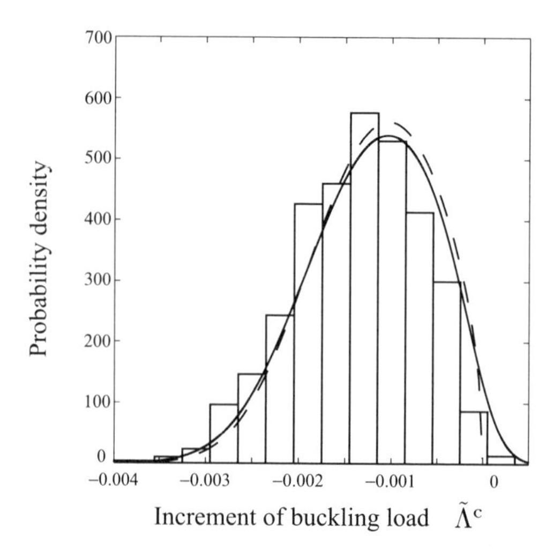

Fig. 16.5 Probabilistic variation of critical loads. Dashed curve: first-order approximation (16.27); solid curve: second-order approximation (16.29).

dashed curve is in a fair agreement with the histogram, but failed to express the critical loads in $\widetilde{\Lambda}^{\rm c} > 0$. The second-order asymptotic approximation by (16.29) shown by the solid curve enjoys a better agreement for $\widetilde{\Lambda}^{\rm c} > 0$.

16.5 Truss Tower Structure

As an example of a large-scale structure, we employ the truss tower structure as shown in Fig. 16.6, which is subjected to the z-directional concentrated load Λ at the crown.

16.5.1 Perfect system

For the perfect system, all members have the same cross-sectional area $A = 1.0$ and the same modulus of elasticity $E = 2.0 \times 10^6$. In the following, the units of length and force are omitted for brevity. The equilibrium paths of the truss tower obtained by path-tracing analysis are shown in Fig. 16.7. The critical load is governed by the unstable-symmetric bifurcation point on the fundamental path shown by $\circ$, and the bifurcation mode triggers the y-directional sway of the tower. The bifurcation load is $\Lambda^{\rm c0} = 1.2505 \times 10^4$.

16.5.2 Generalized imperfection sensitivity law

We employ initial imperfections $\widetilde{v}_i$ $(i = 1, \ldots, 37)$ of the nodal coordinates and cross-sectional areas of the top two layers of the truss tower as shown in Fig. 16.8. The imperfections $\widetilde{v}_i$ $(i = 1, \ldots, 37)$ are defined by

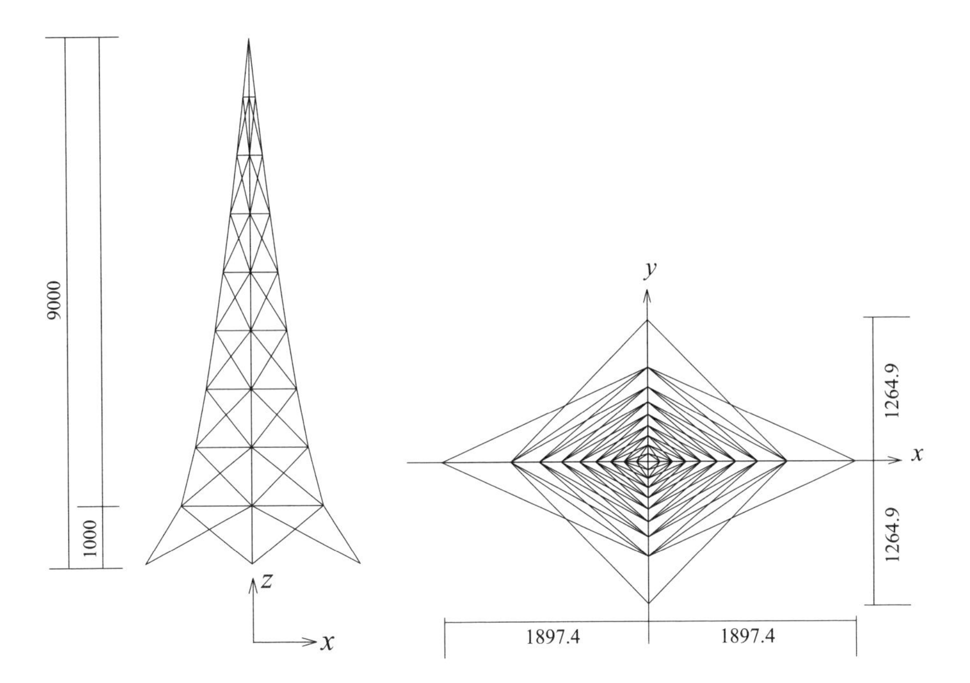

Fig. 16.6 Truss tower.

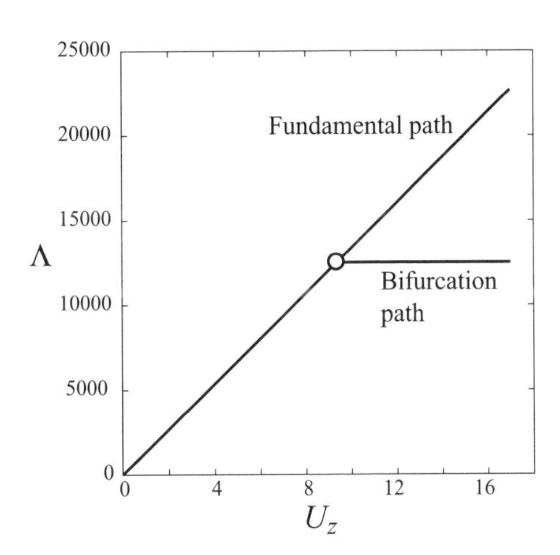

Fig. 16.7 Equilibrium paths of the truss tower. U_z: negative z-directional displacement of node 1; ∘: bifurcation point.

- $\widetilde{v}_{1+3i}$, $\widetilde{v}_{2+3i}$, $\widetilde{v}_{3+3i}$: x-, y- and z-directional movements of nodes $i+1$ ($i = 0, 1, 2, 3, 4$), respectively, and
- $\widetilde{v}_{A_i}$: increase of the cross-sectional area of member i ($i = 1, \ldots, 22$).

Following the same procedure as for the four-bar truss tent in Section 16.4.2, we obtained the generalized imperfection sensitivity law

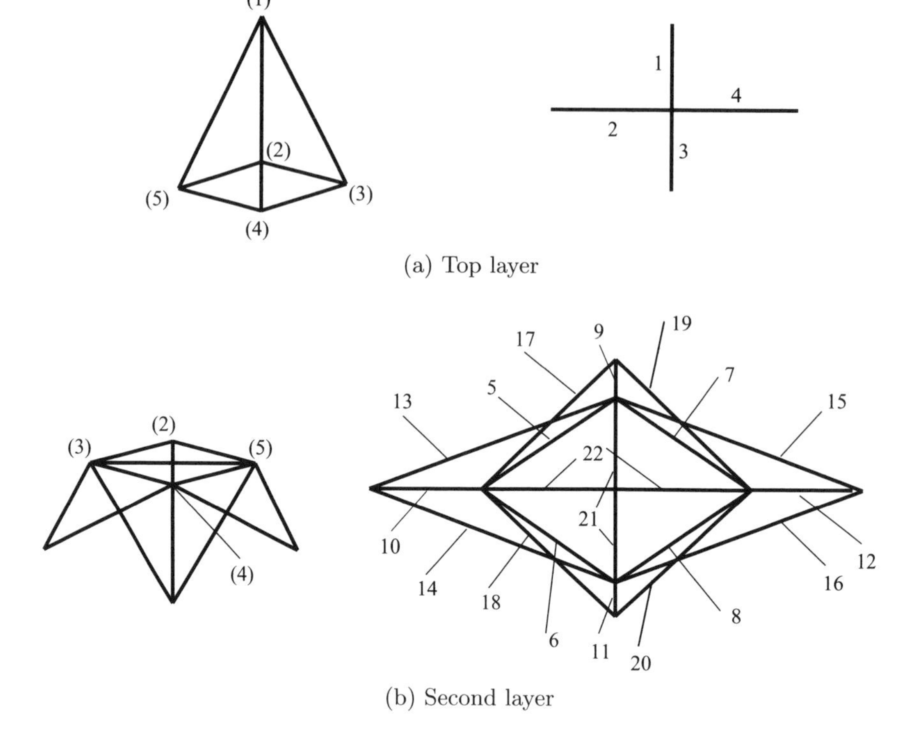

Fig. 16.8 Top two layers of the truss tower. (1),...,(5) denote node numbers; 1,...,22 denote member numbers.

$$
\begin{aligned}
\widetilde{\Lambda}^{\mathrm{c}} \simeq\ & -[551.15\widetilde{v}_2 - 108.40(\widetilde{v}_5 + \widetilde{v}_{11}) \\
& +9.5218(\widetilde{v}_6 - \widetilde{v}_{12}) - 109.62(\widetilde{v}_8 + \widetilde{v}_{14}) - 0.0(\widetilde{v}_9 - \widetilde{v}_{15}) \\
& -6024.0(\widetilde{v}_{A_1} - \widetilde{v}_{A_3}) - 1.9254(\widetilde{v}_{A_5} - \widetilde{v}_{A_6} + \widetilde{v}_{A_7} - \widetilde{v}_{A_8}) \\
& -860.28(\widetilde{v}_{A_9} - \widetilde{v}_{A_{11}}) - 624.40(\widetilde{v}_{A_{13}} - \widetilde{v}_{A_{14}} + \widetilde{v}_{A_{15}} - \widetilde{v}_{A_{16}}) \\
& -178.10(\widetilde{v}_{A_{17}} - \widetilde{v}_{A_{18}} + \widetilde{v}_{A_{19}} - \widetilde{v}_{A_{20}})]^{\frac{2}{3}} \\
& +18.637\widetilde{v}_3 + 127.44(\widetilde{v}_5 - \widetilde{v}_{11}) + 8.0765(\widetilde{v}_6 + \widetilde{v}_{12}) \\
& +6.1196 \times 10^{-2}(\widetilde{v}_7 - \widetilde{v}_{13}) - 0.19304(\widetilde{v}_9 + \widetilde{v}_{15}) \\
& +3508.2(\widetilde{v}_{A_1} + \widetilde{v}_{A_3}) - 6.7274(\widetilde{v}_{A_2} + \widetilde{v}_{A_4}) \\
& +0.94381(\widetilde{v}_{A_5} + \widetilde{v}_{A_6} + \widetilde{v}_{A_7} + \widetilde{v}_{A_8}) + 596.67(\widetilde{v}_{A_9} + \widetilde{v}_{A_{11}}) \\
& -1.1279(\widetilde{v}_{A_{10}} + \widetilde{v}_{A_{12}}) + 329.13(\widetilde{v}_{A_{13}} + \widetilde{v}_{A_{14}} + \widetilde{v}_{A_{15}} + \widetilde{v}_{A_{16}}) \\
& +27.186(\widetilde{v}_{A_{17}} + \widetilde{v}_{A_{18}} + \widetilde{v}_{A_{19}} + \widetilde{v}_{A_{20}}) - 0.93645\widetilde{v}_{A_{21}} + 0.49113\widetilde{v}_{A_{22}}
\end{aligned}
\tag{16.30}
$$

From (16.30), influential imperfections, such as $\widetilde{v}_{A_1}$ and $\widetilde{v}_{A_3}$ with large coefficient values, can be clearly identified.

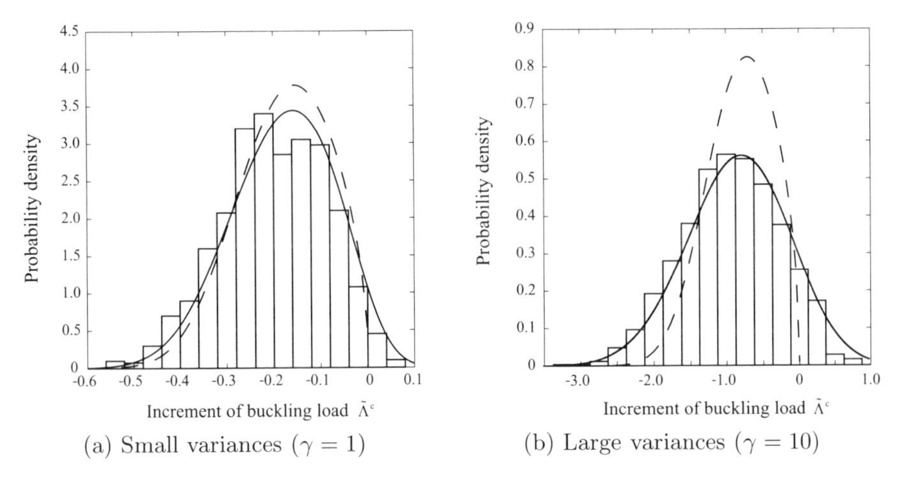

Fig. 16.9 Probabilistic variation of critical loads. Dashed curve: first-order approximation; solid curve: second-order approximation.

16.5.3 Probabilistic variation of critical loads

The probabilistic variation of critical loads for the truss tower is investigated. Initial imperfections $\widetilde{v}_i$ ($i = 1, \ldots, 15$) are assumed to follow a normal distribution $\mathrm{N}(0, 10^{-8}\gamma^2)$, and $\widetilde{v}_{A_i}$ ($i = 1, \ldots, 22$) to $\mathrm{N}(0, 10^{-10}\gamma^2)$ with some scaling constant $\gamma > 0$. The variances in (16.10) and (16.11) for this case are

$$(\sigma_a)^2 = 1.1089 \times 10^{-2}\gamma^2, \quad (\sigma_b)^2 = 2.9059 \times 10^{-3}\gamma^2 \tag{16.31}$$

and the probability density functions in (16.12)–(16.14) are obtained as

$$\phi_a = (1.1365 \times 10/\gamma)|a|^{1/2} \exp[-(4.5090 \times 10/\gamma^2)|a|^3], \qquad (-\infty < a < 0) \tag{16.32}$$

$$\phi_b = (7.4007/\gamma) \exp[-(1.7206 \times 10^2/\gamma^2)b^2], \quad (-\infty < b < \infty) \tag{16.33}$$

$$\phi_{\widetilde{\Lambda}^c} = \int_{-\infty}^{0} \phi_a(a)\phi_b(\widetilde{\Lambda}^c - a)\mathrm{d}a, \quad (-\infty < \widetilde{\Lambda}^c < \infty) \tag{16.34}$$

For two cases of $\gamma = 1, 10$, path-tracing analyses of the truss tower were conducted on an ensemble of 1000 sets of initial imperfections following the normal distribution prescribed above to arrive at the histograms of critical loads $\widetilde{\Lambda}^c$ as shown in Fig. 16.9.

- For $\gamma = 1$ with small variances of initial imperfections, the first-order asymptotic approximation by (16.32) plotted by the dashed curve in Fig. 16.9(a) failed to express the critical loads in $\widetilde{\Lambda}^c > 0$. The second-order asymptotic approximation by (16.34) plotted by the solid curve displays a better agreement with the histogram.
- For $\gamma = 10$ with large variances of initial imperfections, the first-order asymptotic approximation by (16.32) plotted by the dashed curve in Fig. 16.9(b) is apparently inaccurate. The second-order asymptotic

approximation by (16.34) plotted by the solid curve displays a drastic improvement.

16.6 Summary

In this chapter,

- the generalized sensitivity law that can represent the influence of a number of initial imperfections in a systematic manner has been developed,
- the formulas for probabilistic variation of critical loads with Gaussian random imperfections have been developed, and
- the numerical procedures to evaluate the generalized law and the probabilistic formulas have been presented.

The major finding of this chapter is as follows.

- The first order law is simple and readily applicable to practice, while the second order law achieves higher accuracy. As a consequence of this, the proposed methodology combines the theoretical completeness and practical applicability.

Appendix

A.1 Introduction

We introduce derivations of several formulas and details of numerical examples in the Appendix. Although some derivations seem set aside at the end of this book, they offer very important ingredients for readers who are interested in stability theory.

Appendix is organized as follows:

- Interpolation approach for coincident critical points is presented in Appendix A.2.
- Derivation of the explicit diagonalization approach is given in Appendix A.3.
- Derivation of the block diagonalization approach is given in Appendix A.4.
- Details of quadratic estimation of critical loads of the braced column are presented in Appendix A.5.
- Differential coefficients of the bar–spring model are provided in Appendix A.6.
- Imperfection sensitivity laws are derived for a semi-symmetric bifurcation point in Appendix A.7 and for degenerate hilltop points in Appendices A.8 and A.9.

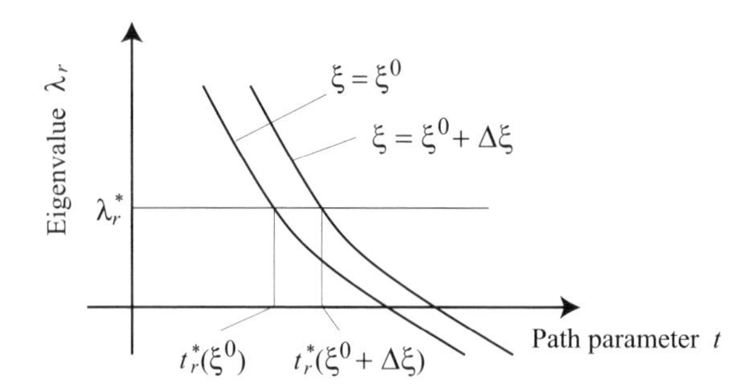

Fig. A.1 Relation between path parameter t and eigenvalues λ_r of the tangent stiffness matrix.

A.2 Interpolation Approach for Coincident Critical Points

The interpolation approach in Section 2.5 for a simple critical point is not applicable to a coincident critical point, because the eigenmodes of the tangent stiffness matrix cannot be determined uniquely from (1.5). In this section, we present an interpolation approach for a coincident critical point, including a hill-top branching point but excluding a group-theoretic double bifurcation point. We consider an m-fold critical point, and assume that imperfections are minor for all bifurcation modes (cf., (2.39) in Section 2.6.2).

The fundamental path is traced by increasing path parameter t, which can represent the load factor, nodal displacement, arc-length parameter, and so on. Suppose, at a regular equilibrium state on the fundamental path, that the design (imperfection) parameter ξ is modified while the eigenvalues of $\mathbf{S}$ are fixed to be $\lambda_i = \lambda_i^*$ $(i = 1, \dots, m)$. The value of path parameter $t = t_r^*$ where $\lambda_r = \lambda_r^*$ is satisfied varies according to modification of design parameter ξ. The relation between t and λ_r is illustrated in Fig. A.1 for two designs corresponding to $\xi = \xi^0$ and $\xi^0 + \Delta\xi$. Other variables corresponding to the equilibrium state with $\lambda_r = \lambda_r^*$ are also indicated by $(\,\cdot\,)^*$ below.

Differentiating (1.1), (1.5) and (1.7) with respect to ξ, and fixing an eigenvalue $\lambda_r = \lambda_r^*$ $(\lambda_r' = 0)$, we obtain the following set of equations

$$\sum_{j=1}^{n} S_{,ij} U_j^{*\prime} + S_{,i\xi} + S_{,i\Lambda} \Lambda^{*\prime} = 0 \tag{A.1}$$

$$\sum_{j=1}^{n} \left(\sum_{k=1}^{n} S_{,ijk} \phi_{rj} U_k^{*\prime} + S_{,ij\xi} \phi_{rj} + S_{,ij\Lambda} \phi_{rj} \Lambda^{*\prime} + S_{,ij} {\phi_{rj}^*}' \right) = \lambda_r {\phi_{ri}^*}' \tag{A.2}$$

$$\sum_{j=1}^{n} \phi_{rj} {\phi_{rj}^*}' = 0 \tag{A.3}$$

The sensitivity coefficients of $\mathbf{U}^*$, Λ^* and $\mathbf{\Phi}_r^*$ for fixed value of $\lambda_r = \lambda_r^* \neq 0$ are found by solving the set of $2n + 1$ linear equations (A.1)–(A.3).

There are several difficulties in the evaluation of the sensitivity coefficients.

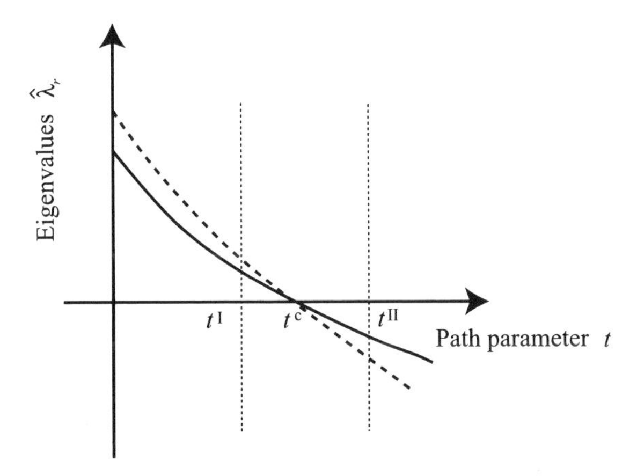

Fig. A.2 Variation of eigenvalues with respect to the path parameter t. Solid curve: $\widehat{\lambda}_1$; dashed curve: $\widehat{\lambda}_2$.

- At the critical point with $\lambda_r = \lambda_r^* = 0$, these sensitivity coefficients cannot be obtained in this way, because the matrix $[S_{,ij}]$ in (A.1) and (A.2) is singular [228] as discussed in Section 2.5 for a simple critical point.
- Nonuniqueness of $\mathbf{\Phi}_r$ is another difficulty in sensitivity analysis at a coincident critical point. Since the eigenmodes of the multiple eigenvalues are expressed as linear combinations of linearly independent modes, the critical mode should be appropriately chosen to obtain the directional derivatives as discussed in Section 2.6.2.
- Since the critical load factor $\Lambda_1^{\rm c}$ is defined as the smallest value of the critical loads, the gradient of $\Lambda_1^{\rm c}$ is discontinuous at a coincident critical point (cf., Fig. 2.7 in Section 2.6.2); therefore, only directional derivatives can be defined.

With the use of the interpolation approach (2.33), nonuniqueness of the eigenmode can be avoided because the eigenvalues λ_r of the tangent stiffness matrix do not exactly coincide at $t = t^{\rm I}$ and $t^{\rm II}$ near the critical point, respectively, at which $\lambda_1 > 0$ and $\lambda_1 < 0$ are satisfied. Let $\widehat{\lambda}_r$ for some r denote the eigenvalue of $\mathbf{S}$, for which the associated mode shape is continuous in the range $t^{\rm I} \leq t \leq t^{\rm II}$. Fig. A.2 illustrates the variation of the two lowest eigenvalues $\widehat{\lambda}_1$ and $\widehat{\lambda}_2$ which are continuously differentiable functions of t. The load factors at $t = t^{\rm I}$ and $t^{\rm II}$ are denoted by $\Lambda^{\rm I}$ and $\Lambda^{\rm II}$, respectively. The sensitivity coefficients $\widehat{\Lambda}_r^{\rm I\prime}$ and $\widehat{\Lambda}_r^{\rm II\prime}$ for fixed values of $\widehat{\lambda}_r = \lambda_r^* = \widehat{\lambda}_r^{\rm I}$ and $\widehat{\lambda}_r^{\rm II}$, respectively, are calculated from (A.1)–(A.3). Then the sensitivity coefficients at the critical load corresponding to $\widehat{\lambda}_r = 0$ $(r = 1, \dots, m)$ are obtained by the interpolation as

$$\widehat{\Lambda}_r^{\rm c\prime} = \frac{\widehat{\lambda}_r^{\rm I}\widehat{\Lambda}_r^{\rm II\prime} - \widehat{\lambda}_r^{\rm II}\widehat{\Lambda}_r^{\rm I\prime}}{\widehat{\lambda}_r^{\rm I} - \widehat{\lambda}_r^{\rm II}}, \quad (r = 1, \dots, m) \tag{A.4}$$

The directional derivatives of the critical load factor can be found as the minimum and maximum values among $\widehat{\Lambda}_r^{\rm c\prime}$ $(r = 1, \dots, m)$. In this way, the directional derivatives at the critical point can be obtained by solving m-times the linear equations of the size $2n+1$.

In Section 4.4, we presented optimization problem P3 incorporating the constraints $\lambda_r^{\mathrm{c}} > 0$ $(r \geq m+1)$. The sensitivity coefficients of λ_r^{c} are obtained as follows also by interpolation.

Suppose $\widehat{\Lambda}_\alpha^{\mathrm{c}\,\prime}$ is the maximum or minimum value among $\widehat{\Lambda}_r^{\mathrm{c}\,\prime}$ $(r = 1,\dots,m)$, which corresponds to the directional derivative of the critical load factor. Differentiation of (1.5) for $r = m+1,\dots,n$ at $t = t^{\mathrm{I}}$ or t^{II} under the condition $\lambda_\alpha = \lambda_\alpha^*$ leads to the following equation:

$$\sum_{j=1}^{n}\Bigl(\sum_{k=1}^{n} S_{,ijk}\phi_{rj}U_k^{*\prime} + S_{,ij\xi}\phi_{rj} + S_{,ij\Lambda}\phi_{rj}\Lambda^{*\prime} + S_{,ij}\phi_{rj}^{*\;\prime}\Bigr) \\ = \lambda_r\phi_{ri}^{*\;\prime} + \lambda_r^{*\prime}\phi_{ri} \tag{A.5}$$

Note that $U_k^{*\prime}$ and $\Lambda^{*\prime}$ have already been obtained from (A.1)–(A.3) at $t = t^{\mathrm{I}}$ and t^{II}. By multiplying ϕ_{ri} to both sides of (A.5), taking summation over i, and incorporating (1.5) and (1.7), we can derive

$$\lambda_r^{*\prime} = \sum_{i=1}^{n}\sum_{j=1}^{n}\Bigl(\sum_{k=1}^{n} S_{,ijk}\phi_{ri}\phi_{rj}U_k^{*\prime} + S_{,ij\xi}\phi_{ri}\phi_{rj} + S_{,ij\Lambda}\phi_{ri}\phi_{rj}\Lambda^{*\prime}\Bigr) \tag{A.6}$$

Therefore, $\phi_{ri}^{*\;\prime}$ is not needed in evaluating $\lambda_r^{*\prime}$, which is similar to the case of sensitivity analysis of linear buckling load in Section 1.4. The sensitivity coefficients $\widehat{\lambda}_r^{\mathrm{I}\,\prime}$ and $\widehat{\lambda}_r^{\mathrm{II}\,\prime}$ are found by evaluating (A.6) at $t = t^{\mathrm{I}}$ and t^{II}, respectively.

Then $\widehat{\lambda}_r^{\mathrm{c}\,\prime}$ at the critical point is calculated by the following interpolation equation:

$$\widehat{\lambda}_r^{\mathrm{c}\,\prime} = \frac{\widehat{\lambda}_\alpha^{\mathrm{I}}\widehat{\lambda}_r^{\mathrm{II}\,\prime} - \widehat{\lambda}_\alpha^{\mathrm{II}}\widehat{\lambda}_r^{\mathrm{I}\,\prime}}{\widehat{\lambda}_\alpha^{\mathrm{I}} - \widehat{\lambda}_\alpha^{\mathrm{II}}} \tag{A.7}$$

A.3 Derivation of Explicit Diagonalization Approach

Explicit diagonalization approach for a minor imperfection is derived for a simple unstable-symmetric bifurcation point and a coincident bifurcation point of a symmetric system. To be precise, the details of the formula (2.37) with (2.38) for the explicit diagonalization are presented.

A.3.1 Simple unstable-symmetric bifurcation point

We consider a minor imperfection at a simple unstable-symmetric bifurcation point, for which $\Lambda^{\mathrm{c}\,\prime}$ exists (cf., Section 2.3).

Differentiation of (1.1) and (1.5) for $r = 1$ with respect to ξ with the use of $\lambda_1 = \lambda_1{}' = 0$ and evaluation at the critical point lead, respectively, to

$$\sum_{j=1}^{n} S_{,ij} U_j^{\mathrm{c}\prime} + S_{,i\xi} + S_{,i\Lambda} \Lambda^{\mathrm{c}\prime} = 0 \tag{A.8}$$

$$\sum_{j=1}^{n} \Bigl(\sum_{k=1}^{n} S_{,ijk} \phi_{1j}^{\mathrm{c}} U_k^{\mathrm{c}\prime} + S_{,ij\xi} \phi_{1j}^{\mathrm{c}} + S_{,ij\Lambda} \phi_{1j}^{\mathrm{c}} \Lambda^{\mathrm{c}\prime} + S_{,ij} \phi_{1j}^{\mathrm{c}}{}' \Bigr) = 0 \tag{A.9}$$

Multiplying ϕ_{ri}^{c} to (A.8) and summing up for i lead to

$$\sum_{i=1}^{n} (\phi_{ri}^{\mathrm{c}} U_i^{\mathrm{c}\prime} \lambda_r^{\mathrm{c}} + S_{,i\xi} \phi_{ri}^{\mathrm{c}} + S_{,i\Lambda} \phi_{ri}^{\mathrm{c}} \Lambda^{\mathrm{c}\prime}) = 0 \tag{A.10}$$

By multiplying ϕ_{1i}^{c} to (A.9), summing up for i and by using (1.5) and $\lambda_1^{\mathrm{c}} = 0$, we obtain

$$\sum_{i=1}^{n} \sum_{j=1}^{n} \Bigl(\sum_{k=1}^{n} S_{,ijk} \phi_{1i}^{\mathrm{c}} \phi_{1j}^{\mathrm{c}} U_k^{\mathrm{c}\prime} + S_{,ij\xi} \phi_{1i}^{\mathrm{c}} \phi_{1j}^{\mathrm{c}} + S_{,ij\Lambda} \phi_{1i}^{\mathrm{c}} \phi_{1j}^{\mathrm{c}} \Lambda^{\mathrm{c}\prime} \Bigr) = 0 \tag{A.11}$$

Since the bifurcation point in question is symmetric, we have (cf., Section 3.3)

$$V_{,111} = \sum_{i=1}^{n} \sum_{j=1}^{n} \sum_{k=1}^{n} S_{,ijk} \phi_{1i}^{\mathrm{c}} \phi_{1j}^{\mathrm{c}} \phi_{1k}^{\mathrm{c}} = 0 \tag{A.12}$$

The relation (3.2) between the displacements U_i^{c} and Q_j^{c}, and their derivatives, at the critical point are written as

$$U_i^{\mathrm{c}} = \sum_{r=1}^{n} \phi_{ri}^{\mathrm{c}} Q_r^{\mathrm{c}}, \quad U_i^{\mathrm{c}\prime} = \sum_{r=1}^{n} \phi_{ri}^{\mathrm{c}} Q_r^{\mathrm{c}\prime} \tag{A.13}$$

Note that ϕ_{ri}^{c} is used only for transformation of displacements, and is fixed during differentiation.

Incorporating (A.13) into (A.10) and using (1.7), we obtain

$$\lambda_r^{\mathrm{c}} Q_r^{\mathrm{c}\prime} + \sum_{i=1}^{n} (S_{,i\xi} \phi_{ri}^{\mathrm{c}} + S_{,i\Lambda} \phi_{ri}^{\mathrm{c}} \Lambda^{\mathrm{c}\prime}) = 0, \quad (r = 2, \ldots, n) \tag{A.14}$$

and incorporating (A.13) into (A.11) and using (A.12), we obtain

$$\sum_{i=1}^{n} \sum_{j=1}^{n} \Bigl(\sum_{k=1}^{n} \sum_{r=2}^{n} S_{,ijk} \phi_{1i}^{\mathrm{c}} \phi_{1j}^{\mathrm{c}} \phi_{rk}^{\mathrm{c}} Q_r^{\mathrm{c}\prime} + S_{,ij\xi} \phi_{1i}^{\mathrm{c}} \phi_{1j}^{\mathrm{c}} + S_{,ij\Lambda} \phi_{1i}^{\mathrm{c}} \phi_{1j}^{\mathrm{c}} \Lambda^{\mathrm{c}\prime} \Bigr) = 0 \tag{A.15}$$

Since $\lambda_r > \lambda_1^{\mathrm{c}} = 0$, $(r = 2, \ldots, n)$, (A.14) yields the expression for passive coordinates

$$Q_r^{\mathrm{c}\prime} = -\frac{1}{\lambda_r^{\mathrm{c}}} \sum_{i=1}^{n} \left(S_{,i\xi} \phi_{ri}^{\mathrm{c}} + S_{,i\Lambda} \phi_{ri}^{\mathrm{c}} \Lambda^{\mathrm{c}\prime} \right), \quad (r = 2, \ldots, n) \tag{A.16}$$

Incorporating (A.16) into (A.15), we can eliminate the passive coordinates to obtain the following formula of sensitivity coefficient $\Lambda^{c\prime} = C_2$ in (2.13) as

$$\Lambda^{c\prime} = \frac{1}{\mu}\Big[\sum_{i=1}^{n}\sum_{j=1}^{n} S_{,ij\xi}\phi_{1i}^{c}\phi_{1j}^{c} - \sum_{r=2}^{n}\frac{1}{\lambda_r}\Big(\sum_{i=1}^{n}\sum_{j=1}^{n}\sum_{k=1}^{n} S_{,ijk}\phi_{1i}^{c}\phi_{1j}^{c}\phi_{rk}^{c}\sum_{i=1}^{n} S_{,i\xi}\phi_{ri}^{c}\Big)\Big] \tag{A.17}$$

where μ is independent of ξ and is given as

$$\mu = -\sum_{i=1}^{n}\sum_{j=1}^{n} S_{,ij\Lambda}\phi_{1i}^{c}\phi_{1j}^{c} + \sum_{r=2}^{n}\frac{1}{\lambda_r}\Big(\sum_{i=1}^{n}\sum_{j=1}^{n}\sum_{k=1}^{n} S_{,ijk}\phi_{1i}^{c}\phi_{1j}^{c}\phi_{rk}^{c}\sum_{i=1}^{n} S_{,i\Lambda}\phi_{ri}^{c}\Big) \tag{A.18}$$

It is noted that $\Lambda^{c\prime}$ in (A.17) permits a simple expression

$$\Lambda^{c\prime} = -\frac{V_{,11\xi}}{V_{,11\Lambda}} \tag{A.19}$$

in the V-formulation (cf., (3.20) and Section 3.2.3).

Remark A.3.1 Note that $Q_1^{c\prime}$ cannot be defined by (A.16) because $\lambda_1^c = 0$. For a symmetric bifurcation point, the term for $Q_1^{c\prime}$ is absent in (A.15) by virtue of (A.12). Such absence is also the case for an asymmetric bifurcation point if the prebuckling deformation $\mathbf{U}^c$ does not have the component of Q_1^c. □

A.3.2 Coincident critical point of symmetric system

The formulas (2.43) and (2.44) for the explicit diagonalization approach are derived below.

We assume Λ^c is differentiable with respect to a minor imperfection ξ. Consider the following linear combination of the eigenvectors corresponding to active coordinates

$$\boldsymbol{\Psi}^c = \sum_{\alpha=1}^{m} a_\alpha \boldsymbol{\Phi}_\alpha^c \tag{A.20}$$

Differentiation of (1.1) and (1.5) for vector $\boldsymbol{\Psi}^c$ in (A.20) with respect to ξ leads to

$$\sum_{j=1}^{n} S_{,ij}U_j^{c\prime} + S_{,i\xi} + S_{,i\Lambda}\Lambda^{c\prime} = 0 \tag{A.21}$$

$$\sum_{j=1}^{n}\Big(\sum_{k=1}^{n} S_{,ijk}\psi_j^c U_k^{c\prime} + S_{,ij\xi}\psi_j^c + S_{,ij\Lambda}\psi_j^c\Lambda^{c\prime} + S_{,ij}\psi_j^{c\prime}\Big) = 0 \tag{A.22}$$

Eq. (A.21) is divided into two parts in preparation for the elimination of passive coordinates (cf., Section 3.2.3). By multiplying $\phi_{\alpha i}^c$ $(\alpha = 1,\ldots,m)$ to (A.21) and summing over i, we obtain the first part

$$\sum_{i=1}^{n}(S_{,i\xi}\phi_{\alpha i}^c + S_{,i\Lambda}\phi_{\alpha i}^c\Lambda^{c\prime}) = 0, \quad (\alpha = 1,\ldots,m) \tag{A.23}$$

by (1.5) and $\lambda_\alpha = 0$ $(\alpha = 1, \dots, m)$. The second part is also derived by multiplying $\phi^{\rm c}_{ri}$ $(r = m+1, \dots, n)$ to (A.21):

$$\sum_{i=1}^{n} (\phi^{\rm c}_{ri} U^{\rm c}_i{}' \lambda^{\rm c}_r + S_{,i\xi}\phi^{\rm c}_{ri} + S_{,i\Lambda}\phi^{\rm c}_{ri}\Lambda^{\rm c}{}') = 0, \quad (r = m+1, \dots, n) \tag{A.24}$$

Consider a completely minor imperfection that satisfies (2.39):

$$\sum_{i=1}^{n} S_{,i\xi}\phi^{\rm c}_{\alpha i} = 0, \quad (\alpha = 1, \dots, m) \tag{A.25}$$

Then it may be observed from (A.23) that $\Lambda^{\rm c}{}'$ can possibly be bounded for a bifurcation load factor satisfying $\sum_{i=1}^{n} S_{,i\Lambda}\phi^{\rm c}_{\alpha i} = 0$.

By multiplying $\psi^{\rm c}_i$ to (A.22), we can derive

$$\sum_{i=1}^{n}\sum_{j=1}^{n}\Big(\sum_{k=1}^{n} S_{,ijk}\psi^{\rm c}_i\psi^{\rm c}_j U^{\rm c}_k{}' + S_{,ij\xi}\psi^{\rm c}_i\psi^{\rm c}_j + S_{,ij\Lambda}\psi^{\rm c}_i\psi^{\rm c}_j\Lambda^{\rm c}{}'\Big) = 0 \tag{A.26}$$

where $S_{,ij}\psi^{\rm c}_j = 0$, since $\mathbf{\Psi}^{\rm c}$ is a critical eigenvector.

The displacements $U^{\rm c}_i$ are expressed by $Q^{\rm c}_j$ as (A.13). By incorporating (A.13) into (A.24), we obtain

$$Q^{\rm c}_r{}' = -\frac{1}{\lambda^{\rm c}_r}(S_{,i\xi}\phi^{\rm c}_{ri} + S_{,i\Lambda}\phi^{\rm c}_{ri}\Lambda^{\rm c}{}'), \quad (r = m+1, \dots, n) \tag{A.27}$$

For symmetric systems, $Q^{\rm c}_\alpha{}' = 0$ $(\alpha = 1, \dots, m)$ is satisfied; i.e., the prebuckling deformation does not have components of the buckling modes of the perfect system for any symmetric system defined by the parameter ξ. Then the following relation is derived from (A.26) with (A.13):

$$\sum_{i=1}^{n}\sum_{j=1}^{n}\Big(\sum_{k=1}^{n}\sum_{r=m+1}^{n} S_{,ijk}\psi^{\rm c}_i\psi^{\rm c}_j\phi^{\rm c}_{rk}Q^{\rm c}_r{}' + S_{,ij\xi}\psi^{\rm c}_i\psi^{\rm c}_j + S_{ij\Lambda}\psi^{\rm c}_i\psi^{\rm c}_j\Lambda^{\rm c}{}'\Big) = 0 \tag{A.28}$$

The elimination of passive coordinates by incorporating (A.27) into (A.28) and using (A.20) leads to the formula of sensitivity coefficient

$$\Lambda^{\rm c}{}' = \frac{\sum_{\alpha=1}^{m}\sum_{\beta=1}^{m} a_\alpha a_\beta \Xi_{\alpha\beta}}{\sum_{\alpha=1}^{m}\sum_{\beta=1}^{m} a_\alpha a_\beta \Theta_{\alpha\beta}} \tag{A.29}$$

where

$$\Xi_{\alpha\beta} = \sum_{i=1}^{n}\sum_{j=1}^{n} S_{,ij\xi}\phi^{\rm c}_{\alpha i}\phi^{\rm c}_{\beta j} - \sum_{r=m+1}^{n}\frac{1}{\lambda^{\rm c}_r}\Big(\sum_{k=i}^{n}\sum_{j=1}^{n}\sum_{k=1}^{n} S_{,ijk}\phi^{\rm c}_{\alpha i}\phi^{\rm c}_{\beta j}\phi^{\rm c}_{rk}\sum_{i=1}^{n} S_{,i\xi}\phi^{\rm c}_{ri}\Big) \tag{A.30}$$

and $\Theta_{\alpha\beta}$ is independent of ξ as

$$\Theta_{\alpha\beta} = -\sum_{i=1}^{n}\sum_{j=1}^{n} S_{,ij\Lambda}\phi^{\rm c}_{\alpha i}\phi^{\rm c}_{\beta j} + \sum_{r=m+1}^{n}\frac{1}{\lambda^{\rm c}_r}\Big(\sum_{i=1}^{n}\sum_{j=1}^{n}\sum_{k=1}^{n} S_{,ijk}\phi^{\rm c}_{\alpha i}\phi^{\rm c}_{\beta j}\phi^{\rm c}_{rk}\sum_{i=1}^{n} S_{,i\Lambda}\phi^{\rm c}_{ri}\Big) \tag{A.31}$$

In the V-formulation in Section 3.2.3, we have alternative expressions of (A.30) and (A.31)

$$\Xi_{\alpha\beta} = V_{,\alpha\beta\xi}, \quad \Theta_{\alpha\beta} = -V_{,\alpha\beta\Lambda}, \quad (\alpha, \beta = 1, \dots, m) \tag{A.32}$$

A.4 Block Diagonalization Approach for Symmetric System

A block diagonalization approach to evaluate the sensitivity coefficient of a bifurcation load of a symmetric system is proposed for a simple symmetric bifurcation point with an antisymmetric bifurcation mode. Note that bifurcation usually takes place for a symmetric system, and block diagonalization method [118, 204] is an established means to exploit symmetry for a structure [138].

A.4.1 Symmetry condition

The symmetry of a structure in general is classified to *reflection symmetry* and *rotation symmetry*. Suppose that the action of the reflection or rotation on nodal displacement vector $\mathbf{U}$ is represented by the orthogonal representation matrix $\mathbf{R} \in \mathbb{R}^{n \times n}$. Then symmetry of $\mathbf{U}$ is expressed as

$$\mathbf{RU} = \mathbf{U} \tag{A.33}$$

The symmetry condition for the total potential energy reads (cf., Section 3.6.5 for more detailed definition of symmetry):

$$\Pi(\mathbf{RU}, \Lambda^{\mathrm{c}}) = \Pi(\mathbf{U}, \Lambda^{\mathrm{c}}) \tag{A.34}$$

Then the tangent stiffness matrix $\mathbf{S}$ satisfies the symmetry condition

$$\mathbf{RS} = \mathbf{SR} \tag{A.35}$$

The eigenmodes $\mathbf{\Phi}_i$ of $\mathbf{S}$ are classified to

$$\begin{cases} \text{symmetric mode :} & \mathbf{R\Phi}_i = \mathbf{\Phi}_i \\ \text{antisymmetric mode :} & \mathbf{R\Phi}_i = -\mathbf{\Phi}_i \end{cases} \tag{A.36}$$

Note that a bifurcation mode of a symmetric bifurcation point is antisymmetric.

A.4.2 Block diagonalization

Let $\mathbf{h}_i^- = (h_{ij}^-) \in \mathbb{R}^n$ and $\mathbf{h}_\alpha^+ = (h_{\alpha j}^+) \in \mathbb{R}^n$ denote the symmetric and antisymmetric sets of basis vectors satisfying

$$\begin{aligned} \mathbf{h}_i^- &= \mathbf{Rh}_i^-, \quad (i = 1, \dots, n^-) \\ \mathbf{h}_\alpha^+ &= -\mathbf{Rh}_\alpha^+, \quad (\alpha = 1, \dots, n^+) \end{aligned} \tag{A.37}$$

where $n^- + n^+ = n$. The symmetric and antisymmetric eigenvectors of $\mathbf{S}$ at the initial undeformed state, e.g., can be chosen as the basis vectors $\mathbf{h}_i^-$ and $\mathbf{h}_\alpha^+$, respectively. Expand the displacement vector into

$$\mathbf{U} = \sum_{i=1}^{n^-} Q_i^- \mathbf{h}_i^- + \sum_{\alpha=1}^{n^+} Q_\alpha^+ \mathbf{h}_\alpha^+ \tag{A.38}$$

and define the vectors $\mathbf{Q}^- = (Q_i^-) \in \mathbb{R}^{n^-}$ and $\mathbf{Q}^+ = (Q_\alpha^+) \in \mathbb{R}^{n^+}$. From (A.37) and (A.38), we have

$$\begin{aligned} \mathbf{RU} &= \sum_{i=1}^{n^-} Q_i^- \mathbf{Rh}_i^- + \sum_{\alpha=1}^{n^+} Q_\alpha^+ \mathbf{Rh}_\alpha^+ \\ &= \sum_{i=1}^{n^-} Q_i^- \mathbf{h}_i^- + \sum_{\alpha=1}^{n^+} (-Q_\alpha^+) \mathbf{h}_\alpha^+ \end{aligned} \tag{A.39}$$

Thus we have the correspondence

$$\mathbf{U} \to \mathbf{RU} \iff (\mathbf{Q}^-, \mathbf{Q}^+) \to (\mathbf{Q}^-, -\mathbf{Q}^+) \tag{A.40}$$

The total potential energy $\Pi^{\mathrm{G}}(\mathbf{Q}^-, \mathbf{Q}^+, \Lambda)$ is symmetric in the sense that

$$\Pi^{\mathrm{G}}(\mathbf{Q}^-, -\mathbf{Q}^+, \Lambda) = \Pi^{\mathrm{G}}(\mathbf{Q}^-, \mathbf{Q}^+, \Lambda) \tag{A.41}$$

Then the equilibrium equations satisfy

$$G_{,i}^-(\mathbf{Q}^-, -\mathbf{Q}^+, \Lambda) = G_{,i}^-(\mathbf{Q}^-, \mathbf{Q}^+, \Lambda) = 0, \quad (i = 1, \dots, n^-) \tag{A.42a}$$

$$G_{,\alpha}^+(\mathbf{Q}^-, -\mathbf{Q}^+, \Lambda) = -G_{,\alpha}^+(\mathbf{Q}^-, \mathbf{Q}^+, \Lambda) = 0, \quad (\alpha = 1, \dots, n^+) \tag{A.42b}$$

Here differentiations of Π^{G} with respect to Q_i^- and Q_α^+ are denoted by $G_{,i}^-$ and $G_{,\alpha}^+$, respectively. On the fundamental path of a symmetric perfect system, (A.42b) is satisfied by $\mathbf{Q}^+ = \mathbf{0}$.

Owing to the symmetry (A.41), the tangent stiffness matrix $\mathbf{G}$ with respect to $\mathbf{Q} = (\mathbf{Q}^{-\top}, \mathbf{Q}^{+\top})^\top$ is block diagonalized as [135, 138]

$$\mathbf{G} = \begin{bmatrix} [G_{,ij}^-] & \mathbf{O} \\ \mathbf{O} & [G_{,\alpha\beta}^+] \end{bmatrix} \tag{A.43}$$

For a minor modification preserving symmetry, the system retains the symmetry in the sense of (A.34) and the bifurcation mode $\mathbf{\Phi}_1^{\mathrm{c}+}(\xi) \in \mathbb{R}^n$ of a modified system is antisymmetric by (A.36). Therefore, $\mathbf{\Phi}_1^{\mathrm{c}+}(\xi)$ can be expressed by a linear combination of antisymmetric basis vectors $\mathbf{h}_\alpha^+$ as

$$\mathbf{\Phi}_1^{\mathrm{c}+}(\xi) = \sum_{\alpha=1}^{n^+} C_\alpha^+(\xi) \mathbf{h}_\alpha^+ \tag{A.44}$$

where C_α^+ is a coefficient.

The conditions for the bifurcation point are written as the reduced form of the extended system

$$G_{,i}^- = 0, \quad (i = 1, \dots, n^-) \tag{A.45}$$

$$\sum_{\beta=1}^{n^+} G_{,\alpha\beta}^+ C_\beta^+ = 0, \quad (\alpha = 1, \dots, n^+) \tag{A.46}$$

with normalization condition

$$\sum_{\alpha=1}^{n^+}(C_\alpha^+)^2 = 1 \tag{A.47}$$

Differentiation of (A.45)–(A.47) with respect to ξ at the bifurcation point leads to the set of $n^- + n^+ + 1$ $(= n+1)$ linear equations

$$\sum_{j=1}^{n^-} G^-_{,ij} Q_j^{-\mathrm{c}\prime} + G^-_{,i\xi} + G^-_{,i\Lambda}\Lambda^{\mathrm{c}\prime} = 0, \quad (i = 1,\ldots,n^-) \tag{A.48}$$

$$\sum_{\beta=1}^{n^+}\left[\left(\sum_{i=1}^{n^-}\frac{\partial G^+_{,\alpha\beta}}{\partial Q_i^-}Q_i^{-\mathrm{c}\prime} + G^+_{,\alpha\beta\xi} + G^+_{,\alpha\beta\Lambda}\Lambda^{\mathrm{c}\prime}\right)C_\beta^+ + G^+_{,\alpha\beta}C_\beta^{+\prime}\right] = 0,$$
$$(\alpha = 1,\ldots,n^+) \tag{A.49}$$

$$\sum_{\alpha=1}^{n^+} C_\alpha^+ C_\alpha^{+\prime} = 0 \tag{A.50}$$

Note that $Q_\alpha^{+\prime} = 0$ holds for the minor modification that preserves symmetry, and $(\partial G^-_{,i}/\partial Q_\alpha^+)Q_\alpha^{+\prime} = 0$ by (A.43).

The sensitivity coefficients $\Lambda^{\mathrm{c}\prime}$ and $Q_i^{-\mathrm{c}\prime}$ of the load factor and the displacements at the critical point can be found by solving the set of equations (A.48)–(A.50). The singularity of the tangent stiffness matrix has been successfully avoided by block diagonalization, as $[G^-_{,ij}]$ in (A.48) is nonsingular.

Remark A.4.1 In order to accurately compute sensitivity coefficients, the critical load has to be computed accurately. In the conventional incremental response analysis, an adoption of a small increment does not always improve the accuracy but often ends up with the divergence of the incremental quantities because the tangent stiffness matrix is singular at the bifurcation point. As a remedy of this, the reduced form of an extended system (A.45)–(A.47) is solved directly for $Q_j^{-\mathrm{c}}$, C_α and Λ^{c} by the Newton–Raphson method [311], as conducted in the numerical examples in Section 2.7. □

A.5 Details of Quadratic Estimation of Critical Loads

The Step 1-ii in Section 12.2.4 for obtaining the critical loads of imperfect systems corresponding to the worst imperfection consists of the following substeps:

Step 1-ii-a: Assign appropriate values for Λ^* and $\Delta\Lambda$, and calculate displacements $\mathbf{U}$ of the perfect system by linear approximation at three load levels Λ^-, Λ^*, Λ^+ $(\Lambda^\pm = \Lambda^* \pm \Delta\Lambda)$.

Step 1-ii-b: Find three sets of eigenvalues and eigenmodes $(\lambda_r^{-0}, \boldsymbol{\Phi}_r^{-0})$, $(\lambda_r^{*0}, \boldsymbol{\Phi}_r^{*0})$, $(\lambda_r^{+0}, \boldsymbol{\Phi}_r^{+0})$ of $\mathbf{S}$ from (1.5), respectively, at $\Lambda = \Lambda^-$, Λ^* and Λ^+.

Step 1-ii-c: Find the worst imperfection mode for each eigenvalue at the three load levels by LP or QP formulation.

Step 1-ii-d: Calculate the increments $\delta\lambda_r^-$, $\delta\lambda_r^*$ and $\delta\lambda_r^+$ of the eigenvalues at Λ^-, $\Lambda = \Lambda^*$ and Λ^+, respectively.

Step 1-ii-e: Estimate λ_r^k of the imperfect system using the following linear relation at the three load levels:

$$\lambda_r^* = \lambda_r^{*0} + \delta\lambda_r^*, \quad \lambda_r^\pm = \lambda_r^{\pm 0} + \delta\lambda_r^\pm \tag{A.51}$$

Step 1-ii-f: Compute the first- and second-order differential coefficients of the eigenvalues of the imperfect system with respect to Λ based on the finite difference approach as

$$\Delta\lambda_r^* = \frac{\lambda_r^+ - \lambda_r^-}{2\Delta\Lambda}, \quad \Delta^2\lambda_r^* = \frac{\Delta\lambda_r^+ - \Delta\lambda_r^-}{\Delta\Lambda} \tag{A.52}$$

where $\Delta\lambda_r^+ = (\lambda_r^+ - \lambda_r^*)/\Delta\Lambda$ and $\Delta\lambda_r^- = (\lambda_r^* - \lambda_r^-)/\Delta\Lambda$.

Step 1-ii-g: Find the critical load factor Λ_r^{c} of the imperfect system from

$$\Lambda_r^{\mathrm{c}} = \Lambda^* + \delta\Lambda_r^* \tag{A.53}$$

with the use of $\delta\Lambda_r^*$ that is calculated from the quadratic approximation

$$\lambda_r^* + \Delta\lambda_r^*\delta\Lambda_r^* + \frac{1}{2}\Delta^2\lambda_r^*(\delta\Lambda_r^*)^2 = 0 \tag{A.54}$$

A.6 Differential Coefficients of Bar–Spring Model

The higher-order differential coefficients of Π of the bar–spring model in Section 8.3 are shown below. The symmetric perfect system in Section 4.6 satisfies $k_{\mathrm{h}2} = 0$, $\overline{\eta} = 0$, $\overline{\theta} = \text{const.}$

The second order differential coefficients of Π are obtained as

$$\begin{aligned} S_{,11} &= k_{h1}H^2\cos^2(\overline{\eta}+\eta) - k_{h1}H^2 a\sin(\overline{\eta}+\eta) + k_{h2}H^3 a\cos^2(\overline{\eta}+\eta) \\ &\quad -\frac{1}{2}k_{h2}H^3a^2\sin(\overline{\eta}+\eta) + k_{\mathrm{r}1}^{\mathrm{L}} + k_{\mathrm{r}2}^{\mathrm{L}}(\theta-\eta) + k_{\mathrm{r}1}^{\mathrm{R}} + k_{\mathrm{r}2}^{\mathrm{R}}(\theta+\eta) \\ &\quad -\Lambda pH\cos(\overline{\eta}+\eta) \end{aligned} \tag{A.55a}$$

$$\begin{aligned} S_{,22} &= 4k_{\mathrm{t}}L^2\sin^2(\overline{\theta}-\theta) - 4k_{\mathrm{t}}L^2[\cos(\overline{\theta}-\theta) - \cos\overline{\theta}]\cos(\overline{\theta}-\theta) \\ &\quad +k_{\mathrm{r}1}^{\mathrm{L}} + k_{\mathrm{r}2}^{\mathrm{L}}(\theta-\eta) + k_{\mathrm{r}1}^{\mathrm{R}} + k_{\mathrm{r}2}^{\mathrm{R}}(\theta+\eta) - \Lambda pL\sin(\overline{\theta}-\theta) \end{aligned} \tag{A.55b}$$

$$S_{,12} = -k_{\mathrm{r}1}^{\mathrm{L}} - k_{\mathrm{r}2}^{\mathrm{L}}(\theta-\eta) + k_{\mathrm{r}1}^{\mathrm{R}} + k_{\mathrm{r}2}^{\mathrm{R}}(\theta+\eta) \tag{A.55c}$$

$$S_{,1\Lambda} = -H\sin(\overline{\eta}+\eta) \tag{A.55d}$$

$$S_{,2\Lambda} = -L\cos(\overline{\theta}-\theta) \tag{A.55e}$$

Higher-order differential coefficients of Π are written as

$$\begin{aligned} S_{,111} &= -3k_{h1}H^2\cos(\overline{\eta}+\eta)\sin(\overline{\eta}+\eta) - k_{h1}H^2 a\cos(\overline{\eta}+\eta) \\ &\quad +k_{h2}H^3\cos^3(\overline{\eta}+\eta) - 3k_{h2}H^3 a\cos(\overline{\eta}+\eta)\sin(\overline{\eta}+\eta) \\ &\quad -\frac{1}{2}k_{h2}H^3a^2\cos(\overline{\eta}+\eta) - k_{r2}^L + k_{r2}^R + \Lambda pH\sin(\overline{\eta}+\eta) \quad \text{(A.56a)} \\ S_{,222} &= -12k_tL^2\sin(\overline{\theta}-\theta)\cos(\overline{\theta}-\theta) \\ &\quad -4k_tL^2[\cos(\overline{\theta}-\theta)-\cos\overline{\theta}]\sin(\overline{\theta}-\theta) \\ &\quad +k_{r2}^L + k_{r2}^R + \Lambda pL\cos(\overline{\theta}-\theta) \quad \text{(A.56b)} \\ S_{,112} &= k_{r2}^L + k_{r2}^R \quad \text{(A.56c)} \\ S_{,122} &= -k_{r2}^L + k_{r2}^R \quad \text{(A.56d)} \end{aligned}$$

Consider a major imperfection $\xi = \overline{\eta}$. Differential coefficients with respect to ξ are obtained as

$$\begin{aligned} S_{,1\xi} &= k_{h1}H^2[\cos(\overline{\eta}+\eta)-\cos\overline{\eta}]\cos(\overline{\eta}+\eta) - k_{h1}H^2a\sin(\overline{\eta}+\eta) \\ &\quad +k_{h2}H^3a\cos(\overline{\eta}+\eta)[\cos(\overline{\eta}+\eta)-\cos\overline{\eta}] - \frac{1}{2}k_{h2}H^3a^2\sin(\overline{\eta}+\eta) \\ &\quad +\frac{1}{(\overline{\theta}+\overline{\eta}+\pi/2)^2}\left[\,\widehat{k}_{r1}(\theta-\eta)+\frac{1}{2}\widehat{k}_{r2}(\theta-\eta)^2\right] \\ &\quad +\frac{1}{(\overline{\theta}-\overline{\eta}+\pi/2)^2}\left[\,\widehat{k}_{r1}(\theta+\eta)+\frac{1}{2}\widehat{k}_{r2}(\theta+\eta)^2\right] \\ &\quad -\Lambda pH\cos(\overline{\eta}+\eta) \quad \text{(A.57a)} \\ S_{,2\xi} &= \frac{1}{(\overline{\theta}+\overline{\eta}+\pi/2)^2}\left[-\widehat{k}_{r1}(\theta-\eta)-\frac{1}{2}\widehat{k}_{r2}(\theta-\eta)^2\right] \\ &\quad +\frac{1}{(\overline{\theta}-\overline{\eta}+\pi/2)^2}\left[\,\widehat{k}_{r1}(\theta+\eta)+\frac{1}{2}\widehat{k}_{r2}(\theta+\eta)^2\right] \quad \text{(A.57b)} \end{aligned}$$

A.7 Imperfection Sensitivity Law of a Semi-Symmetric Bifurcation Point

The imperfection sensitivity law of a semi-symmetric double bifurcation point is derived. As was done in Section 3.6.2, the critical eigenmodes $\boldsymbol{\Phi}_1^{\rm c}$ and $\boldsymbol{\Phi}_2^{\rm c}$ are assumed to be symmetric and asymmetric bifurcation modes, respectively. We focus on the imperfection mode in the direction of $\boldsymbol{\Phi}_2^{\rm c}$ and assume [129, 284]

$$\begin{aligned} &V_{,1\xi} = 0, \quad V_{,2\xi} < 0 \\ &V_{,111} = V_{,122} = 0, \quad V_{,222} < 0, \quad V_{,112} \neq 0 \end{aligned} \quad \text{(A.58)}$$

Employ the further assumption that the fundamental path is stable until reaching this point, i.e.,

$$V_{,11\Lambda} < 0, \quad V_{,22\Lambda} < 0 \quad \text{(A.59)}$$

and define structural parameter

$$\kappa = \frac{V_{,22\Lambda} V_{,112}}{V_{,222} V_{,11\Lambda}} \tag{A.60}$$

From (A.58)–(A.60), the following relation holds

$$\operatorname{sign} V_{,112} = -\operatorname{sign} \kappa \tag{A.61}$$

The total potential energy at the critical point of the perfect system $(\mathbf{q}^{\mathrm{a}}, \widetilde{\Lambda}, \xi) = (\mathbf{0}, 0, 0)$ is expanded as

$$\begin{aligned}\Pi^{\mathrm{V}}(\mathbf{q}^{\mathrm{a}}, \widetilde{\Lambda}, \xi) = \Pi^{\mathrm{V}}(\mathbf{0}, 0, 0) + V_{,2\xi} q_2 \xi + \frac{1}{2} V_{,112} (q_1)^2 q_2 + \frac{1}{6} V_{,222} (q_2)^3 \\ + \frac{1}{2} V_{,11\Lambda} (q_1)^2 \widetilde{\Lambda} + \frac{1}{2} V_{,22\Lambda} (q_2)^2 \widetilde{\Lambda} + \frac{1}{24} V_{,1111} (q_1)^4 + \text{h.o.t.}\end{aligned} \tag{A.62}$$

The bifurcation equations are obtained as

$$\frac{\partial \Pi^{\mathrm{V}}}{\partial q_1} = V_{,112} q_1 q_2 + V_{,11\Lambda} q_1 \widetilde{\Lambda} + \frac{1}{6} V_{,1111} (q_1)^3 + \text{h.o.t.} = 0 \tag{A.63}$$

$$\frac{\partial \Pi^{\mathrm{V}}}{\partial q_2} = V_{,2\xi} \xi + \frac{1}{2} V_{,112} (q_1)^2 + \frac{1}{2} V_{,222} (q_2)^2 + V_{,22\Lambda} q_2 \widetilde{\Lambda} + \text{h.o.t.} = 0 \tag{A.64}$$

The stability (tangent stiffness) matrix of the bifurcation equation becomes

$$\mathbf{S}^{\mathrm{V}} = \begin{pmatrix} V_{,112} q_2 + V_{,11\Lambda} \widetilde{\Lambda} + \frac{1}{2} V_{,1111} (q_1)^2 & V_{,112} q_1 \\ V_{,112} q_1 & V_{,222} q_2 + V_{,22\Lambda} \widetilde{\Lambda} \end{pmatrix} \tag{A.65}$$

From (A.63), we have

$$\begin{cases} q_1 = 0 : & \text{fundamental path} \\ V_{,112} q_2 + V_{,11\Lambda} \widetilde{\Lambda} + \frac{1}{6} V_{,1111} (q_1)^2 + \text{h.o.t.} = 0 : & \text{bifurcation path} \end{cases} \tag{A.66}$$

A.7.1 Limit point load

Consider the fundamental path, on which

$$q_1 = 0, \quad V_{,112} q_2 + V_{,11\Lambda} \widetilde{\Lambda} + \text{h.o.t.} \neq 0 \tag{A.67}$$

are satisfied by putting $\xi = 0$ in (A.63). Then $\det \mathbf{S}^{\mathrm{V}} = 0$ with (A.65) and (A.67) becomes

$$V_{,222} q_2 + V_{,22\Lambda} \widetilde{\Lambda} = 0 \tag{A.68}$$

which yields

$$q_2 = -\frac{V_{,22\Lambda}}{V_{,222}} \widetilde{\Lambda} \tag{A.69}$$

The simultaneous solution of (A.64) and (A.69) gives the locations of the two limit points of an imperfect system

$$\widetilde{\Lambda}^{\mathrm{c}}_{\mathrm{Lim}}(\xi) = \begin{cases} \pm \dfrac{|2 V_{,222} V_{,2\xi} \xi|^{\frac{1}{2}}}{V_{,22\Lambda}} : & \text{for } \xi > 0 \\ \text{non-existent} : & \text{for } \xi < 0 \end{cases} \tag{A.70}$$

$$(q_2)^{\mathrm{c}}_{\mathrm{Lim}}(\xi) = \begin{cases} \pm\left|\dfrac{2V_{,2\xi}\xi}{V_{,222}}\right|^{\frac{1}{2}} > 0 : & \text{for } \xi > 0 \\ \text{non-existent} : & \text{for } \xi < 0 \end{cases} \tag{A.71}$$

Here the double signs take the same order. With the use of (A.59), the imperfection sensitivity law for the limit point on the fundamental path reads

$$\widetilde{\Lambda}^{\mathrm{c}}_{\mathrm{Lim}}(\xi) = \frac{|2V_{,222}V_{,2\xi}\xi|^{\frac{1}{2}}}{V_{,22\Lambda}} \tag{A.72}$$

that exists for $\xi > 0$ (cf., Fig. A.3).

A.7.2 Bifurcation load

Consider the bifurcation path with $V_{,112}q_2 + V_{,11\Lambda}\widetilde{\Lambda} = 0$, namely

$$q_2 = -\frac{V_{,11\Lambda}}{V_{,112}}\widetilde{\Lambda} \tag{A.73}$$

The simultaneous solution of (A.64), (A.73) and $\det \mathbf{S}^{\mathrm{V}} = 0$ leads to the locations of the two bifurcation points of an imperfect system

$$\widetilde{\Lambda}^{\mathrm{c}}_{\mathrm{Bif}}(\xi) = \begin{cases} \pm\dfrac{V_{,112}}{V_{,11\Lambda}}\left|\dfrac{2V_{,2\xi}\xi}{V_{,222}(2\kappa-1)}\right|^{\frac{1}{2}} : & \text{for } (2\kappa-1)\xi > 0 \\ \text{non-existent} : & \text{for } (2\kappa-1)\xi < 0 \end{cases} \tag{A.74}$$

$$(q_2)^{\mathrm{c}}_{\mathrm{Bif}}(\xi) = \begin{cases} \mp\left|\dfrac{2V_{,2\xi}\xi}{V_{,222}(2\kappa-1)}\right|^{\frac{1}{2}} : & \text{for } (2\kappa-1)\xi > 0 \\ \text{non-existent} : & \text{for } (2\kappa-1)\xi < 0 \end{cases} \tag{A.75}$$

A.7.3 Imperfect behaviors

Imperfect behaviors are investigated below. The limit point exists for $\xi > 0$ for any value of κ from (A.70). The locations of the bifurcation points change according to the values of κ and ξ.

From (A.74), for $(2\kappa - 1)\xi > 0$, we can see the presence of a pair of bifurcation points at (q_2^*, Λ^*) and $(-q_2^*, -\Lambda^*)$ with

$$\Lambda^* = -\frac{V_{,112}}{V_{,11\Lambda}}\left|\frac{2V_{,2\xi}\xi}{V_{,222}(2\kappa-1)}\right|^{\frac{1}{2}}, \quad q_2^* = \left|\frac{2V_{,2\xi}\xi}{V_{,222}(2\kappa-1)}\right|^{\frac{1}{2}} > 0 \tag{A.76}$$

Note that

$$\begin{cases} \Lambda^* > 0 : & \text{for } \kappa < 0 \ (V_{,112} > 0) \\ \Lambda^* < 0 : & \text{for } \kappa > 0 \ (V_{,112} < 0) \end{cases} \tag{A.77}$$

In view of the condition $(2\kappa - 1)\xi > 0$, we have further classification:

- For $\kappa < 1/2$, the bifurcation points are existent for $\xi < 0$, as shown in Fig. A.3(a). On the fundamental path of an imperfect system, a limit point exists for $\xi > 0$ and a bifurcation point exists for $\xi < 0$ at $(-q_2^*, -\Lambda^*)$.

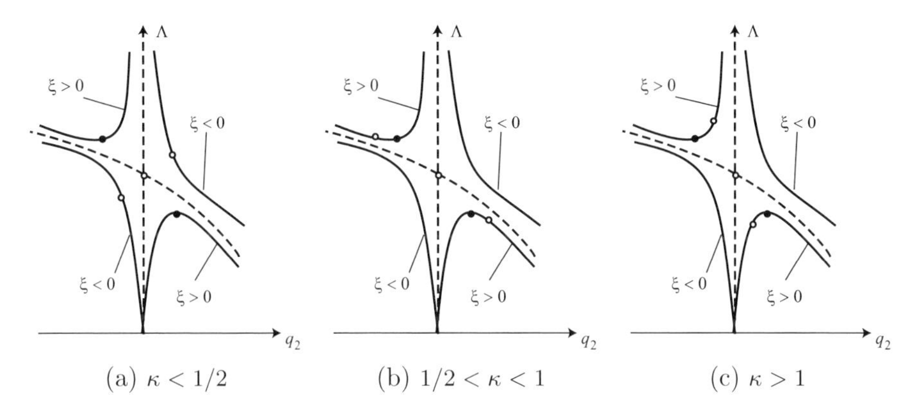

Fig. A.3 Critical points of imperfect semi-symmetric systems. •: limit point; ○: bifurcation point.

- For $\kappa > 1/2$, the bifurcation points are existent for $\xi > 0$. We investigate the relative location of the limit and bifurcation point on the fundamental path. The following relation is obtained from (A.72) and (A.74) with (A.60):

$$|\widetilde{\Lambda}^{\mathrm{c}}_{\mathrm{Lim}}| < |\widetilde{\Lambda}^{\mathrm{c}}_{\mathrm{Bif}}| \tag{A.78}$$

From (A.71) and (A.75), we have

$$(q_2)^{\mathrm{c}}_{\mathrm{Lim}}/(q_2)^{\mathrm{c}}_{\mathrm{Bif}} = \sqrt{2\kappa - 1} \tag{A.79}$$

and hence

$$\begin{cases} (q_2)^{\mathrm{c}}_{\mathrm{Lim}} > (q_2)^{\mathrm{c}}_{\mathrm{Bif}} > 0: & \text{for } \kappa > 1 \\ (q_2)^{\mathrm{c}}_{\mathrm{Bif}} > (q_2)^{\mathrm{c}}_{\mathrm{Lim}} > 0: & \text{for } 1/2 < \kappa < 1 \end{cases} \tag{A.80}$$

In view of (A.80), the imperfect imperfect behaviors are classified as shown in Figs. A.3(b) and (c).

To sum up, the imperfection sensitivity law for the bifurcation load on the fundamental path is given by

$$\widetilde{\Lambda}^{\mathrm{c}}_{\mathrm{Bif}}(\xi) = -\operatorname{sign}\xi \frac{V_{,112}}{V_{,11\Lambda}} \left| \frac{2V_{,2\xi}\xi}{V_{,222}(2\kappa - 1)} \right|^{\frac{1}{2}} \tag{A.81}$$

A.8 Imperfection Sensitivity Laws of Degenerate Hilltop Point I: Asymmetric Bifurcation

The imperfection sensitivity laws of degenerate hilltop points with simple asymmetric bifurcation are derived.

A.8.1 General formulation

We consider the degenerate case of

$$V_{,112} = 0, \quad V_{,1122} \neq 0 \tag{A.82}$$

and focus on the most customary case in practice where the system has a rising fundamental path and becomes unstable at the hilltop point, and we have (cf., (8.19), (8.20) and Remark 8.2.1)

$$V_{,2\Lambda} < 0, \quad V_{,222} < 0 \tag{A.83}$$

$$C_\alpha = \frac{1}{2V_{,2\Lambda}}(V_{,2\Lambda}V_{,1122} - V_{,222}V_{,11\Lambda}) > 0 \tag{A.84}$$

With the use of (A.82), the bifurcation equations (8.10) and (8.11) become

$$\begin{aligned}\frac{\partial \Pi^{\mathrm{V}}}{\partial q_1} &= V_{,1\xi}\xi + \frac{1}{2}V_{,111}(q_1)^2 + V_{,11\Lambda}q_1\widetilde{\Lambda} + V_{,11\xi}q_1\xi + V_{,12\xi}q_2\xi \\ &+\frac{1}{2}V_{,1122}q_1(q_2)^2 = 0\end{aligned} \tag{A.85}$$

$$\begin{aligned}\frac{\partial \Pi^{\mathrm{V}}}{\partial q_2} &= V_{,2\Lambda}\widetilde{\Lambda} + V_{,2\xi}\xi + \frac{1}{2}V_{,222}(q_2)^2 + V_{,12\xi}q_1\xi \\ &+\frac{1}{2}V_{,1122}(q_1)^2 q_2 = 0\end{aligned} \tag{A.86}$$

and the criticality condition (8.13) becomes

$$\det\begin{pmatrix} S^{\mathrm{V}}_{11} & S^{\mathrm{V}}_{12} \\ S^{\mathrm{V}}_{21} & S^{\mathrm{V}}_{22}\end{pmatrix} = 0 \tag{A.87}$$

with

$$S^{\mathrm{V}}_{11} = V_{,111}q_1 + V_{,11\Lambda}\widetilde{\Lambda} + V_{,11\xi}\xi + \frac{1}{2}V_{,1122}(q_2)^2 \tag{A.88a}$$

$$S^{\mathrm{V}}_{12} = S^{\mathrm{V}}_{21} = V_{,12\xi}\xi + V_{,1122}q_1q_2 + V_{,112\Lambda}q_1\widetilde{\Lambda} + \frac{1}{2}V_{,1112}(q_1)^2 \tag{A.88b}$$

$$S^{\mathrm{V}}_{22} = V_{,222}q_2 + V_{,22\Lambda}\widetilde{\Lambda} + V_{,22\xi}\xi + \frac{1}{2}V_{,1122}(q_1)^2 \tag{A.88c}$$

A.8.2 Perfect behavior

Consider the perfect system with $\xi = 0$. From (A.85), we have two solutions:

$$\begin{cases} q_1 = 0: & \text{trivial fundamental path} \\ \frac{1}{2}V_{,111}q_1 + V_{,11\Lambda}\widetilde{\Lambda} + \frac{1}{2}V_{,1122}(q_2)^2 = 0: & \text{bifurcation path}\end{cases} \tag{A.89}$$

The equations (A.86) with $\xi = 0$ and (A.89) give a bifurcation path parameterized by q_2

$$\widetilde{\Lambda} = \widetilde{\Lambda}(q_2) = -\frac{V_{,222}}{2V_{,2\Lambda}}(q_2)^2 \leq 0 \tag{A.90a}$$

$$q_1 = q_1(q_2) = -\frac{2C_\alpha}{V_{,111}}(q_2)^2 \tag{A.90b}$$

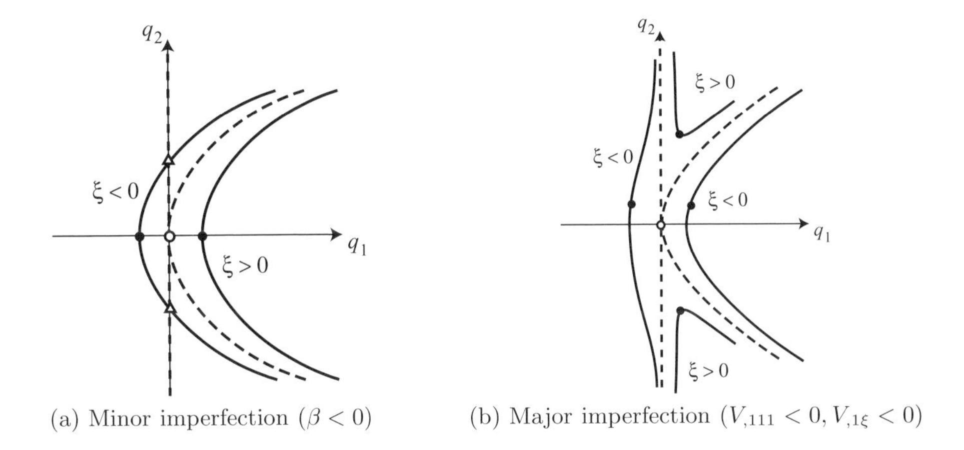

(a) Minor imperfection ($\beta < 0$) (b) Major imperfection ($V_{,111} < 0, V_{,1\xi} < 0$)

Fig. A.4 Asymptotic behaviors in the vicinity of a degenerate hilltop branching point with asymmetric bifurcation. ∘: hilltop point; •: limit point; △: unstable-symmetric bifurcation point; dashed curve: perfect equilibrium path; solid curve: imperfect equilibrium path.

that is always existent. The perfect behavior expressed by (A.90b) is depicted in Fig. A.4(a) by the dashed curves.

A.8.3 Imperfection sensitivity: minor symmetric

Consider a minor symmetric imperfection, for which

$$\begin{aligned}&V_{,1\xi} = 0, \quad V_{,12\xi} = 0, \quad \dots \\ &V_{,11\xi}, \ V_{,2\xi}, \ \dots : \ \text{possibly nonzero}\end{aligned} \tag{A.91}$$

are satisfied by (3.64). The imperfect curves expressed by (A.85) and (A.86) are shown by the solid curves in Fig. A.4(a). Eq. (A.85) has two solutions:

$$q_1 = 0 : \text{trivial fundamental path} \tag{A.92}$$

$$\frac{1}{2}V_{,111}q_1 + V_{,11\Lambda}\widetilde{\Lambda} + V_{,11\xi}\xi + \frac{1}{2}V_{,1122}(q_2)^2 = 0 : \quad \text{bifurcation or aloof path} \tag{A.93}$$

For the trivial fundamental path $q_1 = 0$, (A.86) becomes

$$V_{,2\Lambda}\widetilde{\Lambda} + V_{,2\xi}\xi + \frac{1}{2}V_{,222}(q_2)^2 = 0 \tag{A.94}$$

The limit point load is given for $q_2 = 0$ as

$$\widetilde{\Lambda}^{\rm c}_{\rm Lim}(\xi) = -\frac{V_{,2\xi}}{V_{,2\Lambda}}\xi \tag{A.95}$$

For the bifurcation or aloof path (A.93), we seek for the solution of the order

$$\widetilde{\Lambda} = O(\xi), \quad q_1 = O(\xi^2), \quad q_2 = O(\xi^{\frac{1}{2}}) \tag{A.96}$$

where $O(\,\cdot\,)$ denotes the order of the term in the parentheses. Then the criticality condition (A.87) is satisfied, and the set of equations (A.86) and (A.93) yields

$$\begin{pmatrix} V_{,2\Lambda} & V_{,222} \\ V_{,11\Lambda} & V_{,1122} \end{pmatrix} \begin{pmatrix} \widetilde{\Lambda} \\ \frac{1}{2}(q_2)^2 \end{pmatrix} = - \begin{pmatrix} V_{,2\xi} \\ V_{,11\xi} \end{pmatrix} \xi \tag{A.97}$$

Its solution gives the bifurcation load

$$\widetilde{\Lambda}^{\rm c}_{\rm Bif}(\xi) = - \frac{V_{,2\xi}V_{,1122} - V_{,222}V_{,11\xi}}{V_{,1122}V_{,2\Lambda} - V_{,222}V_{,11\Lambda}}\xi \tag{A.98}$$

that is existent for

$$(q_2)^2 = \beta\xi > 0 \tag{A.99}$$

or otherwise no bifurcation points exist. Here

$$\beta = 2\frac{V_{,2\xi}V_{,11\Lambda} - V_{,2\Lambda}V_{,11\xi}}{V_{,2\Lambda}V_{,1122} - V_{,222}V_{,11\Lambda}} \tag{A.100}$$

A.8.4 Imperfection sensitivity: major antisymmetric

Consider a major antisymmetric imperfection, which satisfies

$$\begin{aligned} &V_{,11\xi} = 0, \quad V_{,2\xi} = 0, \quad V_{,22\xi} = 0 \\ &V_{,1\xi}, \quad V_{,12\xi} : \text{ possibly nonzero} \end{aligned} \tag{A.101}$$

by (3.66). The imperfect curves expressed by (A.85) and (A.86) are shown by the solid curves in Fig. A.4(b) ($V_{,111} < 0$, $V_{,1\xi} < 0$). Then the set of equations (A.85)–(A.87) has two solutions of different orders

$$\text{I)} \quad \widetilde{\Lambda} = O(\xi^{\frac{1}{2}}), \quad q_1 = O(\xi^{\frac{1}{2}}), \quad q_2 = O(\xi^{\frac{1}{4}}) \tag{A.102}$$

$$\text{II)} \quad \widetilde{\Lambda} = O(\xi^{\frac{3}{2}}), \quad q_1 = O(\xi^{\frac{1}{2}}), \quad q_2 = O(\xi) \tag{A.103}$$

Type I) solution

For (A.102), the set of equations (A.85)–(A.87) reduces to

$$V_{,1\xi}\xi + \frac{1}{2}V_{,111}(q_1)^2 + V_{,11\Lambda}q_1\widetilde{\Lambda} + \frac{1}{2}V_{,1122}q_1(q_2)^2 = 0 \tag{A.104}$$

$$V_{,2\Lambda}\widetilde{\Lambda} + \frac{1}{2}V_{,222}(q_2)^2 = 0 \tag{A.105}$$

$$S^{\rm V}_{11} = V_{,111}q_1 + V_{,11\Lambda}\widetilde{\Lambda} + \frac{1}{2}V_{,1122}(q_2)^2 = 0 \tag{A.106}$$

The solutions of this set of equations (A.104)–(A.106) give the locations of two limit points of an imperfect system as

$$q_1^{\rm c}(\xi) = -\,\text{sign}\, V_{,111}\left(\frac{2V_{,1\xi}\xi}{V_{,111}}\right)^{\frac{1}{2}} \tag{A.107}$$

$$q_2^{\rm c}(\xi) = \pm\frac{(2V_{,111}V_{,1\xi}\xi)^{\frac{1}{4}}}{(C_\alpha)^{\frac{1}{2}}} \tag{A.108}$$

$$\widetilde{\Lambda}^{\rm c}(\xi) = -\frac{V_{,222}}{2V_{,2\Lambda}C_\alpha}(2V_{,111}V_{,1\xi}\xi)^{\frac{1}{2}} < 0 \tag{A.109}$$

that are existent for

$$V_{,111}V_{,1\xi}\xi > 0 \tag{A.110}$$

One of these solutions corresponds to the limit point on an imperfect fundamental path, and another to that on an aloof path.

Type II) solution

For (A.103), the set of equations (A.104)–(A.106) reduces to

$$V_{,1\xi}\xi + \frac{1}{2}V_{,111}(q_1)^2 = 0 \tag{A.111}$$

$$V_{,2\Lambda}\widetilde{\Lambda} + V_{,12\xi}q_1\xi = 0 \tag{A.112}$$

$$S_{22}^{\mathrm{V}} = V_{,222}q_2 + \frac{1}{2}V_{,1122}(q_1)^2 = 0 \tag{A.113}$$

The solutions of this set of equations (A.111)–(A.113) give the locations of two limit points of an imperfect system as

$$q_1^{\mathrm{c}}(\xi) = \pm\left(-\frac{2V_{,1\xi}\xi}{V_{,111}}\right)^{\frac{1}{2}} \tag{A.114}$$

$$q_2^{\mathrm{c}}(\xi) = \frac{V_{,1122}V_{,1\xi}}{V_{,111}V_{,222}}\xi \tag{A.115}$$

$$\widetilde{\Lambda}^{\mathrm{c}}(\xi) = \mp\frac{1}{V_{,2\Lambda}}\left(-\frac{2V_{,1\xi}\xi}{V_{,111}}\right)^{\frac{1}{2}}V_{,12\xi}\xi \tag{A.116}$$

that are existent for

$$V_{,111}V_{,1\xi}\xi < 0 \tag{A.117}$$

Here the double signs take the same order. One of these solutions corresponds to the limit point on an imperfect fundamental path, and another to that on an aloof path.

A.9 Imperfection Sensitivity Laws of Degenerate Hilltop Point II: Unstable-Symmetric Bifurcation

The imperfection sensitivity laws of degenerate hilltop points with simple unstable-symmetric bifurcation are derived.

A.9.1 General formulation

We consider a degenerate hilltop point with simple unstable-symmetric bifurcation with

$$V_{,111} = V_{,112} = 0, \quad V_{,1111} < 0, \quad V_{,1122} \neq 0 \tag{A.118}$$

and focus on the most customary case in practice where the system has a rising fundamental path and becomes unstable at the hilltop point, at which a limit point and a simple unstable-symmetric bifurcation point coincide. Then we have (cf., (8.19) and (8.20))

$$V_{,2\Lambda} < 0, \quad V_{,222} < 0, \quad V_{,1111} < 0, \quad V_{,11\Lambda} < 0, \quad V_{,22\Lambda} < 0 \tag{A.119}$$

$$C_\alpha = \frac{1}{2V_{,2\Lambda}}(V_{,2\Lambda}V_{,1122} - V_{,222}V_{,11\Lambda}) > 0 \tag{A.120}$$

With the use of (A.118), the bifurcation equations (8.10) and (8.11) become

$$\begin{aligned} \frac{\partial \Pi^{\mathrm{V}}}{\partial q_1} &= V_{,1\xi}\xi + V_{,11\Lambda}q_1\widetilde{\Lambda} + V_{,11\xi}q_1\xi + V_{,12\xi}q_2\xi + \frac{1}{6}V_{,1111}(q_1)^3 \\ &\quad + \frac{1}{2}V_{,1122}q_1(q_2)^2 = 0 \end{aligned} \tag{A.121}$$

$$\begin{aligned} \frac{\partial \Pi^{\mathrm{V}}}{\partial q_2} &= V_{,2\Lambda}\widetilde{\Lambda} + V_{,2\xi}\xi + \frac{1}{2}V_{,222}(q_2)^2 + V_{,22\Lambda}q_2\widetilde{\Lambda} + V_{,12\xi}q_1\xi \\ &\quad + \frac{1}{2}V_{,1122}(q_1)^2 q_2 = 0 \end{aligned} \tag{A.122}$$

and the criticality condition (8.13) becomes

$$\det \begin{pmatrix} S_{11}^{\mathrm{V}} & S_{12}^{\mathrm{V}} \\ S_{21}^{\mathrm{V}} & S_{22}^{\mathrm{V}} \end{pmatrix} = 0 \tag{A.123}$$

with

$$S_{11}^{\mathrm{V}} = V_{,11\Lambda}\widetilde{\Lambda} + V_{,11\xi}\xi + \frac{1}{2}V_{,1111}(q_1)^2 + \frac{1}{2}V_{,1122}(q_2)^2 \tag{A.124a}$$

$$S_{12}^{\mathrm{V}} = S_{21}^{\mathrm{V}} = V_{,12\xi}\xi + V_{,1122}q_1q_2 + V_{,112\Lambda}q_1\widetilde{\Lambda} \tag{A.124b}$$

$$S_{22}^{\mathrm{V}} = V_{,222}q_2 + V_{,22\Lambda}\widetilde{\Lambda} + V_{,22\xi}\xi + \frac{1}{2}V_{,1122}(q_1)^2 \tag{A.124c}$$

A.9.2 Perfect behavior

For a hilltop point with symmetric bifurcation with $V_{,111} = 0$ and $V_{,1111} \neq 0$, (A.121) and (A.122) with $\xi = 0$ give two bifurcation paths parameterized by q_2

$$\widetilde{\Lambda} = \widetilde{\Lambda}(q_2) = -\frac{V_{,222}}{2V_{,2\Lambda}}(q_2)^2 < 0, \quad q_1 = q_1(q_2) = \pm\left(-\frac{6C_\alpha}{V_{,1111}}\right)^{\frac{1}{2}} q_2 \tag{A.125}$$

that are conditionally existent for $V_{,1111} < 0$ (cf., (A.119)), i.e., for a declining bifurcation path. The perfect behavior is depicted in Fig. A.5 by the dashed curves.

A.9.3 Imperfection sensitivity: minor symmetric

Consider a minor symmetric imperfection, for which

$$\begin{aligned} &V_{,1\xi} = 0, \quad V_{,12\xi} = 0, \quad \dots \\ &V_{,11\xi}, \ V_{,2\xi}, \quad \dots: \ \text{possibly nonzero} \end{aligned} \tag{A.126}$$

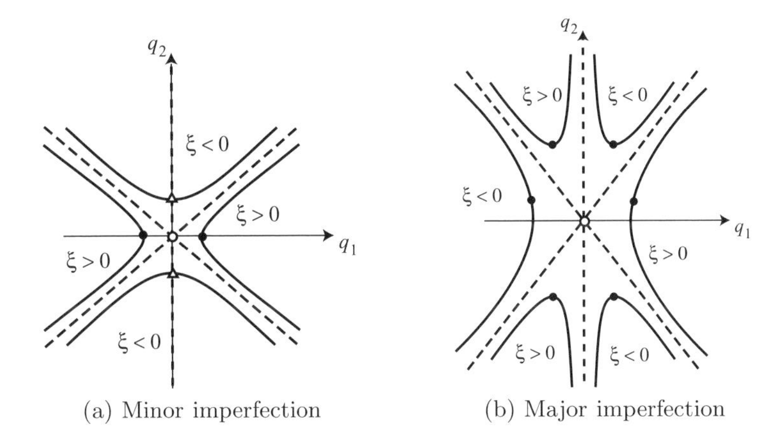

Fig. A.5 Asymptotic behaviors in the vicinity of a degenerate hilltop branching point with symmetric bifurcation ($\gamma < 0$, $V_{,1\xi} > 0$, $V_{,1122}V_{,222} > 0$). ○: hilltop point; •: limit point; △: unstable-symmetric bifurcation point; dashed curve: perfect equilibrium path; solid curve: imperfect equilibrium path.

are satisfied by (3.64). The imperfect curves expresses by (A.121) and (A.122) are shown by the solid curves in Fig. A.5(a) for $\gamma < 0$. Eq. (A.121) has two solutions:

$$q_1 = 0 : \text{trivial fundamental path} \tag{A.127}$$

$$V_{,11\Lambda}\widetilde{\Lambda} + V_{,11\xi}\xi + \frac{1}{6}V_{,1111}(q_1)^2 + \frac{1}{2}V_{,1122}(q_2)^2 = 0 : \quad \text{bifurcation or aloof path} \tag{A.128}$$

For the trivial fundamental path $q_1 = 0$, (A.122) becomes

$$V_{,2\Lambda}\widetilde{\Lambda} + V_{,2\xi}\xi + \frac{1}{2}V_{,222}(q_2)^2 = 0 \tag{A.129}$$

The limit point load is given for $q_2 = 0$ as

$$\widetilde{\Lambda}^{c}_{\text{Lim}}(\xi) = -\frac{V_{,2\xi}}{V_{,2\Lambda}}\xi \tag{A.130}$$

For the bifurcation or aloof path (A.128), we seek for the solution of (A.121)–(A.123) of the order

$$\widetilde{\Lambda} = O(\xi), \quad q_1 = O(\xi^2), \quad q_2 = O(\xi^{\frac{1}{2}}) \tag{A.131}$$

Then we have

$$\widetilde{\Lambda}^{c}_{\text{Bif}}(\xi) = -\frac{V_{,2\xi}V_{,1122} - V_{,222}V_{,11\xi}}{V_{,2\Lambda}V_{,1122} - V_{,222}V_{,11\Lambda}}\xi \tag{A.132}$$

that is existent for

$$(q_2)^2 = 2\gamma\xi > 0 \tag{A.133}$$

otherwise no bifurcation points exist. Here

$$\gamma = \frac{V_{,2\xi}V_{,11\Lambda} - V_{,2\Lambda}V_{,11\xi}}{V_{,2\Lambda}V_{,1122} - V_{,222}V_{,11\Lambda}} \tag{A.134}$$

To sum up,

$$\widetilde{\Lambda}^{\rm c}(\xi) = \begin{cases} \widetilde{\Lambda}^{\rm c}_{\rm Lim}(\xi): & \text{for } \gamma\xi < 0 \\ \widetilde{\Lambda}^{\rm c}_{\rm Bif}(\xi): & \text{for } \gamma\xi > 0 \end{cases} \tag{A.135}$$

Note that

$$\widetilde{\Lambda}^{\rm c}_{\rm Lim} - \widetilde{\Lambda}^{\rm c}_{\rm Bif} = \frac{V_{,222}}{V_{,2\Lambda}}\gamma\xi > 0 \tag{A.136}$$

A.9.4 Imperfection sensitivity: major antisymmetric

Consider a major antisymmetric imperfection, which satisfies

$$\begin{aligned} &V_{,11\xi} = 0, \quad V_{,2\xi} = 0, \quad V_{,22\xi} = 0 \\ &V_{,1\xi}, \quad V_{,12\xi}: \ \text{possibly nonzero} \end{aligned} \tag{A.137}$$

The imperfect curves expressed by (A.121) and (A.122) are shown by the solid curves in Fig. A.5(b) for $V_{,1\xi} > 0$.

We see the presence of two solutions

$$\text{I)} \quad \widetilde{\Lambda} = O(\xi^{\frac{2}{3}}), \quad q_1 = O(\xi^{\frac{1}{3}}), \quad q_2 = O(\xi^{\frac{1}{3}}) \tag{A.138}$$

$$\text{II)} \quad \widetilde{\Lambda} = O(\xi^{\frac{4}{3}}), \quad q_1 = O(\xi^{\frac{1}{3}}), \quad q_2 = O(\xi^{\frac{2}{3}}) \tag{A.139}$$

Type I) solution

For the Type I) solution in (A.138), the bifurcation equations (A.121), (A.122) and the criticality condition (A.123) reduce to

$$V_{,1\xi}\xi + V_{,11\Lambda}q_1\widetilde{\Lambda} + \frac{1}{6}V_{,1111}(q_1)^3 + \frac{1}{2}V_{,1122}q_1(q_2)^2 = 0 \tag{A.140}$$

$$V_{,2\Lambda}\widetilde{\Lambda} + \frac{1}{2}V_{,222}(q_2)^2 = 0 \tag{A.141}$$

$$S_{11}^{\rm V} = V_{,11\Lambda}\widetilde{\Lambda} + \frac{1}{2}V_{,1111}(q_1)^2 + \frac{1}{2}V_{,1122}(q_2)^2 = 0 \tag{A.142}$$

The solutions of this set of equations (A.140)–(A.142) give the locations of two limit points of an imperfect system as

$$q_1^{\rm c}(\xi) = \left(\frac{3}{V_{,1111}}\right)^{\frac{1}{3}}(V_{,1\xi}\xi)^{\frac{1}{3}} \tag{A.143}$$

$$(q_2^{\rm c}(\xi))^2 = -\frac{3^{\frac{2}{3}}(V_{,1111})^{\frac{1}{3}}}{2C_\alpha}(V_{,1\xi}\xi)^{\frac{2}{3}} \tag{A.144}$$

$$\widetilde{\Lambda}^{\rm c}(\xi) = \frac{3^{\frac{2}{3}}V_{,222}(V_{,1111})^{\frac{1}{3}}}{4V_{,2\Lambda}C_\alpha}(V_{,1\xi}\xi)^{\frac{2}{3}} < 0 \tag{A.145}$$

One of these solutions corresponds to the limit point on an imperfect fundamental path, and another to that on an aloof path.

Type II) solution

For the Type II) solution in (A.139), the bifurcation equations (A.121), (A.122) and the criticality condition (A.123) reduce to

$$V_{,1\xi}\xi + \frac{1}{6}V_{,1111}(q_1)^3 = 0 \tag{A.146}$$

$$V_{,2\Lambda}\widetilde{\Lambda} + \frac{1}{2}V_{,222}(q_2)^2 + V_{,12\xi}q_1\xi + \frac{1}{2}V_{,1122}(q_1)^2 q_2 = 0 \tag{A.147}$$

$$S_{22}^{\mathrm{V}} = V_{,222}q_2 + \frac{1}{2}V_{,1122}(q_1)^2 = 0 \tag{A.148}$$

The solution of this set of equations (A.146)–(A.148) gives the locations of a limit point of an imperfect system as

$$q_1^{\mathrm{c}}(\xi) = -\left(\frac{6V_{,1\xi}\xi}{V_{,1111}}\right)^{\frac{1}{3}} \tag{A.149}$$

$$q_2^{\mathrm{c}}(\xi) = \frac{V_{,1122}}{2V_{,222}}\left(\frac{6V_{,1\xi}\xi}{V_{,1111}}\right)^{\frac{2}{3}} \tag{A.150}$$

$$\widetilde{\Lambda}^{\mathrm{c}}(\xi) = -\frac{(V_{,1122})^2}{8V_{,2\Lambda}V_{,222}}\left(\frac{6}{V_{,1111}}\right)^{\frac{4}{3}}(V_{,1\xi}\xi)^{\frac{4}{3}} + \frac{1}{V_{,2\Lambda}}\left(\frac{6V_{,1\xi}\xi}{V_{,1111}}\right)^{\frac{1}{3}}V_{,12\xi}\xi \tag{A.151}$$

This solution corresponds to the limit point on an aloof path.

A.10 Summary

In this chapter,

- the interpolation approach for sensitivity analysis for a minor imperfection has been extended to coincident critical points,
- details of explicit diagonalization approach for sensitivity analysis for minor imperfection have been presented,
- details of quadratic approximation of the critical load of braced frame and higher derivatives of the bar–spring model have been shown, and
- imperfection sensitivity laws for degenerate hilltop points have been derived for minor and major imperfections.

The major findings of the Appendix are as follows.

- Throughout this book, complex bifurcation behaviors to be encountered at coincident critical points have been emphasized. Nonetheless, the systematics of the interpolation approach and explicit diagonalization approach are of great assistance in the numerical computation of sensitivity coefficients at these points.
- The power series expansion method with the assist of the concept of the order of a solution is a strong and insightful tool to derive imperfection sensitivity laws of coincident critical points, including hilltop branching points.

Bibliography

[1] E. Aarts and J. Korst. *Simulated Annealing and Boltzmann Machines: A Stochastic Approach to Combinatorial Optimization and Neural Computing.* John Wiley & Sons, Chichester, U.K., 1989.

[2] H. Abramovich, J. Singer, and R. Yaffe. Imperfection characteristics of stiffened shells: Group 1,. TAE Report, 406, Dept. of Aeronautical Engineering, Technion, Israel Institute of Technology, Haifa, 1981.

[3] B. O. Almroth, F. A. Brogan, E. Miller, F. Zelle, and M. T. Peterson. Collapse analysis of shells of general shape, User's Manual for the STAGS-A computer code. Technical Report AFDL-TR-71(8), Air Force Flight Dynamics Lab., Wright-Patterson Air Force Base, Dayton, OH, 1973.

[4] J. C. Amazigo. Buckling under axial compression of long cylindrical shells with random axisymmetric imperfections. *Quart. Appl. Math.*, 26(4):537–566, 1969.

[5] J. C. Amazigo. Buckling of stochastically imperfect columns on nonlinear elastic foundations. *Quart. Appl. Math.*, 29(3):403–409, 1971.

[6] J. C. Amazigo and B. Budiansky. Asymptotic formulas for the buckling stresses of axially compressed cylinders with localized or random axisymmetric imperfections. *J. Appl. Mech., ASME*, 39(1):179–184, 1972.

[7] A. H-S. Ang and W. H. Tang. *Probability Concepts in Engineering Planning and Design.* John Wiley & Sons, New York, NY, 1975.

[8] J. Arbocz. Present and future of shell stability analysis. *Zeitschrift für Flugwissenschaften und Weltraumforschung*, 5:335–348, 1981.

[9] J. Arbocz and H. Abramovich. The initial imperfection data bank at the Delft University of Technology: Part I. Technical Report LR-290, Dept. of Aeronautical Engineering, Delft University of Technology, Delft, The Netherlands, 1979.

[10] J. Arbocz and J. M. A. M. Hol. Collapse of axially compressed cylindrical shells with random imperfections. *AIAA J.*, 29(12):2247–2256, 1991.

[11] J. Arbocz and C. D. Babcock Jr. Utilization of STAGS to determine knockdown factors from measured initial imperfections. Technical Report LR-275, Dept. of Aeronautical Engineering, Delft University of Technology, Delft, The Netherlands, 1978.

[12] J. Arbocz and J. G. Williams. Imperfection surveys on a 10-ft-diameter shell structure. *AIAA J.*, 15(7):949–956, 1977.

[13] J. S. Arora. *Introduction to Optimum Design*. Academic Press, San Diego, CA, 2nd edition, 2004.

[14] J. S. Arora and C. H. Tseng. IDESIGN User's Manual, Ver. 3.5. Technical report, Optimal Design Laboratory, The University of Iowa, Iowa City, IA, 1987.

[15] J. Astill, C. J. Nosseir, and M. Shinozuka. Impact loading on structures with random properties. *J. Struct. Mech.*, 1(1):63–67, 1972.

[16] G. Augusti, A. Barratta, and F. Casciati. *Probabilistic Methods in Structural Engineering*. Chapman and Hall, New York, NY, 1984.

[17] K. J. Bathe. *Finite Element Procedures*. Prentice-Hall, Englewood Cliffs, NJ, 1996.

[18] K. J. Bathe, E. Ramm, and E. J. Wilson. Finite element formulation for large deformation dynamic analysis. *Int. J. Numer. Meth. Engng.*, 9:353–386, 1974.

[19] Z. P. Bažant and L. Cedolin. Initial postcritical analysis of asymmetric bifurcation in frames. *J. Struct. Engng., ASCE*, 115(11):2845–2857, 1989.

[20] Z. P. Bažant and L. Cedolin. *Stability of Structures*. Oxford University Press, Oxford, U.K., 1991.

[21] Z. P. Bažant and Y. Xiang. Postcritical imperfection-sensitive buckling and optimal bracing of large regular frames. *J. Struct. Engng., ASCE*, 123(4):513–522, 1997.

[22] Y. Ben-Haim. Convex models of uncertainty in radial pulse buckling of shells. *J. Appl. Mech., ASME*, 60(3):683–688, 1993.

[23] Y. Ben-Haim. *Robust Reliability in the Mechanical Sciences*. Springer-Verlag, Berlin, 1996.

[24] Y. Ben-Haim and I. Elishakoff. *Convex Models of Uncertainty in Applied Mechanics*. Elsevier, Amsterdam, 1990.

[25] H. Benaroya and M. Rehak. Finite element methods in probabilistic structural analysis: A selective review. *Appl. Mech. Rev.*, 41(5):201–213, 1988.

[26] M. P. Bendsøe and O. Sigmund. *Topology Optimization: Theory, Methods and Applications*. Springer, Berlin, 2003.

[27] E. S. Bernard, R. Coleman, and R. Q. Bridge. Measurement and assessment of geometric imperfections in thin-walled panels. *Thin-Walled Struct.*, 33(2):103–126, 1999.

[28] M. C. Bernard and J. L. Bogdanoff. Buckling of columns with random initial displacements. *J. Eng. Mech. Div., ASCE*, 97(EM3):755–771, 1971.

[29] E. Bielewicz, J. Górski, R. Schmidt, and H. Walukiewicz. Random fields in the limit analysis of elastic-plastic shell structures. *Comp. & Struct.*, 51(3):267–275, 1994.

[30] J. Bielski. Influence of geometrical imperfections on buckling pressure and post-buckling behavior of elastic toroidal shells. *Mech. Struct. Machines*, 20(2):145–154, 1992.

[31] B. Bochenek. Optimization of geometrically nonlinear structures with respect to both buckling and postbuckling constraints. *Eng. Opt.*, 29:401–415, 1997.

[32] B. Bochenek. Problem of structural optimization for post-buckling behavior. *Struct. Multidisc. Optim.*, 25:423–435, 2003.

[33] V. V. Bolotin. Statistical methods in the nonlinear theory of elastic shells. Otdeleni Tekhnicheskikh Nauk 3 (English translation: NASA, 1962, TTF-85, 1–16), Izvestija Academii Nauk SSSR, 1958.

[34] V. V. Bolotin. *Statistical Methods in Structural Mechanics.* Holden-Day Ser. Math. Phys. 7. Holden-Day, San Francisco, CA, 1969.

[35] V. V. Bolotin. *Random Vibrations of Elastic Systems.* Mech. Elastic Stability 7. Martinus Nijhoff, Hague, The Netherland, 1984.

[36] W. E. Boyce. Buckling of a column with random initial displacement. *J. Aero. Sci.*, 28(4):308–320, 1961.

[37] T. E. Bruns and O. Sigmund. Toward the topology design of mechanisms that exhibit snap-through behavior. *Comp. Meth. Appl. Mech. Engng.*, 193:3973–4000, 2004.

[38] T. E. Bruns, O. Sigmund, and D. A. Tortorelli. Numerical methods for the topology optimization of structures that exhibit snap-through. *Int. J. Numer. Meth. Engng.*, 55:1215–1237, 2002.

[39] T. E. Bruns and D. A. Tortorelli. Topology optimization of non-linear structures and compliant mechanisms. *Comp. Meth. Appl. Mech. Engng.*, 190:3443–3459, 2001.

[40] C. G. Bucher and U. Bourgund. A fast and efficient response surface approach for structural reliability problems. *Struct. Safety*, 7(1):57–66, 1990.

[41] C. G. Bucher and M. Shinozuka. Structural response variability II. *J. Eng. Mech. Div., ASCE*, 114(EM12):2035–2054, 1988.

[42] B. Budiansky and J. C. Amazigo. Initial post-buckling behavior of cylindrical shells under external pressure. *J. Math. Phys.*, 47:223–235, 1968.

[43] M. A. Burke and W. D. Nix. A numerical study of necking in the plane tension test. *Int. J. Solids Struct.*, 15:379–393, 1979.

[44] E. Byskov and J. W. Hutchinson. Mode interaction in axially stiffened cylindrical shells. *AIAA J.*, 15(7):941–948, 1977.

[45] J. B. Cardoso and J. S. Arora. Variational method for design semsitivity analysis in nonlinear structural mechanics. *AIAA J.*, 26(5):595–603, 1988.

[46] R. Casciaro, G. Salerno, and A. D. Lanzo. Finite element asymptotic analysis of slender elastic structures: A simple approach. *Int. J. Numer. Meth. Engng.*, 35:1397–1426, 1992.

[47] G. Cederbaum and J. Arbocz. Reliability of imperfection-sensitive composite shells via the Koiter–Cohen criterion. *Reliability Eng. System Safety*, 56(3):257–263, 1997.

[48] A. H. Chilver. Coupled modes of elastic buckling. *J. Mech. Phys. Solids*, 15:15–28, 1967.

[49] K. K. Choi and N-H. Kim. *Structural Sensitivity Analysis and Optimization 1, Linear Systems.* Springer, New York, NY, 2004.

[50] S. Chow and J. K. Hale. *Methods of Bifurcation Theory.* Grundlehren der mathematischen Wissenschaften 251. Springer-Verlag, New York, NY, 1982.

[51] M. K. Chryssanthopoulos. Probabilistic buckling analysis of plates and shells. *Thin-Walled Struct.*, 30(1–4):135–157, 1998.

[52] M. K. Chryssanthopoulos, M. J. Baker, and P. J. Dowling. Imperfection modeling for buckling analysis of stiffened cylinders. *J. Struct. Eng. Div., ASCE*, 117(ST7):1998–2017, 1991.

[53] M. K. Chryssanthopoulos, M. J. Baker, and P. J. Dowling. Statistical analysis of imperfections in stiffened cylinders. *J. Struct. Eng. Div., ASCE*, 117(ST7):1979–1997, 1991.

[54] M. K. Chryssanthopoulos, V. Giavotto, and C. Poggi. Characterization of manufacturing effects for buckling-sensitive composite cylinders. *Composites Manuf.*, 6(2):93–101, 1995.

[55] F. H. Clarke. *Optimization and Nonconvex Analysis.* John Wiley & Sons, New York, NY, 1983.

[56] M. A. Crisfield. *Non-linear finite Element Analysis of Solids and Structures, Vol. 1: Essentials.* John Wiley & Sons, New York, NY, 1991.

[57] J. G. A. Croll and R. C. Batista. Explicit lower bounds for the buckling of axially loaded cylinders. *Int. J. Mech. Sci.*, 23(6):331–343, 1981.

[58] R. Dancy and D. Jacobs. The initial imperfection data bank at the Delft University of Technology: Part II. Technical Report LR-559, Dept. of Aeronautical Engineering, Delft University of Technology, Delft, The Netherlands, 1988.

[59] P. K. Das and Y. Zheng. Cumulative formation of response surface and its use in reliability analysis. *Prob. Eng. Mech.*, 15(4):309–315, 2000.

[60] W. Day, A. J. Karwowski, and G. C. Papanicolaou. Buckling of randomly imperfect beams. *Acta Applicandae Mathematicae*, 17(3):269–286, 1989.

[61] H. de Boer and F. van Keulen. Refined semi-analytical design sensitivities. *Int. J. Solids Struct.*, 37:6961–6980, 2000.

[62] J. de Vries. Imperfection database. In: *CNES Conf., 3rd European Conf. on Launcher Tech.*, 323–332. Strasbourg, France, 2001.

[63] C-G. Deng. Equations of bifurcation sets of three-parameter catastrophes and the application in imperfection sensitivity analysis. *Int. J. Eng. Sci.*, 32(11):1811–1822, 1994.

[64] A. der Kiureghian. Finite element methods in structural safety studies. In: *Proc. of the Symp. on Struct. Safety Studies*, 40–52. ASCE, Denver, CO, 1985.

[65] A. der Kiureghian and Y. Zhang. Space-variant finite element reliability analysis. *Comp. Meth. Appl. Mech. Engng.*, 168(1-4):173–183, 1999.

[66] E. J. Doedel, H. B. Keller, and J. P. Kernévez. Numerical analysis and control of bifurcation problems I: Bifurcation in finite dimensions. *Int. J. Bifurcation Chaos*, 1:493–520, 1991.

[67] L. H. Donnel and C. C. Wan. Effect of imperfections on the buckling of thin cylinders and columns under axial compression. *J. Appl. Mech., ASME*, 17:73–83, 1950.

[68] M. R. Doup. Probabilistic analysis of the buckling of thin-walled shells using an imperfection database and a two-mode analysis. Technical report, Faculty Aeronautical Engineering, Delft University of Technology, Delft, The Netherlands, 1997.

[69] B. L. O. Edlund and U. L. C. Leopoldson. Computer simulation of the scatter in steel member strength. *Comp. & Struct.*, 5(4):209–224, 1975.

[70] I. Elishakoff. Impact buckling of thin bar via Monte Carlo method. *J. Appl. Mech., ASME*, 45(3):586–590, 1978.

[71] I. Elishakoff. Simulation of an initial imperfection data bank, Part I: Isotropic shells with general imperfections. TAE Report, 500, Dept. of Aeronautical Engineering, Technion, Israel Institute of Technology, Haifa, 1982.

[72] I. Elishakoff. *Probabilistic Methods in the Theory of Structures.* John Wiley & Sons, New York, NY, 1983.

[73] I. Elishakoff. Reliability approach to the initial imperfection sensitivity. *Acta Mech.*, 55(1-2):151–170, 1985.

[74] I. Elishakoff. Convex versus probabilistic modelling of uncertainty in structural dynamics. In: M. Petyt, H. F. Wolfe, and C. Mei, eds., *Struct. Dyna.: Recent Advances*, 3–21. Elsevier, London, 1991.

[75] I. Elishakoff. Possible limitations of probabilistic methods in engineering. *Appl. Mech. Rev.*, 53(2):19–36, 2000.

[76] I. Elishakoff and J. Arbocz. Reliability of axially compressed cylindrical shells with general nonsymmetric imperfections. *J. Appl. Mech., ASME*, 52(1):122–128, 1985.

[77] I. Elishakoff, R. T. Haftka, and J. Fang. Structural design under bounded uncertainty–optimization with anti-optimization. *Comp. & Struct.*, 53(6):1401–1405, 1994.

[78] I. Elishakoff, Y. W. Li, and J. H. Starnes Jr. *Non-Classical Problems in the Theory of Elastic Stability.* Cambridge University Press, Cambridge, U.K., 2001.

[79] I. Elishakoff, Y. K. Lin, and L.P. Zhu. *Probabilistic and Convex Modelling of Acoustically Excited Structures.* Studies Appl. Mech. 39. Elsevier, Amsterdam, 1994.

[80] I. Elishakoff and Y. J. Ren. *Finite Element Methods for Structures with Large Stochastic Variations.* Oxford University Press, Oxford, U.K., 2003.

[81] I. Elishakoff, S. van Manen, P. G. Vermeulen, and J. Arbocz. First-order second-moment analysis of the buckling of shells with random imperfections. *AIAA J.*, 25(8):1113–1117, 1987.

[82] A. G. Erdman, G. N. Sandor, and S. Kota. *Mechanism Design: Analysis and Synthesis.* Prentice-Hall, Englewood Cliffs, NJ, 4th edition, 2001.

[83] L. Euler. *Methodus inveniendi lineas curvas maximi minimive proprietate gaudentes, Appendix, De curvis elasticis.* Marcum Michaelem Bousquet, Lausanne and Geneva, Switzerland, 1744.

[84] G. Falsone and N. Impollonia. About the accuracy of a novel response surface method for the analysis of finite element modeled uncertain structures. *Prob. Eng. Mech.*, 19(1-2):53–63, 2004.

[85] L. Faravelli. Response surface approach for reliability analysis. *J. Eng. Mech. Div., ASCE*, 115(EM2):2763–2781, 1989.

[86] R. S. Fersht. Buckling of cylindrical shells with random imperfections. In: Y. C. Fung and E. E. Sechler, eds., *Thin Shell Structures: Theory, Experiment and Design*, 325–341. Prentice-Hall, Englewood Cliffs, NJ, 1974.

[87] W. Flügge. Die stabilität der Kreiszylinderschale. *Ing.-Arch.*, 3:463–506, 1932.

[88] W. B. Fraser. *Buckling of a Structure with Random Imperfections.* Ph.D. dissertation, Division of Engineering and Applied Physics, Harvard University, Cambridge, MA, 1965.

[89] W. B. Fraser and B. Budiansky. The buckling of a column with random initial deflections. *J. Appl. Mech., ASME*, 36(2):233–240, 1969.

[90] F. Fujii, K. Ikeda, H. Noguchi, and S. Okazawa. Modified stiffness iteration to pinpoint multiple bifurcation points. *Comp. Meth. Appl. Mech. Engng.*, 190:2499–2522, 2001.

[91] H. Fujii, M. Mimura, and Y. Nishiura. A picture of the global bifurcation diagram in ecological interacting and diffusing systems. *Phys. D*, 5(1):1–42, 1982.

[92] N. Gayton, J. M. Bourinet, and M. Lemaire. CQ2RS: A new statistical approach to the response surface method for reliability analysis. *Struct. Safety*, 25(1):99–121, 2003.

[93] R. G. Ghanem and P. D. Spanos. *Stochastic Finite Elements: A Spectral Approach.* Springer-Verlag, New York, NY, 1991.

[94] F. Glover. Tabu search – Part I. *ORSA J. on Computing*, 1:190–206, 1989.

[95] L. A. Godoy. *Theory of Elastic Stability: Analysis and Sensitivity.* Taylor and Francis, Philadelphia, PA, 2000.

[96] L. A. Godoy and D. T. Mook. Higher-order sensitivity to imperfections in bifurcation buckling analysis. *Int. J. Solids Struct.*, 33(4):511–520, 1996.

[97] L. A. Godoy and E. O. Taroco. Design sensitivity of post-buckling states including material constraints. *Comp. Meth. Appl. Mech. Engng.*, 188:665–679, 2000.

[98] D. E. Goldberg. *Genetic Algorithms in Search, Optimization, and Machine Learning.* Addison-Wesley, Reading, MA, 1989.

[99] M. Golubitsky and D. G. Schaeffer. *Singularities and Groups in Bifurcation Theory 1.* Appl. Math. Sci. Ser. 51. Springer-Verlag, New York, NY, 1985.

[100] M. Golubitsky, I. Stewart, and D. G. Schaeffer. *Singularities and Groups in Bifurcation Theory 2.* Appl. Math. Sci. Ser. 69. Springer-Verlag, New York, NY, 1988.

[101] H. M. Gomes and A. M. Awruch. Comparison of response surface and neural network with other methods for structural reliability analysis. *Struct. Safety*, 26(1):49–67, 2004.

[102] S. Gupta and C. S. Manohar. An improved response surface method for the determination of failure probability and importance measures. *Struct. Safety*, 26(2):123–139, 2004.

[103] G. Gusic, A. Combescure, and J. F. Jullien. Influence of circumferential thickness variations on buckling of cylindrical shells under external pressure. *Comp. & Struct.*, 74(4):461–477, 2000.

[104] R. T. Haftka. Simultaneous analysis and design. *AIAA J.*, 23(7):1099–1103, 1985.

[105] R. T. Haftka. Semi-analytical static nonlinear structural sensitivity analysis. *AIAA J.*, 31(7):1307–1312, 1993.

[106] R. T. Haftka, Z. Gürdal, and M. P. Kamat. *Elements of Structural Optimization, Solid Mechanics and Its Applications.* Kluwer Academic Publishers, Dordrecht, The Netherlands, 3rd edition, 1992.

[107] R. T. Haftka, R. H. Mallet, and W. Nachbar. Adaption of Koiter's method to finite element analysis of snap-through buckling behavior. *Int. J. Solids Struct.*, 7(10):1427–1445, 1971.

[108] A. Haldar and S. Mahadevan. *Probability, Reliability and Statistical Methods in Engineering Design.* John Wiley & Sons, New York, NY, 2000.

[109] A. Haldar and S. Mahadevan. *Reliability Assessment using Stochastic Finite Element Analysis.* John Wiley & Sons, New York, NY, 2000.

[110] S. K. Hall, G. E. Cameron, and D. E. Grierson. Least-weight design of steel frameworks accounting for P-Δ effects. *J. Struct. Engng., ASCE,* 115(6):1463–1475, 1988.

[111] J. S. Hansen. Influence of general imperfections in axially loaded cylindrical shells. *Int. J. Solids Struct.*, 11(11):1223–1233, 1975.

[112] J. S. Hansen and J. Roorda. On a probabilistic stability theory for imperfection sensitive structures. *Int. J. Solids Struct.*, 10(3):341–359, 1974.

[113] E. J. Haug and J. Cea, eds. *Optimization of Distributed Parameter Structures, Vol. 1, 2.* Sijthoff and Noordhoff, Alphen aan den Rijn, The Netherlands, 1981.

[114] E. J. Haug and K. K. Choi. Systematic occurrence of repeated eigenvalues in structural optimization. *J. Optimization Theory and Appl.*, 38:251–274, 1982.

[115] E. J. Haug, K. K. Choi, and V. Komkov. *Design Sensitivity Analysis of Structural Systems.* Academic Press, Orlando, FL, 1986.

[116] K. M. Heal, M. L. Hansen, and K. M. Rickard. *Maple V Programming Guide.* Springer-Verlag, New York, NY, 1998.

[117] T. J. Healey. *Symmetry, Bifurcation, and Computational Methods in Nonlinear Structural Mechanics.* Ph.D. dissertation, Dept. of Mechanical. Engineering, University of Illinois, Champaign, IL, 1985.

[118] T. J. Healey and J. A. Treacy. Exact block diagonalization of large eigenvalue problems for structures with symmetry. *Int. J. Numer. Meth. Engng.*, 31(2):265–285, 1991.

[119] R. Hill and J. W. Hutchinson. Bifurcation phenomena in the plane tension test. *J. Mech. Phys. Solids*, 23:239–264, 1975.

[120] T. Hisada and S. Nakagiri. Stochastic finite element method developed for structural safety and reliability. In: *Proc. of the 3rd Int. Conf. on Struct. Safety and Reliability*, 395–408. Elsevier, Amsterdam, 1981.

[121] T. Hisada and H. Noguchi. Development of a nonlinear stochastic FEM and its applications. In: A. H-S. Ang, M. Shinozuka, and G. I. Schuëller, eds., *Struct. Safety and Reliability*, 1097–1104. ASCE Press, New York, NY, 1989.

[122] K. D. Hjelmstad and S. Pezeshk. Optimal design of frames to resist buckling under multiple load cases. *J. Struct. Engng., ASCE,* 117(3):915–935, 1991.

[123] D. Ho. Buckling load of nonlinear systems with multiple eigenvalues. *Int. J. Solids Struct.*, 10:1315–1330, 1974.

[124] M. Hoshiya and I. Yoshida. Conditional stochastic FEM. In: P. D. Spanos, ed., *Comput. Stochastic Mech.* Balkema, Rotterdam, The Netherland, 1995.

[125] L. L. Howell. *Compliant Mechanisms.* John Wiley & Sons, New York, NY, 2001.

[126] G. W. Hunt. Imperfection-sensitivity of semi-symmetric branching. *Proc. Roy. Soc. London, Ser. A*, 357:193–211, 1977.

[127] G. W. Hunt. An algorithm for the nonlinear analysis of compound bifurcation. *Philosophical Transactions of Royal Society of London, Series A*, 300:443–471, 1981.

[128] G. W. Hunt. Hidden (a)symmetries of elastic and plastic bifurcation. *Appl. Mech. Rev.*, 39(8):1165–1186, 1986.

[129] K. Huseyin. *Nonlinear Theory of Elastic Stability.* Noordhoff, Leyden, The Netherland, 1975.

[130] J. W. Hutchinson. Imperfection sensitivity of externally pressurized spherical shells. *J. Appl. Mech., ASME*, 34(1):49–55, 1967.

[131] J. W. Hutchinson. Buckling and initial postbuckling behavior of oval cylindrical shells under axial compression. *J. Appl. Mech., ASME*, 35(1):66–72, 1968.

[132] J. W. Hutchinson and C. Amazigo. Imperfection-sensitivity of eccentrically stiffened cylindrical shells. *AIAA J.*, 5(3):392–401, 1967.

[133] J. W. Hutchinson and W. T. Koiter. Postbuckling theory. *Appl. Mech. Rev.*, 23(12):1353–1366, 1970.

[134] J. W. Hutchinson and J. P. Miles. Bifurcation analysis of the onset of necking in an elastic/plastic cylinder under uniaxial tension. *J. Mech. Phys. Solid*, 22:61–71, 1974.

[135] K. Ikeda, I. Ario, and K. Torii. Block-diagonalization analysis of symmetric plates. *Int. J. Solids Struct.*, 29(22):2779–2793, 1992.

[136] K. Ikeda and K. Murota. Computation of critical initial imperfection of truss structures. *J. Eng. Mech. Div., ASCE*, 116(EM10):2101–2117, 1990.

[137] K. Ikeda and K. Murota. Critical initial imperfection of structures. *Int. J. Solids Struct.*, 26(8):865–886, 1990.

[138] K. Ikeda and K. Murota. Bifurcation analysis of symmetric structures using block-diagonalization. *Comp. Meth. Appl. Mech. Engng.*, 86(2):215–253, 1991.

[139] K. Ikeda and K. Murota. Random initial imperfections of structures. *Int. J. Solids Struct.*, 28(8):1003–1021, 1991.

[140] K. Ikeda and K. Murota. Statistics of normally distributed initial imperfections. *Int. J. Solids Struct.*, 30(18):2445–2467, 1993.

[141] K. Ikeda and K. Murota. Systematic description of imperfect bifurcation behavior of symmetric systems. *Int. J. Solids Struct.*, 36(11):1561–1596, 1999.

[142] K. Ikeda and K. Murota. *Imperfect Bifurcation Phenomena in Structures and Materials – An Engineering Use of Group-theoretic Bifurcation Theory.* Appl. Math. Sci. Ser. 149. Springer, New York, NY, 2002.

[143] K. Ikeda, K. Murota, and I. Elishakoff. Reliability of structures subject to normally distributed initial imperfections. *Comp. & Struct.*, 591(3):463–469, 1996.

[144] K. Ikeda, K. Murota, and H. Fujii. Bifurcation hierarchy of symmetric structures. *Int. J. Solids Struct.*, 27:1551–1573, 1991.

[145] K. Ikeda, M. Ohsaki, and Y. Kanno. Imperfection sensitivity of hilltop branching points of systems with dihedral group symmetry. *Int. J. Nonlinear Mech.*, 40:755–774, 2005.

[146] K. Ikeda, K. Oide, and K. Terada. Imperfection sensitive strength variation of critical loads at hilltop bifurcation point. *Int. J. Eng. Sci.*, 40:743–772, 2002.

[147] K. Ikeda, P. Providéncia, and G. W. Hunt. Multiple equilibria for unlinked and weakly-linked cellular forms. *Int. J. Solids Struct.*, 30(3):371–384, 1993.

[148] K. C. Johns. Imperfection sensitivity of coincident buckling systems. *J. Non-Linear Mech.*, 9:1–21, 1974.

[149] K. C. Johns. Some statistical aspects of coupled buckling structures. In: B. Budiansky, ed., *Buckling of Struct., Proc. IUTAM Symp. on Buckling of Struct.*, 199–207. Springer-Verlag, Berlin, 1976.

[150] M. P. Kamat, N. S. Khot, and V. B. Venkayya. Optimization of shallow trusses against limit point instability. *AIAA J.*, 22(3):403–408, 1984.

[151] M. P. Kamat and P. Ruangsingha. Optimization of space trusses against instability using design sensitivity derivatives. *Eng. Opt.*, 8:177–188, 1985.

[152] Y. Kanno, M. Ohsaki, and N. Katoh. Sequential semidefinite programming for optimization of framed structures under multimodal buckling constraints. *Int. J. Structural Stability and Dynamics*, 1(4):585–602, 2001.

[153] H. Karadeniz, S. van Manen, and A. Vrouwenvelder. *Probabilistic reliability analysis for the fatigue limit state of gravity and jacket type structures, Proc. Third Int. Conf.: BOSS.* McGraw-Hill, London, 1982.

[154] A. Kardara, C. G. Bucher, and M. Shinozuka. Structural response variability III. *J. Eng. Mech. Div., ASCE*, 115(EM8):1726–1747, 1989.

[155] A. Kawamoto, M. P. Bendsøe, and O. Sigmund. Planar articulated mechanism design by graph theoretical enumeration. *Struct. Multidisc. Optim.*, 27:295–299, 2004.

[156] H. B. Keller. Numerical solution of bifurcation and nonlinear eigenvalue problems. In: P. H. Rabinowitz, ed., *Applications of Bifurcation Theory*, 359–384. Academic Press, New York, NY, 1977.

[157] N. S. Khot. Nonlinear analysis of optimized structure with constraints on system stability. *AIAA J.*, 21:1181–1186, 1983.

[158] N. S. Khot and M. P. Kamat. Minimum weight design of truss structures with geometric nonlinear behavior. *AIAA J.*, 23:139–144, 1985.

[159] N. S. Khot, V. B. Venkayya, and L. Berke. Optimum structural design with stability constraints. *Int. J. Numer. Meth. Engng.*, 10:1097–1114, 1976.

[160] H. Kim, R. T. Haftka, W. H. Mason, L. T. Watson, and B. Grossman. Probablistic modeling of errors from structural optimization based on multiple starting points. *Optimization and Eng.*, 3:415–430, 2002.

[161] S-H. Kim and S-W. Na. Response surface method using vector projected sampling points. *Struct. Safety*, 19(1):3–19, 1997.

[162] M. Kleiber. *Parameter Sensitivity in Nonlinear Mechanics.* John Wiley & Sons, New York, NY, 1997.

[163] M. Kleiber and T. D. Hien. *The Stochastic Finite Element Method.* John Wiley & Sons, Chichester, U.K., 1992.

[164] A. W. H. Klompé. Initial imperfection survey on a cylindrical shell, CSE at the Fokker BV. Technical Report LR-495, Dept. of Aeronautical Engineering, Delft University of Technology, Delft, The Netherlands, 1986.

[165] A. W. H. Klompé and den Reyer. The initial imperfection data bank at the Delft University of Technology: Part III. Technical Report LR-568, Dept. of Aeronautical Engineering, Delft University of Technology, Delft, The Netherlands, 1989.

[166] W. T. Koiter. *On the Stability of Elastic Equilibrium.* Ph.D. dissertation, Dept. of Mechanicsl Engineering, Delft University of Technology, Delft, The Netherland, 1945. (English Translation, NASA, TTF-10833, 1967).

[167] W. T. Koiter. The effects of axisymmetric imperfections on the buckling of cylindrical shells under axial compression. *Proc. Kon. Ned. Akad. Wet., Ser. B*, 66:265–279, 1963.

[168] W. T. Koiter, I. Elishakoff, Y. W. Li, and J. H. Starnes Jr. Buckling of an axially compressed cylindrical shells of variable thickness. *Int. J. Solids Struct.*, 31(6):797–805, 1994.

[169] W. T. Koiter and G. D. C. Kuiken. The interaction between local buckling and overall buckling in the behavior of built-up columns. Technical Report WTHD-23, Delft University of Technology, Delft, The Netherland, 1971.

[170] W. T. Koiter and M. Skaloud. Interventions, comportment postcritique des plaques utilisees en construction metallique. *Mém. Soc. Sci. Liege, 5 Serte*, 8(5):64–68, 1963.

[171] L. Kollár, ed. *Structural Stability in Engineering Practice.* E & FN Spon, London, 1999.

[172] A. N. Kounadis. Interaction of the joint and of the lateral bracing stiffnesses for the optimum design of unbraced frames. *Acta Mechanica*, 47:247–262, 1983.

[173] M. Kočvara. On the modelling and solving of the truss design problem with global stability constraints. *Struct. Multidisc. Optim.*, 23:189–203, 2002.

[174] M. Królak, Z. Kołakowski, and M. Kotełko. Influence of load-non-uniformity and eccentricity on the stability and load carrying capacity of orthotropic tubular columns of regular hexagonal cross-section. *Thin-Walled Struct.*, 39(6):483–498, 2001.

[175] T. S. Kwon, B. C. Lee, and W. J. Lee. An approximation technique for design sensitivity analysis of the critical load in non-linear structures. *Int. J. Numer. Meth. Engng.*, 45:1727–1736, 1999.

[176] U. D. Larsen, O. Sigmund, and S. Bouswstra. Design and fabrication of compliant micromechanisms and structures with negative poisson's ratio. In: *Proc. IEEE 9th Annual Int. Workshop on Micro Electro Mech. Sys., An Investigation of Micro Structures, Sensors, Actuators, Machines and Systems*, 365–371. San Diego, California, 1996.

[177] T. H. Lee and J. S. Arora. A computational method for design sensitivity analysis of elastoplastic structures. *Comp. Meth. Appl. Mech. Engng.*, 122:27–50, 1995.

[178] L. Léotoing, S. Drapier, and A. Vautrin. Nonlinear interaction of geometrical and material properties in sandwich beam instabilities. *Int. J. Solids Struct.*, 39:3717–3739, 2002.

[179] R. Levy and H. Perng. Optimization for nonlinear stability. *Comp. & Struct.*, 30(3):529–535, 1988.

[180] L-Y. Li. Influence of loading imperfections on the stability of an axially compressed cylindrical shell. *Thin-Walled Struct.*, 10(3):215–220, 1990.

[181] Y. W. Li, I. Elishakoff, and J. H. Starnes Jr. Axial buckling of composite cylindrical shells with periodic thickness variation. *Comp. & Struct.*, 56(1):65–74, 1995.

[182] Y. W. Li, I. Elishakoff, J. H. Starnes Jr., and M. Shinozuka. Nonlinear buckling of a structure with random imperfection and random axial compression by a conditional simulation technique. *Comp. & Struct.*, 56(1):59–64, 1995.

[183] X. Lignos, G. Ioannidis, and A. N. Kounadis. Interaction of the joint and of the lateral bracing stiffnesses for the optimum design of unbraced frames. *Acta Mechanica*, 47:247–262, 2003.

[184] C-C. Lin and I-W Liu. Optimal design based on optimality criterion for frame structures including buckling constraints. *Comp. & Struct.*, 31(4):535–544, 1989.

[185] X. Lin and J. G. Teng. Iterative Fourier decomposition of imperfection measurements at non-uniformly distributed sampling points. *Thin-Walled Struct.*, 41(10):901–924, 2003.

[186] H. E. Lindberg. Convex models of uncertain imperfection control in multimode dynamic buckling. *J. Appl. Mech., ASME*, 59(4):937–945, 1992.

[187] W. K. Liu, T. Belytschko, and A. Mani. Probabilistic finite elements for nonlinear structural dynamics. *Comp. Meth. Appl. Mech. Engng.*, 56(1):61–81, 1986.

[188] Y-W. Liu and F. Moses. A sequential response surface method and its application in the reliability analysis of aircraft structural systems. *Struct. Safety*, 16(1-2):39–46, 1994.

[189] P. Lokkas and G. A. Croll. Theory of combined sway and nonsway frames buckling. *J. Engng. Mech., ASCE*, 126(1):84–92, 2000.

[190] M. Lombardi and R. T. Haftka. Anti-optimization technique for strutural design under load uncertainties. *Comp. Meth. Appl. Mech. Engng.*, 157(1-2):19–31, 1998.

[191] D. G. Luenberger. *Linear and Nonlinear Programming.* Kluwer Academic Publishers, Boston, MA, 2003.

[192] B. P. Makarov. Statistical analysis of deformation of imperfect cylindrical shells. In: *Raschety na Prochnost (Strength Analysis) 15*, 240–256. Mashinostroenie Publishing House, Moscow, 1971. (in Russian).

[193] Maplesoft. *Maple 9 Programming Guide.* 2004.

[194] N. D. Masters and L. L. Howell. A self-retracting fully compliant bistable micromechanism. *J. MEMS*, 12:273–280, 2003.

[195] E. F. Masur. Buckling, post-buckling and limit analysis of completely symmetric elastic structures. *Int. J. Solids Struct.*, 6:587–604, 1970.

[196] E. F. Masur. Optimal structural design under multiple eigenvalue constraints. *Int. J. Solids Struct.*, 20:211–231, 1984.

[197] B. J. Matkowsky and E. L. Reiss. Singular perturbations of bifurcations. *SIAM J. Appl. Math.*, 33(2):230–255, 1977.

[198] I. C. Medland. A basis for the design of column bracing. *J. Struct. Eng.*, 55(7):301–307, 1977.

[199] T. H. G. Megson and G. Hallak. Measurement of the geometric initial imperfections in diaphragms. *Thin-Walled Struct.*, 14(5):381–394, 1992.

[200] R. Miller and J. M. Hedgepeth. The buckling of lattice columns with stochastic imperfections. *Int. J. Solids Struct.*, 15(1):73–84, 1979.

[201] E. S. Mistakidis and G. E. Stavroulakis. *Nonconvex Optimization in Mechanics: Algorithms, Heuristics and Engineering Applications by the F.E.M.* Nonconvex Optimization and Its Applications 21. Kluwer Academic Publishers, Dordrecht, The Netherlands, 1998.

[202] Z. Mróz and R. T. Haftka. Design sensitivity analysis of non-linear structures in regular and critical states. *Int. J. Solids Struct.*, 31(15):2071–2098, 1994.

[203] Z. Mróz and J. Piekarski. Sensitivity analysis and optimal design of non-linear structures. *Int. J. Numer. Meth. Engng.*, 42:1231–1262, 1998.

[204] K. Murota and K. Ikeda. Computational use of group theory in bifurcation analysis of symmetric structures. *SIAM J. Sci. Statist. Comput.*, 12(2):273–297, 1991.

[205] K. Murota and K. Ikeda. Critical imperfection of symmetric structures. *SIAM J. Appl. Math.*, 51(5):1222–1254, 1991.

[206] K. Murota and K. Ikeda. On random imperfections for structures of regular-polygonal symmetry. *SIAM J. Appl. Math.*, 52(6):1780–1803, 1992.

[207] R. H. Myers and D. C. Montgomery. *Response Surface Methodology: Process and Product Optimization using Design Experiments.* John Wiley & Sons, New York, NY, 1995.

[208] S. Nakagiri and T. Hisada. *An Introduction to Stochastic Finite Element Method: Analysis of Uncertain Structures.* Baifukan, Tokyo, 1985. (in Japanese).

[209] M. S. El Naschie. *Stress, Stability and Chaos in Structural Engineering: An Energy Approach.* McGraw-Hill, London, 1990.

[210] A. Needleman. A numerical study of necking in circular cylindrical bars. *J. Mech. Phys. Solids*, 20(2):111–127, 1972.

[211] F. Nishino and W. Hartono. Influential mode of imperfection on carrying capacity of structures. *J. Eng. Mech. Div., ASCE*, 115(EM10):2150–2165, 1989.

[212] S. Nishiwaki, S. Min, J. Yoo, and N. Kikuchi. Optimal structural design considering flexibility. *Comp. Meth. Appl. Mech. Engng.*, 190:4457–4504, 2001.

[213] H. Noguchi and T. Hisada. Sensitivity analysis in post-buckling problems of shell structures. *Comp. & Struct.*, 47(4):669–710, 1993.

[214] H. Noguchi and T. Hisada. Developement of a sensitivity analysis method for nonlinear buckling load. *JSME Int. J. Ser. A*, 38(3):311–317, 1995.

[215] M. Ohsaki. Discontinuity in sensitivity of elastoplastic bifurcation loads of finite dimensional symmetric structures. *Comp. Mech.*, 23:404–410, 1999.

[216] M. Ohsaki. Optimization of geometrically nonlinear symmetric systems with coincident critical points. *Int. J. Numer. Meth. Engng.*, 48:1345–1357, 2000.

[217] M. Ohsaki. Sensitivity analysis and optimization corresponding to a degenerate critical point. *Int. J. Solids Struct.*, 38:4955–4967, 2001.

[218] M. Ohsaki. Sensitivity analysis of coincident critical loads with respect to minor imperfection. *Int. J. Solids Struct.*, 38:4571–4583, 2001.

[219] M. Ohsaki. Imperfection sensitivity of optimal symmetric braced frames against buckling. *Int. J. Non-Linear Mech.*, 38:1103–1117, 2002.

[220] M. Ohsaki. Maximum loads of imperfect systems corresponding to stable bifurcation. *Int. J. Solids Struct.*, 39:927–941, 2002.

[221] M. Ohsaki. Strcutural optimization for specified nonlinear bunckling load factor. *Japan J. of Industrual and Appl. Math.*, 19(2):163–179, 2002.

[222] M. Ohsaki. Sensitivity analysis of an optimized bar-spring model with hill-top branching. *Archive of Applied Mechanics*, 73:241–251, 2003.

[223] M. Ohsaki. Design sensitivity analysis and optimization for nonlinear buckling of finite-dimensional elastic conservative structures. *Comp. Meth. Appl. Mech. Engng.*, 194:3331–3358, 2005.

[224] M. Ohsaki and K. Ikeda. Imperfection sensitivity analysis of hill-top branching with many symmetric bifurcation points. *Int. J. Solids Struct.*, 43(16):4704–4719, 2006.

[225] M. Ohsaki and T. Nakamura. Optimum design with imperfection sensitivity coefficients for limit point loads. *Struct. Opt.*, 8:131–137, 1994.

[226] M. Ohsaki and S. Nishiwaki. Shape design of pin-jointed multistable compliant mechanism using snapthrough behavior. *Struct. Multidisc. Optim.*, 30:327–334, 2005.

[227] M. Ohsaki and K. Uetani. Sensitivity analysis of bifurcation load of finite dimensional symmetric systems. *Int. J. Numer. Meth. Engng.*, 39:1707–1720, 1996.

[228] M. Ohsaki, K. Uetani, and M. Takeuchi. Optimization of imperfection-sensitive symmetric systems for specified maximum load factor. *Comp. Meth. Appl. Mech. Engng.*, 166:349–362, 1998.

[229] S. Okazawa, K. Oide, K. Ikeda, and K. Terada. Imperfection sensitivity and probabilistic variation of tensile strength of steel members. *Int. J. Solids Struct.*, 39(6):1651–1671, 2002.

[230] N. Olhoff. Optimal design with respect to structural eigenvalues. In: *Proc. 15th Int. IUTAM Congress*, 133–149, 1980.

[231] N. Olhoff and S. H. Rasmussen. On single and bimodal optimum buckling loads of clamped columns. *Int. J. Solids Struct.*, 13:605–614, 1977.

[232] G. V. Palassopoulos. Optimization of imperfection-sensitive structures. *J. Eng. Mech. Div., ASCE*, 115(EM8):1663–1682, 1989.

[233] G. V. Palassopoulos. On the optimization of imperfection-sensitive structures with buckling constraints. *Eng. Opt.*, 17:219–227, 1991.

[234] G. V. Palassopoulos. Reliability-based design of imperfection sensitive structures. *J. Eng. Mech. Div., ASCE*, 117(EM6):1220–1240, 1991.

[235] G. V. Palassopoulos. Response variability of structures subjected to bifurcation buckling. *J. Eng. Mech. Div., ASCE*, 118(EM6):1164–1183, 1992.

[236] G. V. Palassopoulos. New approach to buckling of imperfection-sensitive structures. *J. Engng. Mech., ASCE*, 119(4):850–869, 1993.

[237] C. Papadimitriou, J. L. Beck, and L. S. Katafygiotis. Updating robust reliability using structural test data. *Prob. Eng. Mech.*, 16(2):103–113, 2001.

[238] V. Papadopoulos and M. Papadrakakis. Stochastic finite element-based reliability analysis of space frames. *Prob. Eng. Mech.*, 13(1):53–65, 1998.

[239] J. S. Park and K. K. Choi. Design sensitivity analysis of critical load factor for nonlinear strutural systems. *Comp. & Struct.*, 36(5):823–838, 1990.

[240] C. B. W. Pedersen, T. Buhl, and O. Sigmund. Topology optimization of large-displacement compliant mechanisms. *Int. J. Numer. Meth. Engng.*, 44:1215–1237, 2002.

[241] R. Peek. Worst shapes of imperfections for space trusses with multiple global and local buckling modes. *Int. J. Solids Struct.*, 30:2243–2260, 1993.

[242] R. Peek and N. Triantafyllidis. Worst shapes of imperfections for space trusses with many simultaneously buckling modes. *Int. J. Solids Struct.*, 29(19):2385–2402, 1992.

[243] S. Pellegrino. *Deployable Structures.* CISM International Centre for Mechanical Sciences Courses and Lectures. Springer, Wien, 2002.

[244] H. Petryk. Plastic instability: criteria and computational approaches. *Arch. Comp. Methods Engrg.*, 4:111–151, 1997.

[245] D. A. Pierre and M. J. Lowe. *Mathematical Programming via Augmented Lagrangians.* Addison-Wesley, Reading, MA, 1975.

[246] J. Pietrzak. An alternative approach to optimization of structures prone to instability. *Struct. Multidisc. Optim.*, 11:88–94, 1996.

[247] M. Pircher, P. A. Berry, X. Ding, and R. Q. Bridge. The shape of circumferential weld-induced imperfections in thin-walled steel silos and tanks. *Thin-Walled Struct.*, 39(12):999–1014, 2001.

[248] M. Pircher and A. Wheeler. The measurement of imperfections in cylindrical thin-walled members. *Thin-Walled Struct.*, 41(5):419–433, 2003.

[249] R. H. Plaut, P. Ruangsilasingha, and M. P. Kamat. Optimization of an asymmetric two-bar truss against instability. *J. Struct. Mech.*, 12(4):465–470, 1984.

[250] W. Prager and J. E. Taylor. Problem of optimal structural design. *J. Appl. Mech.*, 35(1):102–106, 1968.

[251] M. R. Rajashekhar and B. R. Ellingwood. A new look at the response surface approach for reliability analysis. *Struct. Safety*, 12(3):205–220, 1993.

[252] S. A. Ramu and R. Ganesan. Parametric instability of stochastic columns. *Int. J. Solids Struct.*, 30(10):1339–1354, 1993.

[253] R. Reitinger, K.-U. Bletzinger, and E. Ramm. Shape optimization of buckling sensitive structures. *Computing Sys. Engng.*, 5:65–75, 1994.

[254] R. Reitinger and E. Ramm. Buckling and imperfection sensitivity in the optimization of shell structures. *Thin-Walled Structures*, 23:159–177, 1995.

[255] E. Riks. Buckling analysis of elastic structures: a computational approach. In: *Advances in Applied Mechanics 34*, 1–76. Academic Press, New York, NY, 1998.

[256] V. J. Romero, L. P. Swiler, and A. A. Giunta. Construction of response surfaces based on progressive-lattice-sampling experimental designs with application to uncertainty propagation. *Struct. Safety*, 26(2):201–219, 2004.

[257] J. Roorda. Stability of structures with small imperfections. *J. Eng. Mech. Div., ASCE*, 91(EM1):87–106, 1965.

[258] J. Roorda. On the buckling of symmetric structural systems with first and second order imperfections. *Int. J. Solids Struct.*, 4:1137–1148, 1968.

[259] J. Roorda. Some statistical aspects of the buckling of imperfection-sensitive structures. *J. Mech. Phys. Solids*, 17(2):111–123, 1969.

[260] J. Roorda. *Buckling of Elastic Structures.* University of Waterloo Press, Waterloo, Canada, 1980.

[261] J. Roorda and J. S. Hansen. Random buckling behavior in axially loaded cylindrical shells with axisymmetric imperfections. *J. Spacecraft*, 9(2):88–91, 1972.

[262] Y. S. Ryu, M. Harrian, C. C. Wu, and J. S. Arora. Structural design sensitivity analysis of nonlinear response. *Comp. & Struct.*, 21(1):245–255, 1985.

[263] Z. Sadovsky. A theoretical approach to the problem of the most dangerous initial deflection shape in stability type structural problems. *Aplikace Mathematiky*, 23(4):248–266, 1978.

[264] M. P. Saka and M. Ulker. Optimum design of geometrically nonlinear space trusses. *Comp. & Struct.*, 42(3):289–299, 1991.

[265] P. Samuels. Bifurcation and limit point instability of dual eigenvalue third order system. *Int. J. Solids Struct.*, 16(8):743–756, 1980.

[266] D. H. Sattinger. *Group Theoretic Methods in Bifurcation Theory.* Lecture Notes in Mathematics 762. Springer-Verlag, Berlin, 1979.

[267] C. A. Schenk, G. I. Schuëller, and J. Arbocz. Buckling analysis of cylindrical shells with random imperfections. In: *Proc. Int. Conf. on Monte Carlo Simulation*, 18–21. Monte Carlo, 2000.

[268] G. I. Schuëller. A state-of-the-art report on computational stochastic mechanics. *Prob. Eng. Mech.*, 12(4):197–321, 1997.

[269] N. D. Scott, J. E. Harding, and P. J. Dowling. Fabrication of small scale stiffened cylindrical shells. *J. Strain Anal.*, 22(2):97–106, 1987.

[270] T. Sekimoto and H. Noguchi. Homologous topology optimization in large displacement and buckling problems. *JSME Int. J., Series A*, 44:610–615, 2001.

[271] A. P. Seyranian, E. Lund, and N. Olhoff. Multiple eigenvalues in structural optimization problem. *Struct. Opt.*, 8:207–227, 1994.

[272] M. Shinozuka. Structural response variability. *J. Eng. Mech. Div., ASCE*, 113(EM6):825–843, 1987.

[273] M. Shinozuka and F. Yamazaki. Stochastic finite element analysis: An introduction. In: S. T. Ariaratnama, I. Schuëller, and I. Elishakoff, eds., *Stochastic Structural Dynamics: Progress in Theory and Applications*, 271–291. Elsevier, London, 1988.

[274] O. Sigmund. On the design of compliant mechanisms using topology optimization. *Mech. Struct. & Mach.*, 25(4):493–524, 1997.

[275] J. Singer, H. Abramovich, and R. Yaffe. Initial imperfection measurements of integrally stringer-stiffened cylindrical shells. TAE Report, 330, Dept. of Aeronautical Engineering, Technion, Israel Institute of Technology, Haifa, 1978.

[276] L. Spunt. *Optimum Structural Design*. Englewood Cliffs, NJ, 1971.

[277] W. J. Supple. Coupled branching configurations in the elastic buckling of symmetric structural system. *Int J. Mech. Sci.*, 9:97–112, 1967.

[278] J. Takagi and M. Ohsaki. Design of lateral braces for columns considering critical imperfection of buckling. *Int. J. Structural Stability and Dynamics*, 4(1):69–88, 2004.

[279] R. C. Tennyson. The effect of shape imperfections and stiffening on the buckling of circular cylinders. In: B. Budiansky, ed., *Buckling of Struct., Proc. IUTAM Symp. on Buckling of Struct.*, 251–273. Harvard University, Cambridge, MA, Springer-Verlag, Berlin, 1976.

[280] R. C. Tennyson, D. B. Muggeridge, and R. D. Caswell. Buckling of circular cylindrical shells having axisymmetric imperfection distributions. *AIAA J.*, 9(5):924–930, 1971.

[281] P. Thoft-Christensen and M. J. Baker. *Structural Reliability Theory and its Applications*. Springer-Verlag, Berlin, 1982.

[282] J. M. T. Thompson. Towards a general statistical theory of imperfection-sensitivity in elastic post-buckling. *J. Mech. Phys. Solids*, 15(6):413–417, 1967.

[283] J. M. T. Thompson. A general theory for the equilibrium and stability of discrete conservative systems. *J. Appl. Math. Phys. (ZAMP)*, 20:797–846, 1969.

[284] J. M. T. Thompson. *Instabilities and Catastrophes in Science and Engineering*. John Wiley & Sons, Chichester, U.K., 1982.

[285] J. M. T. Thompson and G. W. Hunt. *A General Theory of Elastic Stability*. John Wiley & Sons, New York, NY, 1973.

[286] J. M. T. Thompson and G. W. Hunt. Dangers of structural optimization. *Eng. Opt.*, 1:99–110, 1974.

[287] J. M. T. Thompson and G. W. Hunt. *Elastic Instability Phenomena.* John Wiley & Sons, Chichester, U.K., 1984.

[288] J. M. T. Thompson and G. M. Lewis. On the optimum design of thin-walled compression members. *J. Mech. Phys. Solids*, 20(2):101–109, 1972.

[289] J. M. T. Thompson and P. A. Schorrock. Bifurcation instability of an atomic lattice. *J. Mech. Phys. Solids*, 23(1):21–37, 1975.

[290] J. M. T. Thompson and W. J. Supple. Erosion of optimum designs by compound branching phenomena. *J. Mech. Phys. Solids*, 21(3):135–144, 1973.

[291] J. M. T. Thompson, J. K. Y. Tan, and K. C. Lim. On the topological classification of postbuckling phenomena. *J. Struct. Mech.*, 6:383–414, 1978.

[292] Y. Tomita. Simulation of plastic instabilities in solid mechanics. *Appl. Mech. Rev.*, 47(6):171–205, 1994.

[293] I. Trendafilova and J. Ivanova. Loss of stability of thin, elastic, strongly convex shells of revolution with initial imperfections, subjected to uniform pressure: A probabilistic approach. *Thin-Walled Struct.*, 23(1-4):201–214, 1995.

[294] N. Triantafyllidis and R. Peek R. On stability and the worst imperfection shape in solids with nearly simultaneous eigenmodes. *Int. J. Solids Struct.*, 29(18):2281–2299, 1992.

[295] F. Turčić. Resistance of axially compressed cylindrical shells determined for measured geometrical imperfections. *J. Construct. Steel Res.*, 19(3):225–234, 1991.

[296] V. Tvergaard. Imperfection-sensitivity of a wide integrally stiffened panel under compression. *Int. J. Solids Struct.*, 9(1):177–192, 1973.

[297] V. Tvergaard. Effect of thickness inhomogeneities in internally pressurized elastic-plastic spherical shells. *J. Mech. Phys. Solids*, 24(5):291–304, 1976.

[298] V. Tvergaard. Studies of elastic-plastic instabilities. *J. Appl. Mech.*, 66:3–9, 1999.

[299] Tvergaard V and A. Needleman. On the development of localized buckling patterns. In: J. M. T. Thompson and G. W. Hunt, eds., *Collapse: The Buckling of Struct. in Theory and Practice, Proc. IUTAM Symp.*, 299–332. Cambridge University Press, Cambridge, U.K., 1982.

[300] R. A. van Slooten and T. T. Soong. Buckling of a long, axially compressed, thin cylindrical shell with random initial imperfections. *J. Appl. Mech., ASME*, 39(4):1066–1071, 1972.

[301] G. N. Vanderplaats. *Numerical Optimization Techniques for Engineering Design.* Vanderplaats Research & Development Inc., 1999.

[302] E. Vanmarcke, M. Shinozuka, S. Nakagiri, G. I.Schuëller, and M. Grigoriu. Random fields and stochastic finite elements. *Struct. Safety*, 3:143–166, 1986.

[303] G. Venter, R. T. Haftka, and J. H. Starnes Jr. Construction of response surface approximation for design optimization. *AIAA J.*, 36(12):2242–2249, 1998.

[304] W. D. Verduyn and I. Elishakoff. A testing machine for statistical analysis of small imperfect shells: Part I. Technical Report LR-357, Dept. of Aeronautical Engineering, Delft University of Technology, Delft, The Netherlands, 1987.

[305] C. A. Vidal and R. B. Haber. Design sensitivity analysis for rate-independent elastoplasticity. *Comp. Meth. Appl. Mech. Engng.*, 107:393–431, 1993.

[306] B. P. Videc and J. L. Sanders Jr. Application of Khasminskii's limit theorem to the buckling problem of a column with random initial deflection. *Quart. Appl. Math.*, 33:422–428, 1976.

[307] T. von Kármán and H-S. Tsien. The buckling of spherical shells by external pressure. *J. Aero. Sci.*, 1:43–50, 1939.

[308] VR&D. *DOT User's Manual, Ver 5.0*, 1999.

[309] F-Y. Wang. Monte Carlo analysis of nonlinear vibration of rectangular plates with random geometric imperfections. *Int. J. Solids Struct.*, 26(1):99–109, 1990.

[310] V. I. Weingarten, E. J. Morgan, and P. Seide. Elastic stability of thin-walled cylindrical and conical shells under axial compression. *AIAA J.*, 3(3):500–505, 1965.

[311] P. Wriggers and J. C. Simo. A general procedure for the direct computation of turning and bifurcation points. *Int. J. Numer. Meth. Engng.*, 30:155–176, 1990.

[312] P. Wriggers, W. Wagner, and C. Miehe. A quadratically convergent procedure for the calculation of stability points in finite element analysis. *Comp. Meth. Appl. Mech. Engng.*, 70:329–347, 1988.

[313] B. Wu. Buckling mode interaction in fixed-end column with central brace. *J. Engng. Mech., ASCE*, 125(3):316–322, 1999.

[314] C. C. Wu and J. S. Arora. Design sensitivity analysis and optimization of nonlinear structural response using incremental procedure. *AIAA J.*, 25:1118–1125, 1987.

[315] C. C. Wu and J. S. Arora. Simultaneous analysis and design optimization of nonlinear response. *Engineering with Computers*, 2:53–63, 1987.

[316] C. C. Wu and J. S. Arora. Design sensitivity analysis of non-linear buckling load. *Comp. Mech.*, 3:129–140, 1988.

[317] W. Wunderlich and U. Albertin. Analysis and load carrying behaviour of imperfection sensitive shells. *Int. J. Numer. Meth. Engng.*, 47(1-3):255–273, 2000.

[318] Y. Zhang, S-H. Chen, Q-L. Liu, and T-Q. Liu. Stochastic perturbation finite elements. *Comp. & Struct.*, 59(3):425–429, 1996.

Index

Mechanical Engineering Series *(continued from page ii)*

J. García de Jalón and E. Bayo, **Kinematic and Dynamic Simulation of Multibody Systems: The Real-Time Challenge**

W.K. Gawronski, **Advanced Structural Dynamics and Active Control of Structures**

W.K. Gawronski, **Dynamics and Control of Structures: A Modal Approach**

G. Genta, **Dynamics of Rotating Systems**

D. Gross and T. Seelig, **Fracture Mechanics with Introduction to Micro-mechanics**

K.C. Gupta, **Mechanics and Control of Robots**

R. A. Howland, **Intermediate Dynamics: A Linear Algebraic Approach**

D. G. Hull, **Optimal Control Theory for Applications**

J. Ida and J.P.A. Bastos, **Electromagnetics and Calculations of Fields**

M. Kaviany, **Principles of Convective Heat Transfer, 2nd ed.**

M. Kaviany, **Principles of Heat Transfer in Porous Media, 2nd ed.**

E.N. Kuznetsov, **Underconstrained Structural Systems**

P. Ladevèze, **Nonlinear Computational Structural Mechanics: New Approaches and Non-Incremental Methods of Calculation**

P. Ladevèze and J.-P. Pelle, **Mastering Calculations in Linear and Nonlinear Mechanics**

A. Lawrence, **Modern Inertial Technology: Navigation, Guidance, and Control, 2nd ed.**

R.A. Layton, **Principles of Analytical System Dynamics**

F.F. Ling, W.M. Lai, D.A. Lucca, **Fundamentals of Surface Mechanics: With Applications, 2nd ed.**

C.V. Madhusudana, **Thermal Contact Conductance**

D.P. Miannay, **Fracture Mechanics**

D.P. Miannay, **Time-Dependent Fracture Mechanics**

D.K. Miu, **Mechatronics: Electromechanics and Contromechanics**

D. Post, B. Han, and P. Ifju, **High Sensitivity and Moiré: Experimental Analysis for Mechanics and Materials**

R. Rajamani, **Vehicle Dynamics and Control**

F.P. Rimrott, **Introductory Attitude Dynamics**

S.S. Sadhal, P.S. Ayyaswamy, and J.N. Chung, **Transport Phenomena with Drops and Bubbles**

A.A. Shabana, **Theory of Vibration: An Introduction, 2nd ed.**

A.A. Shabana, **Theory of Vibration: Discrete and Continuous Systems, 2nd ed.**

Y. Tseytlin, **Structural Synthesis in Precision Elasticity**

Printed in the United States of America.